MW00379986

The Greatest Space Generation

As Inspired by Wernher von Braun

Foreword by Harrison H. Schmitt,
Apollo 17 Lunar Module Pilot and First Scientist on the Moon.
Edited by Ed Buckbee

Acclaim Press
MORLEY, MISSOURI

P.O. Box 238
Morley, MO 63767
(573) 472-9800
www.acclaimpress.com

Editor: Ed Buckbee
Book Design: Rodney Atchley
Cover Design: Ed Stewart II

ISBN-13: 978-1-942613-22-0
ISBN-10: 1-942613-22-9

Library of Congress Control Number: 2015916032

First Printing 2016
Printed in the United States of America
10 9 8 7 6 5 4 3 2 1

This publication was produced using available information.
The publisher regrets it cannot assume responsibility for errors or omissions.

Contents

Acknowledgments 5

Foreword 6

Part I: Beginning 9

Part II: Organization 39

Part III: Management 131

Part IV: Marketing. 237

Part V: Reflections 251

Index 310

Acknowledgments

A special thanks to the following listed
contributors who made this book possible:

Robert Altenkirch
Deborah Barnhart
Charles L. Bradshaw
Heidi Weber Collier
Jack Conner
Carol Foster
Jay Foster
Ray Garner
Nancy Guire
Tim Hall
Uwe Hueter
David W. Hughes
Rein Ise
Thomas J. "Jack" Lee
William R. Lucas
Charles A. Lundquist
Brooks Moore
Robert "Bob" Naumann
Brian C. Odom
Holly Ralston
Stanley R. "Stan" Reinartz
Harrison H. Schmitt
Bill Sneed
Mike Sneed
Christoph Stuhlinger
Jim Splawn
Ruth von Saurma
J.R. Thompson
Annette Tingle
Margrit von Braun
Bob Ward
Frank Williams
Zack Wilson

Edited by *Lee Roop*
Graphics and Cover by *Edward C. Stewart II*
Photography by *Edward C. Stewart II* and *Brian C. Odom*

Foreword

Many technologies enabled exploration of the Moon between 1969 and 1972. These ranged from rockets to fuel cells to water-cooled underwear. No integrated technology, however, was more critical than the Saturn V launch vehicle. The scientists, engineers, managers and skilled workers who created this magnificent space launch vehicle richly deserve the designation of The Greatest Space Generation given to them by historian and editor Ed Buckbee.

As with the 1940s generation that built the weapons and fought the battles that defeated totalitarian regimes in World War II, the 1960s space generation convinced the Soviet Union that it could not win the Cold War and rule the world under communism. Émigrés of that time from the Soviet Union's space program said the first successful test launch of a Saturn V convinced Soviet leaders that they could not win the race to the Moon and that President Reagan's 1980s Strategic Defense Initiative would be successful. It was a remarkable and historic result of the patriotism, dedication, stamina and courage of those who responded to President Kennedy's challenge to go to the Moon.

Young Americans, largely in their 20s, made up the vast majority of the generation that designed, engineered, built and helped astronauts operate the spacecraft of Apollo, as well as the Saturn V that carried those astronauts to the Moon. Astronauts like me merely occupied the tip of the spear held by these remarkable people. My experience working with them from Apollo 8 through Apollo 17 always will remain one of the highlights of my career.

These young men and women, led by only slightly older men experienced in the technology development efforts of World War II, the Cold War and the National Advisory Committee on Aeronautics, made it possible for my colleagues and me to view and explore first-hand the wonders of a small planet we call the Moon. In the valley of Taurus-Littrow, explored by Apollo 17, the Massif walls of that valley stand higher than the cliffs of the Grand Canyon. The Massifs rise in reverse silhouette against a black sky containing a brilliant sun and a blue and white globe we call Earth. We 12 experienced scenes on a Moon that another generation of young people will settle while their progeny prepare to explore and settle Mars. In the rocks of these mountains we found recorded the early history of our own planet from a time when life began.

The leaders of the Apollo program were each unique in their experiences and abilities but they meshed in ways that we can hope will happen again as another generation prepares to return to the Moon on the way to Mars: James Webb, Robert Seamans, Samuel Phillips, and George Mueller at NASA Headquarters; Robert Gilruth, George Low, Max Faget, and Chris Kraft of the Manned Spacecraft Center; Kurt Debus, Walt Kapryan, and Rocco Petrone at the Kennedy Space Center; and, of course, Wernher von Braun of the Marshall Space Flight Center.

This book focuses appropriately on von Braun. Without von Braun's guidance and confidence, it is not clear that the portion of the Greatest Space Generation at Marshall Space Flight Center in Huntsville, Alabama would have successfully integrated, tested and flown the vast spectrum of technology contained within a Saturn V: five, 1,500,000-pound-thrust, F-1 kerosene-oxygen first stage engines; six, 250,000-pound-thrust, J-2 hydrogen-oxygen second and third stage engines; tanks, piping and pumps that fed fuel and oxygen to these engines; an inertial guidance system that controlled flight to orbit and to the Moon; electronic systems for monitoring and sequencing; and a ground complex that successfully launched a Saturn rocket every two months between November 1968 and November 1969! Imagine the skill and courage required to commit to such a schedule.

But the Greatest Space Generation did not reside just in NASA. In fact, of the approximately 450,000 Americans actively involved in Apollo between 1961 and 1972, some 400,000 worked for contractors and subcontractors. Industry leaders in these numerous private companies and laboratories played an equally important role to their NASA counterparts in implementing the design, manufacture and construction of the procured hardware and software, and doing so at the level of quality required for successful Apollo missions.

Personnel in NASA and industry possessed the imagination, science and engineering, and manufacturing skill to create the massive, 364-foot-tall, 6,800,000 pound Saturn Vs. Prior to each mission, Apollo astronauts had complete confidence in their design, development and manufacturing quality control. When they said the rocket was ready, it was ready. Attention to detail was intense. Greater performance and increased reliability were their mantra. They exceeded expectations in every way.

Ed Buckbee has woven the tale of the Greatest Space Generation and Wernher von Braun into a readable and important historical narrative. He has added significant new insights and guidance for a following generation that will win the next geopolitical battle in space and then take Americans to Mars.

Astronaut Harrison H. Schmitt
Apollo 17 Lunar Module Pilot
First scientist and last person to step on the Moon

Part I
Beginning

The Visionaries–Setting the Stage for the Greatest Space Generation

By David W. Hughes and Charles A. Lundquist

The Stuff of Dreamers and a Prelude to Achievement

It is hard to imagine a paradigm-shattering human endeavor that did not begin with a dream. Three of the seminal dreamers for what became the Space Age were Konstantin E. Tsiolkovskiy, Robert H. Goddard, and Hermann J. Oberth. Each of these men was an exceptional dreamer from startlingly disparate backgrounds. They came from different countries, from different sorts of families, and from different educational systems. But each man penned a seminal publication that secured, in large measure, his respective place in history. Moreover, each man intuited that a rocket would be the vehicle of choice for futuristic space missions. And one dream that they all had in common was the vision of a future involving the exploration of space and man's eventual travel through the cosmos.

Arguably, the eventual accomplishments of the Greatest Space Generation were heralded a half century before by the visionary works of Tsiolkovskiy, Goddard, and Oberth. Admittedly, there were other individuals who, in early writings, speculated about mankind leaving the Earth. But these three men legitimized their dreams with valid analyses. Along the way, they demonstrated that space travel was possible using rockets that could be constructed and flown with reasonably achievable technologies. And in doing so, they elevated their standing from simply dreamers to the higher echelon of visionaries.

During the active years of the Greatest Space Generation, *circa* 1950-1975, advances in space operations became legion. But for simplicity here, three particularly seminal achievements will be focused upon:

(i) utilization of, and investigations with, unmanned instrumented spacecraft in near-Earth, in lunar, in planetary, and in interplanetary environments;
(ii) establishment of a human-occupied space station in Earth orbit; and
(iii) human travel to, and productive operations on, another body in our solar system.

Of course, looking forward, it is recognized that numerous variations and extensions of these three types of missions are possible. However, it is likely that the basic technologies underpinning the three incredible achievements enumerated above also will be fundamental to future ventures in space.

In contemplating the Greatest Space Generation, it will be instructive to enumerate when and where the trio of visionaries discussed what became the three major categories of advancement in space operations. In the meantime, it seems appropriate to recall that certain milestones reached by the Greatest Space Generation were international in nature. As is now well known, achievements (i) and (ii) were pioneered by more than one nation. In contrast, achievement (iii) was accomplished solely by the United States, albeit with a significant German heritage.

Through the clear lens of history, the honor of being referred to as the Greatest Space Generation is justified because the engineers, scientists, and managers working from 1950 until 1975 successfully demonstrated all three of the major categories of space achievements. Not only did they do it first, they also did it in a remarkably short period of time. From the launching of the first satellite in October 1957 until the lunar landing in July 1969, fewer than 12 years had passed—about half the time span often allotted to a "generation." This generation did it first, they did it all, and they did it fast. They are truly the Greatest Space Generation.

Konstantin E. Tsiolkovskiy

Much of what is known about Konstantin E. Tsiolkovskiy comes from a massive project sponsored by the Academy of Sciences of the USSR to publish his *Collected Works*. The three-volume result, printed in Russian in 1954, was later translated into English by the United States National Aeronautics and Space Administration.

In the introduction to Volume II, containing Tsiolkovskiy's rocket writings, the Russian editors relate that he was born in a remote village in Russia and that his father was a forester. As a child, Tsiolkovskiy contracted scarlet fever, which left him almost deaf and ended his formal schooling. Thereafter, he was self-educated from reading available books.

In 1879, he passed an examination for teachers, and in 1880 became an instructor of arithmetic and geometry in the Kaluga district. He is quoted as loving teaching. But even while doing so, he used much of his free time and his inventive genius to study the technologies of airplanes, all-metal dirigibles, and long-range rockets.

In 1903, Tsiolkovskiy published his first article, "Exploring the Universe with Reactive Machines," presented in a periodical with the translated title *Science Review*. A "reactive machine" was his phrase for what is now commonly called a liquid-propellant rocket. In this document, he formulates rocket mechanics in the gravity field of a planet and also illustrates his concept of a manned rocket.

In 1911, Tsiolkovskiy published another article with the extended title, "The Exploration of the Universe with Reaction Machines, The Rocket Reaction Machine" that appeared in a journal translated as *Herald of Aeronautics*. In this paper, he describes how to launch an artificial satellite of the Earth and remarks on what a spectacular view of Earth would be obtained. He also calculates the escape velocity from Earth and explains the mechanics of reaching the Moon. Subsequently he suggests occupied satellites of Earth as "a base for further dissemination of humanity." In a penultimate section of his 1911 paper, Tsiolkovskiy says, "The moment reaction machines come into use, we shall witness the beginning of a great new era in astronomy."

Tsiolkovskiy continued his studies and writing until his death in 1935, but it is abundantly clear from his 1911 publication that he discussed satellites of the Earth and of other planets, considered space stations orbiting the Earth, and anticipated travel to the Moon.

Robert H. Goddard

When Robert H. Goddard began serious research into rocketry, he soon realized that to hurl useful payloads to extreme altitudes by rocket power the combustion gases should be ejected from the rocket motor at a very high velocity and that most of the mass of the rocket be comprised of the propellant material itself. Goddard settled on the de Laval nozzle, a geometry that consisted of a converging element near the combustion chamber, followed by a narrow "throat," and then a diverging exhaust section. Ultimately, Goddard favored liquid fuel.

Goddard wrote a lengthy paper entitled "A Method of Reaching Extreme Altitudes," published by the Smithsonian Institution in 1919. In his milestone report, Goddard covered the basic background theory of rocketry, his extensive experimental testing, and some forward-looking calculations.

Goddard's initial investigations helped to establish that solid-propellant rockets alone could never be used to reach outer space. Indeed, at least as early as a patent application filed on 15 May 1914, Goddard saw the possibilities of liquid-fuel rockets. In fact, on a chilly 16 March 1926, Robert Goddard successfully launched the first liquid-fuel rocket. Thereafter, further developing and perfecting the liquid-fuel rocket was a quest that would occupy the remainder of Goddard's life.

Before his death in 1945, Goddard proposed—and, in most cases, tested—numerous technologies that later became standard fare on rockets employed to send astronauts to Earth orbit, to space stations, and to the Moon. These developments include the use of a de Laval nozzle, the selection of liquid propellants consisting of both a fuel component and an oxidizer component, and the inclusion of turbine-powered centrifugal pumps for both the fuel and the oxidizer. In his efforts to enhance the stability of a rocket in flight, Goddard tested retractable air vanes to help with directional control while the rocket was still in the atmosphere, and exhaust jet deflector vanes to help control the direction when powered flight occurs beyond a sufficiently dense atmosphere. Moreover, Goddard flew gimbal-mounted rocket motors, including an on-board autopilot incorporating a three-axis, gyro-stabilized reference platform—essentially a precursor to an inertial guidance system. During his lifetime, Goddard performed pioneering work with automatic launch and flight sequence control comprising timed actuation of tank pressurization, ignition, umbilical release, vehicle release, thrust determination, controlled engine shut-down, guidance, parachute deployment, and payload recovery.

As his legacy, Robert Goddard converted liquid-fuel rockets from simply ideas on paper into actual, working mechanical vehicles.

Hermann J. Oberth

Hermann J. Oberth was born in Austria-Hungary to a Transylvanian Saxon family. An avid theoretician, Oberth studied physics in Munich and Göttingen. During the Depression of the 1930s, Oberth supported his family by teaching mathematics and physics at a high school. After initially having his ideas about rocketry rejected as too forward thinking, Oberth privately published *Die Rakete zu den Planetenräumen* (*The Rocket into Planetary Space*) in 1923. This book, and the ideas that it contained, helped to catalyze the formation and gestation of the *Verein für Raumschiffahrt* (*VfR*) (Society for Space Travel). Certainly, the VfR was the premier amateur rocketry group in all of Europe, providing an inspirational home to the likes of Willy Ley and Wernher von Braun (WvB).

Hermann Oberth was a visionary and a theoretician, but in possession of only meager engineering experience. As such, he set out to validate his futuristic assertions simply with formulas, with sketches, and with well-reasoned arguments. Along the way, he laid down the mathematical laws governing rocketry and spaceflight and showed that rockets appropriate for space travel were entirely feasible. Oberth was honored as a VIP guest on-site at the Kennedy Space Center during the launch of Apollo 11 on 16 July 1969.

Among his contributions, Oberth pointed out that a change from solid to liquid fuels removed whatever size limitation is inherent in solid fuel rockets. He calculated that a rocket could leave the Earth's atmosphere and, indeed, proposed the use of a rocket sufficient to escape the gravitational field of the Earth. In addition, Oberth advocated a rocket designed such that it could carry people. He assessed the likely effects of both microgravity and high acceleration on human beings, and he explained that rocket passengers could be strapped into their seats to avoid floating in weightlessness. Finally, Herman Oberth even considered the practical utility of satellites and of space stations.

Setting the Stage

By 1950, global circumstances were auspicious for dramatic advancements in space. World War II had been over for five years and both the United States and the USSR had wartime-developed industrial capabilities that were now available for other uses. The deployment of the V-2 rocket during the war, plus subsequent V-2 flights conducted at White Sands, New Mexico, had clearly demonstrated that large liquid-propellant rockets were practical for both military and scientific purposes.

International communications were stimulated by peace and by new technologies. The revolutionary ideas of space visionaries Tsiolkovskiy, Goddard, and Oberth became more widely circulated. Their predictions for space travel were now thought to be not only possible, but inevitable.

In the scientific community, international cooperation flourished and a globally supported International Geophysical Year (IGY) was planned for July 1957 through December 1958. Instrumented rocket flights directed at geophysical research were considered by several nations, and the United States and the USSR both pledged to launch an artificial satellite as part of their respective participation in the IGY.

Think tanks and the associated industries in various locales were examining the possibility of communication satellites, meteorological satellites, and navigation satellites. Air forces of advanced nations were replacing propeller-drive planes with jet aircraft. And, by military test pilots it was believed that the skills honed in experimental jet airplanes would be somehow transferable to a flight into space on a rocket.

The United States and the USSR both initiated major rocket development projects. The scale—and the funding—of these ambitious programs were influenced by a competitive environment spawned by the Cold War. Fears of not being first in space ensured that the necessary monetary support was in place for such urgent space activities.

Thus, the stage was set for the achievements of the Greatest Space Generation.

Achievements

When Tom Brokaw in his landmark 1998 book, *The Greatest Generation,* describes his highlighted generation members, he opines, "At every stage of their lives they were part of historic challenges and achievements of a magnitude the world had never seen before." Similarly, with the Greatest Space Generation, their own historic challenges and achievements attest to the adjective *greatest.*

For Brokaw's Greatest Generation, their eventual challenges and ultimate achievements were largely unpredictable. In contrast, for the Greatest Space Generation, the challenges and potential achievements were predicted by the earlier visionaries.

All three visionaries anticipated instrumented satellites of the Earth. The Soviet Union met its IGY commitments by launching Sputnik I on 4 October 1957 and Sputnik II on 3 November 1957; and the United States launched Explorer I on 31 January 1958. A multitude of satellites with diverse objectives quickly followed. The Space Age was off to a vigorous start.

Rocket flights by humans were likewise expected by all three visionaries. Here also the Soviets were the first to achieve this goal with the flight of Yuri Gagarin on 12 April 1961. The first United States manned satellite flight—carrying astronaut Alan B. Shepard Jr.—followed on 5 May 1961. A rudimentary space station named Salyut I was launched by the USSR on 19 April 1971. The crew completed a 21-day mission to the station, but the voyage turned to tragedy when the crew perished during reentry. The United States launched a large space station, Skylab, in 1973, with many different scientific investigations. Skylab was populated by three different crews, with increasing occupation times, through 1974.

But, by far, the most ambitious human space venture was the Apollo program of the United States. The rocket development needed to send men to the Moon was led by Wernher von Braun, director of Marshall Space Flight Center. The first manned lunar landing voyage was in July 1969. It was followed by five more missions that each put two men on the surface of the Moon to perform a substantial range of investigations.

Thus, the United States complement of the Greatest Space Generation accomplished each of the three prime space operations expounded by Tsiolkovskiy, Goddard, and Oberth. The specific accomplishments mentioned above are but the barest representation of the much more extensive list of activities actually executed. Further explanation is unnecessary here because other parts of the present book relate stories of many of these activities. In particular, other sections of this book will describe in detail exactly *how* the historic achievements of the Greatest Space Generation were actually accomplished.

WHERE IT ALL STARTED

By Ed Buckbee

When you arrive at Huntsville's International Airport and view the diorama of famous space accomplishments, you wonder, "How did this happen in Alabama?" Driving through Cummings Research Park, passing the University of Alabama-Huntsville campus and observing the signs of numerous Fortune 500 companies such as Boeing, Lockheed Martin, Raytheon—not to mention home grown companies like Dynetics, ADTRAN and Intergraph—a visitor may wonder, "Why are these companies here? Who was the mover and shaker who created this technological oasis in the Heart of Dixie?"

Some credit must go to a brilliant young German, Wernher von Braun, who at the age of 25 began building rockets. Von Braun was born into an aristocratic family. His father was a baron and high-level government official. His mother was the motivator in the family, spoke six languages, loved music and traveled a great deal.

She passed these traits on to her son and consequently, WvB attended the best schools. He was not a perfect student, however, as initial failures in math can attest. Imagination and leadership were qualities that coalesced in the young WvB, with the former contributing significantly to his passion for space and science fiction. He quickly and prudently realized if his dream of space flight was to become reality, then he would need to cultivate technical and engineering skills. Knowledge and understanding of math are crucial to succeed in the aforementioned endeavors, and WvB swiftly mastered his past failure.

He ultimately joined the Germany Army as an engineer and worked at a place on the Baltic Sea called Peenemunde. WvB became the technical director of this highly secret, research and development rocket center under his first mentor, General Walter Dornberger. In 1939 at the age of 27, WvB had 10,000 people working for him. Construction began in the late 1930s and by 1942, WvB and his team of rocketeers were building and launching the monumental and notorious V-2 rocket, the first rocket to be launched into space. On Oct. 3, 1942, the WvB team successfully launched the first V-2. To WvB, Dornberger is reported to have said, "Do you know what we have done? Today, we launched the spaceship!"

The combined facilities at Peenemunde appear to have covered approximately 40,000 acres—about the size of Redstone Arsenal. The majority of the facilities were destroyed either by the Royal Air Force or U.S. bombings in 1943-44. What wasn't taken out by aerial bombings, the Soviets moved to Russia or destroyed during their occupation of East Germany after WWII. The only structure untouched is the power plant that today is the home of the Peenemunde Historic-Technical Museum. The museum features an impressive collection of artifacts from the V-1 and V-2 programs. Von Braun is presented along with several other Germans—including Dornberger and Oberth—but nothing of his work in the U.S. is displayed.

At the young age of 25, WvB began to form a team of rocketeers that within one organization possessed the capability and skill to conceive, design, develop, fabricate, test and launch a revolutionary new rocket system. Peenemunde East, the research and development complex, included a power plant, liquid oxygen plant, wind tunnels, laboratories for rocket fabrication and assembly, quality control, guidance and control, blockhouse and nine rocket test stands. It also included an airfield and on-site housing for engineers and scientists. The German Air Force conducted extensive research and development at Peenemunde East where the V-I buzz bomb, manned rocket powered aircraft and cruise missile work was conducted.

Bombing runs by the British and U.S. in 1943-44 killed several of WvB's team members and destroyed a portion of the research and development facilities. After those bombings, the production of V-2 rockets moved to underground facilities. As the war came to a close, WvB and his team of about 500 scientists and engineers made plans to safeguard V-2 documents and components consisting of more than 100 V-2s. They did everything in their power to avoid surrendering to the advancing Russian forces. The team voted unanimously to surrender to the advancing U.S. forces. In 1946, the U.S. Army initiated Operation Paperclip—the plan designed to move captured German rocket scientists and V-2 equipment to the U.S.

U.S. Army Col. Holger Toftoy was the strategist who made Operation Paperclip a reality. He planned it and he convinced the Pentagon and the president to approve it. Major General John B. Medaris—referred to as 'Big M' by many of us—was the General George Patton of Army missiles. He was the only general I ever met who carried a swagger stick, a leather-covered, foot-long stick that resembles a horsewhip. He was the shaker and mover of the program who successfully lobbied for the authorization and funding necessary for the Army to pursue serious research and development related to rockets and eventually, satellites. Both of these men became important mentors to WvB, teaching him how to market the program with the Congress, the press and other governmental agencies.

The WvB team mused that they had become "POPs" prisoners of peace. Toftoy tried his best to get permission to bring 300 German scientists to the U.S. He received approval for 100, but another 20 were added and they arrived in the U.S. in 1946. The von Braun team launched about 100 V-2s and modified V-2s from Whites Sands Missile Range and Cape Canaveral. Additionally, they all enrolled in a crash course to learn English.

In 1950, most of the 120 Germans transferred to Huntsville. During WWII, Huntsville was the home of two operational Army arsenals, one specializing in munitions and the other in chemicals. They were closed after the war and offered for sale. Col. Toftoy began working with the city fathers and came up with a plan to move the WvB team here and establish the Army's first missile research and development (R&D) center at the newly named Redstone Arsenal. The Germans were welcomed as members of the community. As their families arrived, they began purchasing houses, enrolling in schools, joining the Jaycees and Rotary, and signing up for library cards.

Similarities between Peenemunde and Redstone Arsenal were many. Both were government owned and had research and development teams working on advanced technologies. The facilities were isolated and in secured areas. They were accessible by water, rail and air. There was an available workforce in the area. Soldiers were used to support the workforce. Redstone had a special program called Scientific and Professional (S&P's) that permitted soldiers with engineering degrees to work side-by-side with scientists in the laboratories. There was competition between military services for missions and funding. WvB formed a team of rocketeers that within one organization had the unique in-house capability to conceive, design, fabricate, test and launch a completely new missile system.

The Germans taught us how to defend ourselves against a missile attack. We were a generation living under a persistent threat of nuclear war. The Cold War and the fallout shelters associated with it impacted the lives of all Americans. My daughters often came home from school describing how they practiced crawling under their desk in the event of a Soviet nuclear attack. It was real. We were a target for the Soviets, and they could have attacked us during the Cold War. Today, we could very well be speaking Russian if the Nike–X anti-missile defense system hadn't become operational. That's when the Soviets came to the negotiation table. By the way, von Braun had a fallout shelter in the backyard of his Huntsville home.

Our current missile defense system is built on a foundation that was developed by the U.S. Army team mentored by WvB almost six decades ago. Many of the team members who came to Redstone Arsenal for their first job, describe how their German supervisor taught them specialized skills—missile trajectories, propulsion, guidance and control—subjects that were not taught in our universities at that time.

Peenemunde is the place where rocket technology was conceived and developed that enabled two countries—the U.S. and the Soviet Union—to become space faring nations. We built on that technology. Adding German brainpower, American ingenuity and national pride, we beat the Russians to the Moon. Other nations—China and India—have in recent memory used this same technology to gain access to space. Peenemunde truly is the cradle of early rocket science, and could be considered the birthplace of space flight. The same could be said of Huntsville: In the context of American spaceflight, it is also the cradle.

The Arsenal System

The U.S. Army's arsenal system and the early WvB team were a perfect match. The initial WvB team was accustomed to working together in Germany and they transitioned smoothly into the arsenal system, initially at White Sands and subsequently, Redstone Arsenal and Marshall Space Flight Center. If you think about it, industry in general works in an arsenal-like mode. A company does what it can revolving around a specialty that is the heart of its product. Industry goes outside for additional products or services to enhance the product or to hire skills that it doesn't have or are avalable at a lower cost than doing in-house. For Marshall and the WvB team, as the workload expanded it became imperative to look to American industry. There was a number of reasons for this: 1) government entities normally are limited in resources for in-house activity, 2) American industry in aeronautics and related fields had and was acquiring skills that met the requirements that Marshall required, 3) the government desired to utilize and enhance this industry.

The Team

The civil service workforce on Redstone Arsenal grew from 300 in 1955 to 7,000 in 1965. In the early years, recruiting engineers to Huntsville, Alabama was difficult. Industry was limited and there were limited opportunities for engineers. The Army being the Army, with high priority from the Pentagon, waved its arms and 300 young Army engineers from all over the country showed up. Additional young engineers and other skills such as chemists were recruited in the formative years. After President Kennedy announced the Apollo Lunar Landing program, recruiting became much easier.

The WvB team had a number of advisors who were American officials assigned to the U.S. Army. These were Major James P. Hamill, Generals Holger Toftoy and John B. Medaris and public information officer Gordon Harris. These military-based personnel were very instrumental in advising WvB on how to deal with local officials, as well as providing consultation on meeting congressmen and senators at the Washington level.

When the WvB team was transferred to NASA in 1960, von Braun knew he had to find managers from other agencies and the private sector. These men came from the Atomic Energy Commission and other agencies of government and private companies. Some were on loan from the military services.

He formed a new team of personnel, primarily civilians who became his administrative managers and advisors. They included Delmar Morris, Harry Gorman, Chauncey Huth, William Guilian, Dave Newby, Wilbur Davis, Victor Sorensen, Paul Styles, Marion Kent and Foster Haley. These gentlemen who made up Marshall's upper level management brought skills and experience from various fields.

On the technical side, Germans were heading up the laboratories and research departments. Many members of the team were reluctant to leave because of the dedication and loyalty they held for the Army. The U.S. Army had given them a new life. But this is what WvB and his team wanted to do their entire career… explore space for peaceful purposes.

WvB personally selected the following men he brought with him from Germany as the "technical brain trust" of the team: Stuhlinger–Research, Hoelzer–Computation, Heimburg–Testing, Geissler–Aeroballistics, Neubert–Quality, Haeussermann–Guidance and Control, Mrazek–Propulsion and Vehicle Design, Hueter–Systems Support, Rees–Deputy for Technical, Debus–Launching, and Maus–Fabrication and Manufacturing. They were the directors of the laboratories, the heart of WvB's organization. They were referred to as the board of directors. A number of Americans who were on the job training (OJT) would eventually assume key positions within these laboratories.

When considering accomplished scientists, one usually thinks about Galileo, Thomas Edison or Alexander Graham Bell. Seldom do you find a scientist, equally gifted in both science and management. Rocket scientist WvB was considered to be one of the more successful and effective managers of a peacetime technological program. The team was charged by the U.S. Army to build a missile R&D center in a country with little or no rocket experience. General Medaris was once asked, why did you want WvB in this country after WWII? His response:

> "The answer is simple. At the time he headed up the world's only experienced ballistic missile team. He was willing to help us with our embryo missile program. He had impressed me favorably with his knowledge, his progressive thinking and cooperation. Above all, he had a burning desire to continue rocketry and astronautics as his life's work. Besides, I kind of liked the guy."

I saw WvB in both his roles, as a scientist directing a high-tech program and in his public relations role—giving speeches, leading press conferences, appearing before Congress, and hosting corporate and government officials. He asked us to call him Wernher von Braun because as he stated, "I'm an American citizen and I wish my name to be pronounced as you might pronounce an American." In his presence and around visitors, it was Dr. von Braun. However, when we were not in his presence around the office, we referred to him as, V-B.

I studied his weekly notes, 8,000 pages, and reviewed his daily calendars. Over the past five decades, I've had the opportunity to interact and talk with the majority of the German and American team.

WvB recruited college graduates who received their engineering degrees through the GI Bill after WWII and trained them in Rocketry 101. Many of them already stationed at Redstone were transferred to the team. Some were designated S & P's, scientific and professional soldiers already serving in the Army, assigned to the 9330th Technical Services Unit. The WvB team members served as mentors and teachers to the young Americans, who had no training or experience in such subjects as trajectories, rocket propulsion and guidance and control. Together, they developed a capability that never existed in the U.S., a technical team to conceive, design, develop, fabricate, test and launch a completely new missile-rocket system in-house, within one organization.

WvB was the perfect match for the "doer" position: a manager with excellent leadership, and a charismatic organizer with impressive communications skills, technical knowledge, and human resource skills. Without a doubt, his aristocratic provenience, his good looks, and his charming and courteous appearance were valuable assets in reaching his goals. In Germany and later in the U.S., he had access to a network of high-ranking supporters and a loyal team of coworkers. The prerequisites were there and the path to the Space Age was open.

The development of the Redstone ballistic missile received the most attention. It was successful on its first flight in August 1953 and the U.S. had its first operational, nuclear-tipped surface-to-surface missile. WvB began giving speeches and interviews about rockets and Redstone Arsenal. He became recognized by the national press. One reporter named him, "missile man" and that was the beginning of WvB's popularity with the national press.

"We can publish scientific treaties until hell freezes over. But unless we make people understand what space travel is, and unless the man who pays the bills is behind us, nothing is going to happen."—Von Braun [1]

WvB began writing articles about flights into space and men being flown on rockets to the Moon. He wrote a series for *Colliers* magazine that became quite famous during the 1950s. This was the first indication that WvB had another interest—flying people into space.

In October 1957, the Russians shocked the world and certainly got the attention of WvB and his team, with the launch of the satellite, Sputnik. The space race was on. The Russians began their efforts to outstrip the U.S. in space—launching dogs, men and women cosmonauts.

WvB and his team of Germans and newly recruited Americans had a satellite ready to be launched well before Sputnik, but the Pentagon put a halt on that program. The Army was not allowed to engage in space projects. The U.S. Navy had that mission and their Vanguard missile had failed. The WvB team responded in January 31, 1958 with the launch into orbit of the free world's first satellite, Explorer I, on a modified Redstone. The U.S. was finally in space.

The last weapon developed by the WvB team was the Pershing missile, Redstone's replacement. The team's contribution to the development of the Nike X, anti ballistic missile system—a bullet hitting a bullet—laid the foundation for the national missile defense program that protects our nation today.

Congress declared that the general welfare and security of the U.S. required adequate provisions be made for aeronautical and space activities. Congress further declared that such activities shall be the responsibility of and shall be directed by, a civilian agency exercising control over aeronautical and space activities sponsored by the United States.... [2]

In 1958, President Dwight Eisenhower created NASA, the National Aeronautics and Space Administration. Why? Because he knew the competition between the military services would hamper success just as it had done during the Explorer satellite program. The WvB team was moved from the U.S. Army to this new governmental agency, NASA, that has been charged with the mission of peaceful exploration of outer space.

The president came to Huntsville to dedicate the new NASA space center named after WWII General George C. Marshall. This was a significant change for the team and the way they had done business in a classified environment. Now, they were free to talk about their work and show their progress—accomplishments and achievements—to the American public. WvB was very keen on letting the taxpayers know what they were getting for their money.

The First Redstone Ballistic Missle Launch Team

By Annette Tingle

On October 27, 2015, a select group of about a dozen space veterans gathered for a roundtable discussion of their pioneering experiences with Redstone missile development.* Historians and future space generations can be grateful to participant Jack Conner for the idea and for his efforts in convening the venerable group. Most of these men were now approaching the century milestone. Their hair may have been graying, but their recall, passion, and pride about the work they did were as vibrant as yesterday. Fortunately, their reminiscences were videotaped and provide a glimpse into the genesis of the U. S. space program. In many ways the participants were representative of many other employees in the early years of space exploration. Although the Redstone Ballistic Missile Launch Team (hereinafter Redstone BMLT) members are distinct in personality, they shared much in common. They were trailblazers for the Greatest Space Generation.

More than one of the team members remarked that he had been in the "right place at the right time." Indeed, the circumstances on the historical stage were perfectly aligned for success. If the Apollo 11 Moon landing was a defining milestone of the latter half of the 20th century, the Great Depression and World War II were defining milestones for the first half. The Redstone BMLT was of a generation that had spent the formative years of youth in rather austere economic conditions. Then hundreds of thousands, still in their youth, volunteered or were drafted into military service. Like a significant segment of the early NASA workforce, most of the Redstone team had served in WWII. A logical conclusion is that the high percentage of Veterans in the early workforce was not due merely to hiring preference policy but primarily because of skills that were in demand.

Military service, especially in a war of the epic scope of WWII, is a great leveler, giving confidence to the insecure and correction to those too cocky. Many of those who served departed as kids and returned home as significantly matured men. Many had faced down death and terror and extreme hardship. Bill Grafton, one of those sharing his story in the roundtable discussion, spent months as a POW in German stalags after his B-17 flight crew was shot down. He was one of the fortunate POWs surviving "The Long March" in a forced retreat away from the Eastern front in the incredibly harsh winter of 1945. Those who survived the war weren't afraid of much and possessed an ingrained mindset that they could accomplish difficult and risky tasks, something that early missile development had in abundance. It should not be surprising that for the first Redstone launch Grafton would be the last man on the pad, the one who had to plug in the igniter. He may have been confident and unafraid, but he was not foolhardy.

> *When I plugged the plug-in onto the launcher ring, it accidentally shorted out and the LOX valve opened and it dumped all this liquid oxygen down right on my feet and on the deflector. Of course, a big cloud of white smoke* [went up]. *I didn't wait around to see what happened. I took off like a shot. I had a ground power headset on and a hard hat. When I came to the end*

of that 25-foot cord, it nearly took my head off, but I kept running. Finally I couldn't hear anything firing, so I turned around and looked; everything looked perfect so I went back to the blockhouse.[1]

The problem was corrected and eventually the launch occurred and development of the Redstone continued. Grafton's unassuming manner and relaxed humor do not emphasize his varied experience, ranging from the Redstone and Pershing programs with the Army to NASA and the Test Laboratory, Mississippi Test Facility, and the Saturn launch vehicles.

Team members benefited from another aspect of military service. Training has high priority in the military because it is essential to effectiveness and survival. An added element of prominence in WWII was aviation. The Redstone BMLT, like others of the first space generation, entered the workforce in the midst of aviation expansion and on the cusp of the space exploration era. Veterans with technical skills and experience joined others graduating from college with engineering degrees. Engineering experience with the National Advisory Committee on Aeronautics (NACA), such as Bill Grafton's X-plane flight research work at Edwards AFB and Reed Barnett's wind tunnel work at Langley Field further enhanced job prospects. Some, like team member Reed Barnett, had graduated in the relatively new field of aeronautical engineering. Even if novel, his credentials counted for something in his interview with Dr. Kurt Debus for a job in the Firing Lab.

I went to Dr. Debus and told him I was interested in working for him. He asked what my background was, and I told him I was an aeronautical engineer and working for NACA. He said, "Well, Mr. Barnett, I have 30 or 40 openings," whatever it was, "not one aeronautical engineer." So we chatted about that a little bit, but he hired me.

The nation's pent up demand for domestic products, retooling of industry, and emerging new technologies made the late 1940s a propitious time for job opportunities, especially for motivated entrepreneurs. This is evidenced by some familiar Alabama corporations such as Harbert Corporation (known today as Harbert Management Corporation) and Blount Brothers, the construction company that worked on Launch Complex 39A at Cape Canaveral. Both were founded by WWII veterans. Closer to home, the design engineering for the Huntsville International Airport was constructed by a local family company expanded and enhanced by the combat engineering experience a son, Carl T. Jones, had gained in construction and combat planning in the Alaska territory.[2]

New and expanding industries offered various job opportunities for engineers. Nevertheless, their professional interest and passion attracted many of the Redstone BMLT members to the missile work taking place in Huntsville, even if the conditions were comparatively Spartan. Brooks Moore, who came to Huntsville in 1952 after working for three years on torpedo countermeasures at a Navy research facility in Florida, found conditions much more austere in terms of available funding and laboratory equipment.

> *I had to think long and hard to leave what I left down there. I had been there for three years. But I reasoned that maybe a torpedo is just an underwater mine and maybe I had some experience that would be pertinent to the missile program.*

He hastened to add, and later reiterated, how fortunate he was that he made the right decision.

The politics of missile development and inter-service rivalry, no doubt, contributed to the scarcity of resources, not only for the research and development but also extending to the launch facilities. Reed Barnett described conditions for the early Redstone launches:

> *My job was the assembly, test, and repair branch [on the pad launch crew]. We got the missile's major components sent down by plane to Patrick. Then we loaded them up and trailered them up to an old Quonset hut up on the base, weeds grown up all around it, a little dirt road going back about a quarter mile to the hangar – very crude, very severe. We would put the components together, check them out, make sure everything was tight, all the electrical continuity, all the valves worked and everything. We would put it on the transporter and take it over to the pad and put it on the launch frame and turn it over to Bill* [Grafton].

For the most part, the members of the team came from the North Alabama area or nearby states. The goal of these team members was not simply a job in Huntsville. They were highly motivated to find the right fit for their career goals. Alex McCool, who came in 1954, described his interview experience as a gradual revelation. After interviewing in several areas – Launch and Handling, Test, etc. – and being offered three jobs in Launch and Handling that he thought might be interesting, he concluded with an interview with Wilhelm Schulze in Structures and Mechanics Laboratory.

> *He pulls out this drawing; he starts showing me all this stuff; there's a Redstone sitting on a pedestal, all kind of tanks, trucks, and propellant lines, and everything. He looks at me and says, "Now, Mr. McCool, we work on **this**." He pointed to the missile. He knew I had been over talking to Hans Hueter, Ground Support – lines, tanks, propellants. Ground Support, it's important; but he was really trying to get me. Well, the light came on when he said we work on.... Well, I didn't know a rocket ... here's a picture of a drawing. I said well I want to work on a rocket, not on trucks and all that kind of stuff. So I went to work for him.*

And the implication is that he never looked back, an experience shared by most of the team. One of them was Jim Pearson. He described becoming part of the Redstone BMLT as something he "backed into" after first taking a job in the packaging and handling unit. One of his early jobs was working on putting an atomic bomb in the missile nose cone and getting the handling and shipping equipment in response to General MacArthur's order to Ordnance for a missile with atomic capability. When a cool-down in the Korean conflict effort ended that project, he moved to the Redstone BMLT. Pearson's modest manner belies the extraordinary breadth of his career in testing, ranging from Operation Hardtack (a Pacific atomic bomb test using the Redstone) to Jupiter C and Explorer, and to Mercury Redstone. His career progression tracks the history of the U.S. space program, including the one static test of Centaur.

> *We did get into liquid hydrogen, which was used on the second and third stage of Saturn V. That made the difference in the race with the Russians. They couldn't get their kerosene second and third stages operating, and Saturn V did. So we won that race. After that I was involved with the Shuttle program, which was really interesting.*

Bo Cloud, who came out of the Navy in 1951 as a senior electrician on a diesel submarine, first found employment in a research position with Rohm and Haas, one of Redstone Arsenal's major contractors. After less than a year, he landed a job with the Army as a senior electrician, but the lure of the potential new field of space rocketry soon drew him to that area:

> *During that period of time I saw where this area was being built up. We went through that area; you could see the test stands coming up, but you didn't know what was going on – at least we didn't. Then in '56 I got hired in the Test Area.... The first test firing I saw [was] of the Redstone.*

The enduring value behind NASA's extraordinary successes is the knowledge gained – a chief element of the space exploration mission. But the excitement, of course, has always been in the "fire and smoke." Many on the Redstone team spent considerable time in that arena. They were in the vanguard of U. S. space pioneers, building on and improving on the foundation the Germans had established with the V-2. Almost to a man they were unstinting in their respect and praise for the leadership of the German members of the team and the teamwork environment. Jim Rorex and several of the participants alluded to their confidence in the readiness to proceed with missile testing in comparison to the government's lagging commitment to move the program forward. Curley Chandley was candid in expressing the frustration of some: *In fact, I think the Germans should have showed up in Huntsville about 5 years earlier really because I think we lost that much time.*

Focus on the mission, confidence in the leader, and teamwork are essential for survival in the military and represent a cultural norm that would have been familiar to the Redstone team. Brooks Moore, and several others, in commenting on the success of the Redstone and Saturn programs expressed it this way:

> *"...we developed the Saturn all the way through the Saturn program and had it ready to go essentially before we had systems engineering. That's just*

because we literally worked together as a team, and it was all the way from the top to the bottom. I think that had a lot to do with our success, too.

For the most part, the Americans in the early years of the U. S. space program were young engineers following the leadership of more senior and experienced German engineers and scientists, but they were quick learners. More importantly, they responded to and perfected a teamwork culture long before "teamwork" became an industry buzzword. When Moderator Tim Hall posed a question about what the working relationships were like at Redstone Arsenal in the 1950s, Curley Chandley eagerly responded with a description of the cooperation developed within the design group in Huntsville. This served the team well when a field engineering change to a system had to be made once the vehicle was at the Cape.

This would happen during the night maybe, and Bob Aden and the bunch up in Huntsville were asleep; but the next day we would get the engineering changes that coincided with exactly what we had done. It was just logical the way we all thought and the way we worked together. After working with these people up here, it just came natural, no problem at all with field engineering changes. ... The minds of the individuals concerned were like one, very much so.

Jim Rorex was firm in his praise for the German leadership and the teamwork environment:

I think that group was the cream of the crop. They were great people absolutely. They were just fantastic. I've always thought about how lucky I was to get to come in and work with those people – for them, but you didn't work for them, you worked with them. That was the way it was.

Practicality, resourcefulness, and appropriate reaction in risky situations are useful assets in battle or a missile launch, particularly when resources are not abundant. Those living through the 1930s and WWII were skilled at improvisation and making do with what is at hand. Considering today's sophisticated processes and procedures, safety and quality inspectors might be appalled at previous methods of checking for various leaks, as described by Ken Riggs:

The technicians would get up in the bottom of a vehicle to check it for leaks. They would put their hands up to try to find the leaks. The way we checked hydrogen was we would take a broom down there, and it would catch on fire if you ran across a joint that was leaking. Those were the kind of simple minded things that worked.

Things that had not been tried before may have appeared simple after the fact, but they often had been proven by trial and error. Looking back from the 21st century, the Redstone project assembly and launch operation appear extremely simple, and the basics are still discernible. Layers and layers of complexity and bureaucracy have been added to the simplicity of, as Curley Chandley described it, a single instruction to push a button when it was time to go; but the process still comes down to the basic steps perfected by the first Redstone team.

> *We had all our instructions on one page, front and back. They were a bunch of one-liners… And all this is on one page, right on up to the firing day. No test conductors; we didn't have a test conductor for about three or four vehicles before we finally found we needed one, I guess.*
>
> *We didn't have any supervision or events panel or anything like that. All the panels were very well segregated to each individual system, like the mechanical panel, the networks panel, the electrical panel. There was nothing that said, "Here's the sequence where we are." In fact, we didn't even have control of the instrumentation very well; we had to turn on each recorder one at a time. We built the first countdown panel and the first recorder control panel, and all this stuff was built during getting ready for the first launch, not before.*

It was new ground. Reed Barnett also described how they were developing the process instrumentation and the control procedures simultaneously with preparing the missile for launch. However, none of the team members had any doubt about the importance of what they had accomplished. Reed Barnett: *When the fire observer called for ignition switch, pre-stage, main stage, and that first rocket lifted off, that was the beginning of our space program.*

Another common characteristic of the team was their immense pride in the work they had done. Even in 2015 the pride of John Kastanakis was palpable. He appreciated the independence he had been given in his responsibilities for instrumentation in the Marshall east and west test area facilities. Even after the passage of many years, he still maintained a sense of kinship and ownership:

> *I was very proud of that instrumentation. I think we would put it up – the east blockhouse and the west blockhouse with the instrumentation that we had – against anyone in the world. We were very, very fortunate.*

"Fortunate" was echoed many times during the discussion. The unaware person perhaps might regard the attitude of some of the team members as arrogant. However, in most of these men one senses an attitude of humbleness. Most of them had come to Huntsville as young engineers (some "green behind the ears", as one of them said). With the exception of Jack Conner, who was directed to Redstone Arsenal in military service, most actively sought to come because of interest in the work and ended up spending their careers in space activities. Even Connor, who came to work in 1955, stayed on and worked at the Interim Test Stand in the Test Laboratory. They were committed to the mission goal but realistic about themselves. Ken Riggs best sums up the position of many in the early space generation:

I could have probably been a better person in some respects, but I was just like everybody else. We did our best.

All of the participants had a sense of having had a part in history. Many are still passionate about the work they did and have maintained an active interest in the space program. However, they are not fixated on past successes. In closing, the moderator asked the group about going to Mars: *Do you all think we're going to make it there, and what do you say to the next generation of folks interested in space?* Most of the team members were optimistic. Various ones responded with the confidence they had always exhibited, even if they could not be specific about the support and timetable.

If it's there, we'll get there. Might take a while, but we'll get there.
It may not happen as soon as we expect, but it will happen.
Mars is there, and there's a reason eventually that we'll want to explore it, and we will explore it.

Such confidence is indeed a hopeful note for a future generation and a tribute to the unique past one represented by the first Redstone Ballistic Missile Launch Team members of the *Greatest Space Generation.*

Editors Note: Jack Conner initially conceived the idea of assembling former members of the Redstone Ballistic Missile Launch Team for the roundtable and was instrumental in encouraging the U. S. Space & Rocket Center (USSRC) to sponsor it. Tim Hall, Vice-President, Business and Media Initiatives, arranged videotaping and served as Moderator for the discussion. Bennie Jacks, a USSRC Docent, set up the interviews and made arrangements for the meeting. The roundtable video tape was transcribed by Annette Tingle.

David Newby—Marshall's First Employee Transfer of the WvB Rocket Team

By Brian C. Odom

On July 1, 1960, WvB's German rocket team was transferred to NASA, formally establishing the George C. Marshall Space Flight Center. But the transfer of the WvB group was never a forgone conclusion or a simple administrative stroke of a pen. The transfer entailed not only a major shift of manpower, resources and facilities from the Army to the new Marshall organization, but also a tremendous amount of new construction and the implementation of new management responsibilities. Getting NASA's new Marshall Space Flight Center operational would prove to be a monumental challenge. Among the many talented individuals who took up this challenge, David (Dave) H. Newby stands out. Long before he would be known for his role in the establishment of the U.S. Space & Rocket Center and Huntsville's Von Braun Center, Newby was NASA's first employee in Huntsville and a man tasked with creating the administrative infrastructure for the agency's newly minted space organization.

Dave Newby was born on February 12, 1920 in Chickamauga, Georgia, and graduated from the Georgia Institute of Technology in 1942 with a degree in electrical engineering with a communications option. He began his engineering career later that year as a research scientist in the instruments laboratory at Langley Research Center, which was then part of the National Advisory Committee for Aeronautics (NACA). His job at Langley was to develop research instrumentation for wind tunnels, airplanes and rocket test facilities including those at Wallops Island, where he used computers to reduce the high volume of data generated by instruments. It was in August of 1951 that the health of one of his children forced Newby to move his family to Huntsville, where he accepted the job of deputy chief of the rocket test laboratory in the Ordnance Corp. One year later, Newby was named chief of the lab and was tasked with identifying, designing and building test facilities for the Army's solid rocket test range at Redstone Arsenal. This experience would provide Newby with an opportunity to develop skills and insights that would serve him well in the coming years.[1]

Newby's initial involvement with NASA came on May 4, 1959, when he was appointed the agency's representative to the Army at Redstone Arsenal. In this new role, Newby was tasked with overseeing the work done by the Army with NASA funding. A memorandum from NASA Administrator Dr. T. Keith Glennan outlined the functions of the position, which would report to the Director of Space Flight Development as:

> A. Furnishing technical monitoring of the Army Ordnance Missile Command (AOMC) effort on projects funded by NASA, so as to secure compliance with specifications and to assist in timely and economic accomplishment.
> B. Maintaining liaison between AOMC and NASA on problems and developments of mutual interest, including technical and fiscal problems, schedules, and programs.
> C. Coordinating contracts, visits, and exchanges of data and information between NASA and AOMC personnel.[2]

During an October 1959 visit to the Arsenal, Dr. Glennan had stated his desire for the ABMA group to join NASA as the agency's vehicle systems and launching operations branch and for all large booster system activities to be consolidated into one program to be overseen by the new organization.

After President Eisenhower approved the transfer on October 21, 1959, and logistical discussions began in earnest related to the transfer of the ABMA team to NASA, Newby was assigned to the agency's Transfer Team headed by Al Seifert, Dr. Glennan's Business Administration Director. It was this team that was charged with the delicate and demanding job of paving the way for the administrative move of the von Braun Team to NASA and laying the groundwork for a new organization which included everything from test facilities to telephone and printing services. Newby's counterpart on the Army side of the negotiations, Col. Thomas Cook, worked side-by-side with him and other NASA representatives, performing much of the administrative legwork during the negotiations. In the meetings that took place over the last several days of October 1959, Newby was responsible for briefing the task team on issues related to scheduling and staffing, meeting with Gen. John B. Medaris and Eberhard Rees in order to get their thoughts on how the transfer should be effected, and working with the rest of the Task Group Coordinators to determine the most critical problems moving forward. It was during one of the meetings with Siepert and Newby that Rees voiced the most prominent concerns among the ABMA group, which included where Huntsville would fit in among the top NASA organizations, what job von Braun would be expected to fill, how much flexibility there would be in programming, and what Dr. Glennan's attitude was on "whether the U.S. is, or is not, really in a competitive race with the Russians [regarding] space?"[3] Joining Newby on the Task Group Coordinators was Wesley L. Hjornevik, Dr. Glennan's assistant and the future associate director for manned space flight in Houston. Hjornevik was selected to serve as vice chairman for the group and was, himself, tasked with preparing the President's Report.

The agreement for the transfer was approved by President Eisenhower on November 2, 1959, and a memorandum dated November 3, 1959 listed seven critical management problems associated with the transfer. These ranged from basic organizational relationships to the transfer of equipment and supplies. Also included in the list was the question of financial arrangements. According to the memo, it was a "forgone necessity that NASA must establish its own accounting setup in advance" and have it operational by the July 1 transfer.[4] This would give those responsible for the task just over six months to implement an accounting system that would oversee one of the largest non-defense accounting systems in the country. Newby would soon find himself personally involved in finding solutions to many of these problems.

According to the "Summary and Concepts of a Transfer Plan" worked out by the group, Newby was a key member of and dated Dec. 11, 1959, the transfer of Army personnel, facilities, and equipment from the Development Operations Division to NASA would be conducted in a "manner calculated to prevent dislocation or disruption of on-going programs." The plan stated that the new NASA organization at Redstone Arsenal would be "locally self-sufficient" with "functions involving the management control functions" and those "functions involving immediate service to the technical groups" operated by the new NASA center. The plan also provided for the transfer to NASA of a contiguous area at Redstone "encompassing all the facilities now within the Development Operations Division" as well as that of "equipment and inventories as appropriate in the particular case." Arranged to coincide with the issuance of President Eisenhower's executive order for the transfer, the plan called for all "unobligated" and "unexpended" balances on contracts on major Saturn funds to be transferred from the Army to NASA. From a personnel standpoint, "limited numbers of key personnel" would transfer from the Army to NASA after the executive order, while the remainder would transfer on July 1, 1960.[5]

Overall, the transfer from the Army to NASA entailed the shifting of 4,670 people from the Development Operations Division of ABMA along with 1,200 acres at Redstone Arsenal and facilities valued at $100 million. The transfer also included ABMA's Missile Firing Laboratory at Cape Canaveral, which would become the NASA's Launch Operations Directorate and be headed by a member of the WvB team, Kurt Debus.[6] The newly established George C. Marshall Space Flight Center, with its 5,500 civil servants and 1,189 on-site contractors would now be the largest NASA center by far.[7]

What appeared to be relatively straightforward on paper, would prove exceedingly difficult from an administrative standpoint. During Marshall's first year of existence in 1960, Newby was named by Dr. von Braun to be the director of the Technical Services Office. The duties of this position included the construction and maintenance of the facilities and supplies necessary to carry out the new organization's duties. Another of those involved in the process was Woody Bethay, who later remembered that "We had to build every process, every procedure. We didn't even have a budgetary process, or many of the things needed by a major agency. We had to develop everything. We wrote the book as we went along."[8] And from his new position as director of technical services, Newby would be charged with finding solutions for these considerable administrative obstacles.

Kennedy's objective of placing a "man on the Moon and returning him safely to Earth" provided Marshall with a tremendous amount of resources. From 1961 to 1965, Marshall's Administrative Operations obligations were more than double that of the other NASA Centers and in the early years of construction. Marshall's Construction of Facilities obligations, including those in Huntsville, New Orleans and Mississippi were second only to those at the Cape. During its first four and a half years of existence, an investment of more than $125 million was made in facilities at Marshall.[9] With this new stream of funding in place, Newby's first key task was to oversee the design and construction of a central laboratory and office building for Marshall on the Arsenal. When completed, this building would be home to Marshall's administration, which until that time would remain in Army buildings 4488 and 4484. Intense debates broke out between Newby and Hannes Luehrsen, the master planner and original member of the WvB group, who objected to the nine-story design due to the fact that the rest of the buildings on the Arsenal were three-stories. Von Braun intervened on Newby's side of the debate and in August 1961, Marshall accepted a $4 million bid for construction on the 9-story building that would soon become known as the "Von Braun Hilton." The 227,000 square-foot facility was completed and occupation began in July 1963.

Very soon, Newby's responsibilities expanded far beyond NASA property on the Arsenal to encompass facilities at Edwards Air Force Base, Seal Beach, Cal., and Cape Canaveral, Fla. Newby also served on the committee charged with locating facilities for the manufacture and testing of Saturn launch vehicles. This committee recommended the use of facilities at the Michoud plant in New Orleans, La. (now Michoud Assembly Facility) and the development of testing facilities at what would become the Mississippi Test Facility (now Stennis Space Center). Newby assisted NASA Headquarters with arrangements for the Army Corp of Engineers in building the stage test facilities in Mississippi.

In a 1965 interview in the center's newspaper the *Marshall Star*, Newby stated that when speaking to engineering graduates, he liked to point out that while employers were always eager to find employees who could use their heads, "you can't climb this ladder of success with your hands in your pockets."[10] Newby's career in the space program and as Marshall's first employee certainly stands as a testament to this statement.

THE TEAM
INVENTED MAN-RATED MISSILES
Launching an Astronaut on a Missile
By Ed Buckbee

The new commercial space flight industry was faced with designing and building a rocket capable of safely carrying humans into space. WvB and the team tackled that problem early on. Man-rating a missile to carry an astronaut was a major challenge for the U.S. Army in 1958. Alan B. Shepard, Jr. was the first astronaut to ride a man-rated missile. He was proud of the fact that not only was he the first American astronaut to climb aboard a rocket, but the only space flyer to command WvB's Redstone and Saturn V Moon rockets. He was a bit curious about the Redstone and often asked me what the WvB team did to man-rate the Redstone missile. My stock answer was, "We added more fuel and increased the amount of explosives in the destruction package." I don't think that was the answer Shepard was expecting. Actually, I wasn't sure what was done until I discovered a document in WvB's personal papers written by Joachim (Jack) Kuettner, WvB's project manager for the Mercury Redstone program at the Army Ballistic Missile Agency (ABMA). It was entitled, "The Launching of a Manned Missile" written in 1958. It must have been well done, because WvB added a handwritten note, "Very fine paper!" and signed it with his famous, "B".

I showed it to Shepard in 1989 when we were collecting artifacts and memorabilia from the Mercury astronauts for the U.S. Astronaut Hall of Fame near Kennedy Space Center, FL. He had not seen it. Following are some excerpts from the paper and responses by Shepard.

Kuettner begins, "In the launching of a manned missile, conflicts are encountered between aircraft technology and missile technology. These stem from the difference between man-controlled and automatically controlled vehicles. The possible roles of the occupant as a living payload, as a manned backup for automatic controls, or as a true pilot during the powered flight are considered. Reliability standards of missiles versus manned vehicles create problems of pilot safely. Protection of the occupant on the launching pad, after firing command and during the rest of powered flight must be made compatible with existing launching procedures. Finally the possibility is considered of training space travelers through simulated rides during static firings."

He continues, "The difficulties that will be encountered in the launching of manned missiles are primarily founded on the conflict between aircraft and missile technology. This conflict results from the differences between man-controlled and automatically controlled vehicles and is partly of historical and partly of technical origin. By putting man in a missile, the separate avenues on which these technologies have proceeded are made to intersect, bringing into sharp focus the following questions: (1) What is the role of the occupant? (2) To what degree is the safety of the occupant assured? (3) What special developments and procedures are required to integrate man into the missile system?

"The role of the occupant is, of course, that of a living payload which, unfortunately, is very sensitive and positively inexpendable. As such, missile engineers eye him with misgivings. In turn, existing boosters are, in the eyes of the occupant, barbaric machines grazing human tolerance limits in all directions. In this cargo function, the occupant is strictly a problem child."

Watching Shepard's reaction while reading this about the engineers eyeing the occupant with misgivings and being strictly a problem child brought on this comment, "What the hell is he talking about, barbaric machines, occupants, problem child, he's talking about me. Who is this guy? I think I met him one night at Wernher's house. Wasn't he a Luftwaffe fighter pilot?"

I reminded him it was written forty years ago and, more importantly, before a manned flight on a rocket. "Read on, Jose," Shepard's call sign, "it gets better," I said. He did like the part about training the occupant through simulated rides during static firings.

Kuettner continues, "More justice and more satisfaction would be given to the human occupant if he were used at least as a backup for the automatic controls. Taking missiles as they presently are and coming down to facts, we discover that there is little time for intelligent pondering during the powered phase and that, therefore, a detailed information display is of little help to the occupant. Man may be a lightweight computer of amazing versatility, but in missile and computer terms, he is outrageously slow." This brought Shepard out of his chair.

"Look, we were all trained as test pilots," Shepard says. "In aircraft we were expected to be the primary system and we proved over and over again we could react and save the aircraft. Why should it be any different with rockets? If the engineers give us the right tools and controls, we can be more than an 'occupant.' I hate that word and I'm not real fond of 'backup' either. We astronauts should always be considered the primary system when designing a system like an aircraft or rocket. As to his comment about us being 'outrageously slow', well that sounds like an engineer talking. We had many heated discussions with engineers about controlling the spacecraft and this was after Mercury Redstone. At the beginning of Gemini, we got our control in the Gemini spacecraft—over a lot of dead bodies in the ranks of the engineers. Gemini was probably the best designed spacecraft for flying in space that an astronaut would ever want to strap on and that's because we astronauts were involved in the design from the beginning."

Kuettner continues, "Man's value as an intelligent being would come fully into play, if he were at the controls of the missiles in the true sense, determining flight path, thrust, etc. as we are accustomed to seeing in space fiction stories. For existing hardware, such a function is completely out of the question. Whether or not this will change in the future will depend on whether the booster and manned payload can be designed as an integrated system. It is likely that the required large boosters will be used as versatile work horses designed to carry manned as well as unmanned upper stages, and will therefore have to incorporate automatic controls. The X-15 flights on the one side and the Mercury flights on the other side will give the first answers to this question. After separation from the booster, things may be quite different."

Shepard commented, "Now I remember this guy Kuettner. He briefed us on the Redstone. Regarding his comments about the X-15, it certainly gave us an indication of a pilot's value in being the primary system in a highly sophisticated rocket plane. Look at what Scott Crossfield, Jim Irwin and Neil Armstrong were able to do with that machine that featured rocket-powered engines, guidance and control with thrusters and fly-by wire. It didn't fly to the edge of space by itself."

"In introducing human life into missile technology, certain missile standards change entirely," Kuettner continues. "What has been considered a most successful missile may turn out to be a hazardous monster in terms of conventional flight-testing. Let us look first at the reliability angle, being careful to distinguish between mission reliability and survival reliability. As manned flights should be made only with well-tested missiles, one can base the reliability estimates on actual firing records, rather than on a paper prediction according to components. We should clearly state that we could be only 90% sure that the mission will be accomplished in 7 out of 10 cases. Or, to make it sound a little better, we may say that there is only a 5% chance that 3 out of 10 missions will fail, and there is a 5% chance that 9 out of 10 missions will be successful."

Shepard said, "Well, with those odds, I know some guys who might not have taken a ride on those rockets."

"Although this is a somewhat frightening situation from an aviation standpoint, one must realize that the survival probability would be much higher than the mission reliability. In this case, the confidence limits will be quite low for lack of sufficiently large test samples. However, pioneers have taken much greater chances in the past. Due to the nature of their undertaking, the chances have been generally unpredictable."

Shepard said, "Who said we weren't pioneers ourselves."

Kuettner continues, "In an abort system, the space capsule itself will, of course, have to sense its own internal malfunctions, but here the occupant plays more active role. Three questions arise in the design of an abort system: (1) shall the system be located in the booster or in the space capsule? (2) Should it be made redundant? (3) What information shall be displayed to the occupant in order to use him as a backup? Since the malfunctions under question originate in the booster and since manned payloads may be fired on different types of boosters, it appears naturally to locate the abort systems in the booster. As to the last question, the analysis shows that the emergency cases sensed by the automatic systems are of such critical urgency that even the most intelligent occupant will have a hard time trying to interpret displayed information. There is little time to tell him the cause of the emergency; he will just have to push the ejection button if the abort signal comes on and the automatic systems fails. This cannot be helped. Of course, the normal capsule instrumentation will give the occupant sufficient information on the general course of events."

Shepard's response, "Well, if this had been the case, Sky Ray (Wally Schirra) wouldn't be with us today. If you remember on his Gemini-Titan launch, he had an automatic shutdown on the pad and didn't hit the chicken switch (ejection controls) because he knew they didn't lift off. You know why he knew they didn't have a lift-off — because he couldn't feel it in the seat of his pants. How did he know that, because of his previous experience in his Mercury flight. My point is the engineers have to give us some credit for having the experience and skills to make decisions during such critical times. If they leave everything up to the machine, why even put us 'occupants' on board?"

Kuettner continues, "Space-medical training will take care of most of the problems to be faced by the astronauts during the space ride, except of course, prolonged weightlessness. Some special thought has been given to the time before and after fire command when the space traveler must sit on top of a huge fueled missile and wait for the rocket engines to fire, exposing him to powerful noise and vibration. How can he be prepared for this experience? Is it feasible and safe to give the passenger a simulated ride on an actual missile during a static firing?

Shepard read this and said, " Well, that's exactly what I proposed to Wernher in 1960, that we be permitted to stand atop the Redstone tower while you guys were firing it during ground tests."

"Did he let you do it", I asked.

Shepard said, "No, he didn't let me do it. Later, I heard he let some of his engineers and test conductors walk around the top of the tower to show off in front of some Washington politicians. I don't know what that proved."

Kuettner concluded, "No manned missile has as yet been fired. We are just feeling our way into an unknown terrain. The integration of man and rocket will, for sure, take its course until a clear pattern develops. The ideas discussed here originated in the minds of many members of the von Braun team at Huntsville, AL. These ideas have their basis in long experience and are necessary because the Army Ballistic Missile Agency is actually faced with the problem of launching the first man into space in the near future."

"And that first man was me," Shepard added.

How I Became a Rocket Scientist

By Charles Bradshaw

The U.S. Army's Guided Missile Development Division (GMDD) had established its personnel office at the headquarters of the old Redstone Arsenal some twelve miles south of Huntsville near the Tennessee River. I had been directed to report to these offices on a Monday in early June 1951 to be interviewed for a job assignment.

After filling out a few forms, I was escorted to a van and taken to the area where the operating elements of the GMDD were located on the Northeast side of the arsenal. The area was fenced, had a guard gate, and consisted of three two-story office buildings and two or three small buildings. The driver of the van told me that there were about 120 German-born scientists, and one hundred civil service employees, and military personnel employed by the GMDD, and the little complex where I was to work was named "Squirrel Hill."

I reported to an office on the second floor of Building 111. The door to the office was closed, and when I knocked gently on the door, I heard a grunt from inside, which I interpreted as an invitation to enter.

On entering, I saw a very serious man wielding a slide rule and looking at a large sheet of paper on his desk. I immediately assumed I was about to meet my first rocket scientist, who motioned for me to wait as he completed the series of calculations he was doing. I stood more or less at attention with the folder I had brought tucked under my arm, and it was awe-inspiring for me to stand and watch the first rocket scientist I had ever seen hard at work.

When the rocket scientist completed his series of calculations, he wrote down a couple of numbers and turned his attention to me. He asked, "*Was haven sie das unter dem arm.*"

Wow! I thought. These rocket scientist sure have funny accents. I handed him my folder, and he proceeded to look it over and gave it a couple grunts. After a minute or so, he looked at me and asked, "Mr. Bradshaw, do you truncate?"

Now I wasn't quite sure whether that was a technical question or a question about my morals, so I decided to gloss over my answer with a grunt and a mumble. The rocket scientist looked at my paper again, and then asked another question. "Mr. Bradshaw, do you root polynomials?"

I was ready for that question as my mother had once given me some advice that would help. In the mid-1930s while we were living in the zinc mining town long the Holston River in East Tennessee, I observed the dead-end cycle taking placed about me. When a boy reached sixteen, he would marry a girl from a street or two away, and the couple would move in with one of their parents. The boy would apply for a job in the mines, and a company house of his own. The children became copies of their parent, and grandparents. Having read books and stories by Rudyard Kipling, Jack London, and Mark Twain, I had made some big plans for my life, and the cycle I saw then did not fit those plans.

One day I told my mother what I had observed and that I was not going to follow the town's cycle. My plans were to go out in the big world, out beyond Knoxville, and see what was there. Mother set me down and talked rather sternly to me. She said," Son, you are a good boy, and you need to stay with us." She continued: "We have had some pretty hard times here in our town but we have always had enough to eat, and you are growing up strong and healthy. Out there in the big world where you are talking about going, it is *"dog eat dog and root hog or die"*

Remembering the words of my mother, I assured the rocket scientist I could root polynomials or anything else that might need rooting. This seemed to satisfy the rocket scientist, and he led me down the hall to an empty office, and sat me down in front of my first computer. It wasn't really a computer but a Friden electro-mechanical adding machine which could, in addition to its basic functions of adding, subtracting and multiplying, divide if you had time to wait. I sat down and began the process of learning to operate the Friden calculator, and hoped I was on my way to becoming a c rocket scientist. My boss gave me some arithmetic calculations to do on the Friden so the results could be compared with the results he obtained on the slide rule. It took me longer, but my calculations were correct to six decimal places, while only one or two decimal places were good on the slide rule. In a couple of days, my boss told me I was moving to another office at the end of the hall, working for another rocket scientist. My new boss was Dr. Hans Friedrich, who had been a university professor in Germany before joining the WvB group in the rocket development program at Peenemunde.

I had been assigned to him as a student to learn aerodynamics and flight mechanics. It soon became obvious that Dr. Friedrich was an outstanding teacher. I was to be taught in the old-world style of sitting at the professor's feet in one-on-one learning sessions. My desk was placed directly in front of Dr. Friedrich's so we face each other all day.

Every morning I would move my chair beside him, and he would lecture while making notes on a pad. After about an hour or so of this, I would take the notes back to my desk and spend the rest of the morning studying and transcribing them into the form of a technical paper. In the early afternoon, I would move my chair back to Dr. Friedrich, who would go over my paper to determine if I had captured, and fully understood the points of his morning lecture. We would then go on the next subject, and the process would be repeated.

We continued in this manner for two or three weeks. When I had a question, I would raise my hand and ask. This turned out to be the most productive learning situation I was ever to experience.

After I had been with Dr. Friedrich for about three weeks, he suggested I spend a few days with Dr. Adolph Thiel to study the engineering aspects of rockets. The time with Dr. Thiel was less structured than the time with Dr. Friedrich. We first discussed the three essential engineering components of constructing a guided missile, which were in their order of engineering complexity and importance: the rocket motor, the missile guidance system, and the fuel tanks with associated plumbing and controls. Dr. Thiel gave me a few internal documents about the early days of rocket development and asked that I spend some time going through them.

As part of my tutorial with Dr. Thiel, I spent four or five days going through the papers provided on the history of rockets, made an outline of the information, and then went through my notes with Dr. Thiel.

The U. S. Army was in the process of compiling a technical library containing many of the papers brought to the United States from Germany along with a few other books. These proved to be adequate for a cursory study of the history of rockets.

The mentoring I received from these world-renowned German scientists, Friedrich and Thiel inspired me to embark on a career involving rockets, reactors and computers.

Editors Note: Charles Bradshaw served as the Chief of Computation of the Aeroballistics Laboratory of the GMDD. Later, he became Branch Chief of the Computation Laboratory of ABMA. When the transfer to NASA occurred, he served as the Deputy to Director Helmut Hoelzer of the Computation Laboratory.

Part II
Organization

Space Sciences

By Robert Naumann

Editor's Note: The WvB team was more than "fire and smoke"' Robert Naumann spent his career contributing to significant science projects at Marshall. Here he describes projects of the early days and introduces several other science members of the team.

When the WvB team moved to Redstone Arsenal in 1950 as the Guided Missile Development Division (GMDD) of the Ordnance Missile Laboratories (OML), its primary assignment was the development of the Redstone missile, a vehicle with a massive payload capability needed to accommodate the nuclear weapons of that day, but with a range of only 150 miles. Included in this operation was the Technical Feasibility Study Office (TSFO) which, in a sense, was to become the forerunner of the George C. Marshall (MSFC) Space Sciences Laboratory. The TSFO, under the direction of Adolph (Dolph) Thiel, reported to the Director of OML, Col. Miles Chatfield, and was tasked with long range planning for the Army guided missile program. One of the topics being studied was the feasibility of extending the range of such vehicles to 1,500 miles (intermediate range ballistic missiles or IRBMs) and eventually to 3,000 miles (intercontinental ballistic missiles or ICBMs).

Meanwhile, a major effort was under way to develop a nose cone that could survive the searing heat of reentry of a long range missile. This had not been a problem for the V-2 and Redstone missiles, because their reentry speeds had not been high enough to require special materials, but it would become a serious problem for 1,500-mile IRBMs and 3,000-mile ICBMs. It was known that meteors that survived their fiery plunge to Earth were actually cold inside, the heat of the reentry having been carried away by melting and ablating away the material on their surface. So, a search was on for a suitable material from which to fabricate an ablative cooled nose cone.

Gerhard Heller, a Scientific Assistant to WvB, had been using a ramjet test stand at White Sands to test various materials' resistance to high temperature gas streams, both for reentry cones and for materials to use as jet vanes in the nozzles of rocket engines. This was before rocket engines with swivel nozzles were developed, and both the Redstone and V-2 used graphite vanes in the exhaust stream to steer the vehicle. Wolfgang Stuerer's Materials Branch in the GMDD had been using the Redstone engine test stand for similar testing after the team moved to Huntsville. In 1951, a young Army draftee, Pvt. Jim Kingsbury, with an electrical engineering degree from Penn State was assigned to Stuerer's group. After he finished his tour, he returned as a civilian and started testing materials to be used as jet vanes in the exhaust of a Redstone engine. He decided to try Micarta, a pressed composite of linen, canvas, paper, fiberglass, and carbon fibers in thermosetting resin developed by George Westinghouse in 1910. It is widely used as an electrical insulator and in printed circuit boards.

As Kingsbury relates it, the first test completely destroyed the material, but he had more vanes to test so he put another Micarta vane in the test holder just to fill the space. The engine shut down prematurely and he noticed that he could scrape some of the burnt material away and there was undamaged material underneath. Kingsbury realized he had found a material suitable for a reentry cone, and Micarta was used for both the Jupiter-C and the Jupiter nose cones. Apparently, the secret was to find materials that were strong enough to withstand the flow but that had a low enough thermal conductivity to keep the heat from penetrating into the interior.

As a side note, Harry Julian Allen at the National Advisory Committee for Aeronautics (NACA) Ames Research Center (which became part of NASA in 1958) had developed the blunt body theory for a reentry cone. The shock wave from a blunt body reentering the atmosphere detaches itself from the reentry body, thus reducing the heat transferred to the body. The Air Force had adopted this concept for the reentry cones for the Thor, Atlas and Titan, and used a heat sink to absorb the heat that was transferred to the body. All of the Mercury, Gemini, and Apollo designs used a combination of a blunt body with ablative-cooled heat shields on their return capsules.

In 1953, WvB issued a memorandum stating that it was feasible to build a missile with a 1,000-mile range and requested permission to start working on it. However, the Army showed little interest in such a project and Secretary of Defense Charles Wilson imposed a range limit of 200 miles on Army missiles. Anything beyond that belonged to the Air Force or the Navy.

In June 1954, WvB proposed the use of existing Army hardware (an extended Redstone booster with upper stages of clustered 2.5-inch diameter Loki rockets) to launch a satellite. The Army was quite interested in this possibility at the time, because of the potential military applications. Pvt. Charles (Chuck) Lundquist, another Army draftee, and Col. Nickerson wrote a white paper showing satellites could be used for communications. Lundquist had joined the TFSO in the fall of 1954. After completing his dissertation at the University of Kansas in 1953 on a radiative transport problem posed by Chandrasekhar, he accepted a position as an Assistant Professor of Engineering at Penn State University, where he worked on a torpedo guidance systems contract. In the summer of 1954 he was drafted into the Army's Science and Professional program. He was assigned to the TFSO at Redstone, where he developed the orbital equations of motion that were later programmed by the Computational Lab and used to predict the orbits and provide the position data needed to track the early Explorer satellites.

As Chuck describes it, the office was a large bullpen with about a dozen desks in the same building as the old Redstone Library and the Rocket Auditorium. There was only a single phone line, which one of the people was continually using for personal business. Any time the phone would ring, the other workers would scramble to answer it so they could say that the phone hog didn't work there, which they could do with a straight face.

Dolph Thiel left the Army in 1955 and went to the Space Technology Laboratory (STL) (which later became a division of Thompson Ramo Woldridge,

Inc., which eventually became TRW), where he became program manager for the Thor intermediate range ballistic missile (IRBM). Ernst Stuhlinger, who had been working in the Guidance and Control Laboratory on the guidance system for the Redstone missile, became director of the TFSO Office. The famous German rocket scientist Hermann Oberth came to the U.S. later that year and joined the TFSO.

Shortly thereafter, the planning committee for the International Geophysical Year (IGY) called for the launching of a satellite as part of the IGY activities. The Army with the strong support of Commander George W. Hoover from the Office of Naval Research (ONR) offered WvB's proposal as Project Orbiter. In the meantime, two more proposals were received, one from the Naval Research Laboratory (NRL) using the Vanguard (a new rocket design based on the Viking missile), and one from the Air Force based on the Atlas. On August 3, 1955, the Stewart Committee voted to reject Project Orbiter, citing a poor tracking system and inferior scientific value. A factor in this decision may have been President Eisenhower's reluctance to involve the military in any vehicle over-flying the Soviet Union because of Khrushchev's violent objection to his Open Skies Plan (and the fact that we were flying clandestine U-2 flights over the Soviet Union at the time).

The NRL plan was selected and placed under the direction of the National Academy of Sciences (NAS). The civilian managers were not to solicit any data from existing military ballistic missile programs. The Army appealed the ruling, but the appeal was rejected and the Army was directed to cease all work on space activities. The committee chair, Homer Stewart, reportedly told WvB, "We have pulled a great boner."

Now the WvB team had been ground ruled out of both long range missiles and space vehicles. The Air Force had given Convair some money under Project Atlas to study the feasibility of building an ICBM, but most of their money had gone to developing cruise missiles such as the Snark and the Navajo, both of which were dismal failures (the Atlantic Ocean near Cape Canaveral had been described as "snark infested waters"). Part of the problem was that high yield nuclear weapons then were heavy and bulky, and it was by no means certain that rockets could deliver such heavy payloads over intercontinental distances.

However, in February1954, the von Neumann committee reported that with the recent breakthrough in developing smaller and lighter thermonuclear weapons, an ICBM could be built and both this committee and the RAND report recommended acceleration of the Atlas program. In May 1955, the Air Research and Development Command authorized the development of the Titan as a backup to the Atlas Program. Because of the Soviet Union's advances in nuclear technology and heavy lift rockets, the Killian Committee in February 1955 recommended not only that the ICBM program be given top National Priority, but also recommended that 1,500-mile IRBMs be built concurrently. As a result the Air Force was given the responsibility for the Thor and Army was given responsibility for developing the Jupiter IRBM to be used by the Navy as a shipboard or submarine-launched missile.

The Army Ballistic Missile Agency (ABMA)

General John B. Medaris was given full responsibility for the development of the Jupiter Program and the Army Ballistic Missile Agency was created from the GMDD on February 1, 1956 to develop the Jupiter missile. There were two parts to the Jupiter program: Jupiter-A and Jupiter-C (C for composite; there was no Jupiter-B). Jupiter-A focused on the main configuration, engine, thrust control, and guidance system while Jupiter-C focused on the reentry problem. It consisted of an elongated Redstone fitted with clusters of scaled down 6-inch diameter Sergeant solid rockets (instead of the smaller Loki rockets) in a spinning tub for the second and third stages to provide the high speed reentry conditions typical of a 1,500-mile ballistic missile. In September of 1956, a Jupiter-C launched a payload (an inert Sergeant rocket) that traveled 3,300 miles. Legend has it that Secretary Wilson had the dummy fourth stage rocket inspected to make sure it was not live so that WvB could not "accidentally" launch a satellite.

The Jupiter program almost ended in early 1957 when the Navy decided that they didn't want to deal with LOX and RP-1 on their ships or submarines and pulled out of their agreement with ABMA in order to develop their own solid propellant submarine-launched Polaris IRBM. This left the Jupiter with no real mission. However, the Thor as well as the Atlas and Titan were having their problems with frequent explosions on the launch pad or in flight, and it was decided to keep the Jupiter program alive as an alternative to the Thor. The difficulties in the early missile programs were exemplified by a comment made by WvB, "It is our objective to create a weapon that is more dangerous to the enemy than to those launching it."

Given the success of the 3,300-mile Jupiter-C flight, the Army again requested permission to launch six satellites using the Jupiter-C, and the request was again rejected. In August of 1957, a Jupiter-C launched a simulated Jupiter-A nose cone over 1,500-miles and successfully recovered it, thus proving the feasibility of an IRBM with an ablative-cooled nose cone. Following this launch, General Medaris ordered the remaining Jupiter-C vehicles to be placed in protective storage.

The new Secretary of Defense nominee, Neil McElroy, was visiting ABMA and having dinner with WvB and General Medaris on October 4, 1957, when news broke of the successful launch of the Sputnik by the Soviet Union. A plea was made to the new Secretary to give the Army a go-ahead with a satellite launch using the stored Jupiter-C vehicles. WvB said he could be ready in 60 days; Medaris prudently suggested 90 days. On October 30, Eisenhower agreed to an Army attempt as a backup to the Vanguard, whose launch had been moved up to December 1957. Shortly after the launch of Sputnik 2 on November 8, Medaris received an order to "prepare" for a launch with the caveat that the order may be rescinded if Vanguard was successful. WvB, Medaris and William Pickering (Director of the Jet Propulsion Laboratory, JPL), tendered their resignations to Secretary of the Army William Brucker unless they were given clear orders to proceed with a satellite launch regardless of the success of the Vanguard. On November 21, Medaris received the order to proceed and a launch date was set for January 29, 1958.

Right after October 4, Medaris had stuck his neck out and secretly instructed WvB to remove two Jupiter-C vehicles from storage and prepare them for shipment to the Cape. Meanwhile Pickering had JPL busy developing the satellite, which was to become Explorer *I*.

(As a side note, just after the Soviets launched their Sputnik in October 1957, President Eisenhower ordered full production of both the Thor and Jupiter, but the Jupiters were placed under control of the Air Force. The Air Force deployed the Thors in England and the Jupiters in Italy and Turkey. The first Jupiter base in Turkey was declared operational in April 1962. Ironically, it was the deployment of the Jupiter missiles in Turkey, close to the Soviet border, that prompted Soviet Premier Nikita Khrushchev to deploy nuclear tipped missiles in Cuba, which sparked the Cuban missile crisis. The crisis was resolved when President Kennedy secretly agreed to remove the Jupiters from Turkey and deactivate the other IRBMs deployed in Europe, thus avoiding a possible nuclear confrontation with the Soviet Union.)

The Research Projects Office (RPO)

When ABMA was created, WvB set up the Research Projects Office (RPO) to function much as the earlier Technical Feasibility Study Office, and named Stuhlinger as its Director. Lundquist and Oberth came over with Stuhlinger from the old TFSO organization. Joining them from the GMDD were Gerhard Heller, Rudy Schlidt, and Wolfgang Stuerer with his materials group — Spencer Frary, Hank Martin, and Bill Adams. Ruth Christopher was the Secretary and Dick Potter, affectionately known as "Potter the Plotter", provided technical support in the form of graphs and illustrations. George Bucher transferred from the Army Ordnance Depot in St. Louis shortly after ABMA was created. He had a background in nuclear engineering, but was able to take advantage of the Army's education program to go to Oklahoma State University to get his doctorate in management with the revolutionary belief that managers with a technical background should manage engineers and scientists. He became Stuhlinger's assistant and remained so until Stuhlinger's retirement. Stuerer left shortly thereafter to join private industry. Kingsbury elected to stay with the Materials Lab.

The Research Projects Office grew rapidly into a number of diverse fields to cover many aspects of the emerging roles that ABMA and later MSFC were to play in the future. George Bucher had the distinction of having the best looking secretary in the Agency. The RPO was adjacent to the Personnel Office, and one morning he noticed this tall attractive lady applying for a job and instantly hired Landa Thornton. George also hired Jim Downey, who had been trained as a Nuclear Engineer by the Air Force. One of Jim's first tasks was to gather research requirements for all of the labs at the Agency, package them, and go to Washington to try to get money to carry out the research tasks, a skill at which he became very adept.

Shortly after ABMA was formed, Steurer interviewed John Bensko, who had a background in mining, petroleum geology, and metallurgy. John had heard about the potential for research with ABMA from former coworkers Richard Stein and Gene Cataldo, who had worked for Rudy Schlidt and for Wolfgang Stuerer before

ABMA was created. John's interview with Wolfgang Steurer went very well. Steurer shared WvB's zeal for space exploration, despite ABMA's limits. However, Research Projects Office had no laboratory facilities and was small compared to the other administrative Divisions at ABMA. Steurer explained that they were an applied research group and that their effort was in support of missions determined by other elements of ABMA. Personal research would be coupled with designing research ideas for others to be performed by contract. It was a disappointment to Bensko to find that the Army team was not quite ready to shove off for the Moon, despite WvB's public articles on Lunar and Martian exploration. He was surprised to find that ABMA was not really into space exploration at that time but was a rocket development organization. However, Steurer saw his group as having broader objectives, beyond those that were spelled out officially, and there was no problem with thinking beyond their current objectives. There was no NASA space program or Advanced Research Projects Agency (ARPA) at the time, and the only prospect for doing space-related work was right there with Steurer's group. Steurer's personal interest in lunar and planetary environments remained years after leaving ABMA.

Bensko, the only geologist, shared an office with Hank Martin, a metallurgist, along with Bill Adams and Spencer Frary, who were both chemists. Spencer Frary's interest was in the upper atmosphere and in the effects on materials exposed to a high vacuum. Hank Martin was concerned with reentry problems before he shifted his interest to the effects of micrometeorite impact. Bill Adams assisted Steurer, and helped anyone who needed help. They were shortly joined by Gentry Miles and Dan Gates. Gentry served as a liaison with the Lawrence Radiation Laboratory at Livermore (now Lawrence Livermore National Laboratory, LLNL), to provide information on radiation effects. Dan was a ceramics engineer with a specialty in coatings, and was instrumental in the development of the thermal control coatings used on the early satellites, several of which are still used today. Frary had worked for the old chemical warfare group at Redstone during WWII, and recalled that periodically a switch engine would deliver a tank car full of pure ethanol to the building where he worked. After the ethanol was unloaded, the workers persuaded the engineer to move the tank car to a shady part of the siding. By the end of the day the vapors in the tank car would have condensed so that it was possible to draw off several pints of pure grain alcohol for personal consumption. Since Madison County was dry in those days, this amounted to a valuable perk for those working in this group.

Frary and Bensko were part of the High Altitude Committee, an informal group set up at ABMA to discuss space projects. Klaus Schocken from the Thermal group also participated in the meetings. After ARPA was formed in February of 1958, some Research Projects members who were interested in lunar exploration took an active part in project Horizon, a classified ARPA-approved study directed by Hermann Koelle (Structures and Mechanics) on the setting up of a military base on the Moon.

After Steurer left ABMA, Dan Gates was placed in charge of his group. Stuhlinger moved Gentry Miles to the front office and tasked John Bensko to develop a proposal to NASA to soft-land an instrument package on the Moon's surface.

The atmosphere in the Research Projects Office was very collegial. Landa Thornton recalls, "A very sweet memory is of the way we took a few minutes out of the day to learn more about each other. We knew the name of each other's sweetheart, wife, husband, children (by name) who was ill, etc., and we were all friends." Don Cochran, a young engineering summer student from Clemson remembers, "Landa adopted me, and helped me a lot to learn the ropes there and always remained a good friend of mine." Don was working for Wolfgang Steurer and as he tells it, "My primary responsibility that summer was testing nose cone materials in the high temperature burner (a rocket engine). The testing was performed on the outside with just a thick wall barrier between me and the rocket engine, which was very, very loud. I was testing a nose cone one day when someone tapped me on the shoulder as the rocket engine cut off and said, 'Mr. Cochran, you are doing a good job'. The person who tapped me on the shoulder and who knew about my work was none other than Dr. von Braun. It both surprised me and thrilled me to meet von Braun. The testing was so loud, I didn't even know he had come up behind me. He was looking over my shoulder through the same thick portal during the test."

Hermann Oberth is considered along with the American, Robert Goddard, and the Russian, Konstantin Tsiolkovsky, to be one of the Fathers of Rocketry. He was not only a visionary, but also produced calculations to show that his dreams were theoretically possible, if not practical for his day and time. His doctoral dissertation, "Die Rakete zu den Planetenräumen" ("By Rocket into Planetary Space") was rejected as being too futuristic. Later George Bucher was able to arrange an honorary doctorate for him from Iowa Wesleyan University. Oberth expanded on his original dissertation to publish "Menschen im Weltraum" ("Men in Space"), and participated in writing the study, "The Development of Space Technology in the Next Ten Years" while at ABMA. In July 1955, *Life Magazine* published an article on a manned space station written by Oberth and other German scientists. He was offered an endowed Chair at St. Louis University, but he decided to return to Germany to accept a small pension he had earned from teaching before the war.

Professor Oberth was very much a stereotype of an eccentric professor. He wrote an article for *The American Weekly* in 1954 in which he stated his belief that flying saucers are manned by intelligent beings that have been investigating Earth for centuries. When he was working in the old TFSO bullpen, every day at noon he would set his alarm clock for 30 minutes and put his head on his desk to take a 30-minute nap. He was also known to write and draw on brown paper grocery bags. Every morning when he pulled his old car into the parking lot, he would get out and carefully place pieces of cardboard in the wheel wells to cover his tires. No one could ever figure out why. Maybe he thought the sunlight would deteriorate the rubber in his tires. Also, he was a novice driver, having learned to drive after he came to the U.S. In those days there were guard shacks in the middle of the roads leading into the classified plants area of the Arsenal. One morning Oberth was apparently preoccupied with some astronautical problem, and managed to take out one of the guard shacks. Fortunately, the guard saw him coming and bailed in the nick of time. Landa Thornton recalls another amusing incident. The secretaries in the office had been collecting stubs of pencils to be given to the Harris Home for Children. As she recalls, "I briefly explained about collecting the used pencils, asked Professor Oberth if he had any used, short pencils. He said, 'No,' in his heavy accent, reached in the bag, took a handful of pencils, and said, 'Thank you.' What could I say?"

The Jupiter-C Program

Stuhlingers's dream had always been electric propulsion, ranging from automobiles to spacecraft and this new organization gave him an opportunity to actively pursue this dream. However, first things first: the development of the Jupiter-C. The spinning upper stages had to be oriented by the Redstone booster prior to their separation, as the whole vehicle coasted to its apogee after booster cutoff. The time from cutoff to apex was not always predictable in those days and depended on the booster performance. There was only about an 80-second time frame from booster cutoff to the time a command had to be given to fire the upper stages at the apex. Estimates of the time to apex were made from the measured booster performance, the radar tracking data, and the Doppler tracking data. Stuhlinger built a simple, but elegant, analog computer that took the weighted average of these three times with their confidence interval and started a timer that would fire the upper stage. A manual override switch was also provided in case the device failed. This method was used to launch all of the Jupiter-C upper stages.

Tracking the upper stages of the Jupiter-C also presented some new challenges. Remember, this was the first time anything had been put into a ballistic trajectory of this length. To back up the range radar data, these upper stages contained a series of Daisy flashes, pyrotechnics that could be fired in a precise time sequence and recorded on tracking cameras to get a precise measure of the burnout velocity. There was also the need to photograph the reentry cone to verify that it remained intact and to aid in its recovery. To assist in these tasks, Chuck Lunquist, who headed the Physics and Astrophysics Branch, brought in Ray Hembree and Dave Woodbridge. Woodbridge was a Meteorologist with an interest in the link between solar activity and the weather. Hembree had been head of the Photo Lab with a background at the Navy Air Weapons Center at China Lake, where he had been photographing the Navy's missiles.

Ray also served as Lundquist's Deputy. He was not only an able administrator, but was calming influence on upstart budding young scientists such as myself (Bob Naumann), who were always ranting against government regulations and red tape. One day Ray took me aside and said, "Son, when you are walking through the woods and come to a tree, it is much easier to go around it than to try to go through it." I'll never forget those words of wisdom.

Several amusing anecdotes told by Lundquist occurred shortly after Medaris took charge of ABMA when Chuck was still in the Army. The general came looking for Pvt. Lundquist, who was not at his desk. The general was informed that Pvt. Lundquist was on KP that day, but they could go and get him if the general needed to see him. On the day he was discharged, Chuck said, "I changed clothes and got a tenfold increase in salary."

Another time during the planning for a night launch, there was concern that the full Moon might interfere with the optical tracking. Medaris, anxious to be a useful member of the team, asked if there was anything he could do to help, to which WvB replied, "Well, General, could you do something about the Moon?"

Satellite Thermal Control

Gerhard Heller was named as Stuhlinger's Deputy, and was also put in charge of the Thermo Physics Branch. This branch supported Stuhlinger's work on ion and plasma propulsion as well as studies on the thermal and optical properties of lunar soil, but its primary mission was the thermal design of all of the satellites lunched by ABMA. Again, this was a pioneering effort since no one had ever done this before. The design had to be totally passive, as there was no weight to spare for any active thermal control on these early satellites.

The problem is a simple radiation transfer problem where the satellite receives sunlight from the Sun plus the sunlight that is reflected from the Earth plus the infrared radiation from the Earth and whatever internal heat is generated from its electronics. Heat rejection is through infrared radiation from the surface of the satellite. The equilibrium temperature is determined by the optical properties of the satellite's surface, i.e. the ratio of its alpha, the fraction of sunlight absorbed to its epsilon, the fraction absorbed or emitted at thermal infrared wavelengths. The problem is that the satellite goes in and out of the Earth's shadow so the *a/e* has to be chosen so that the internal batteries won't freeze in the shadow portion or overheat in the sunlight portion of the orbit. It was difficult at that time to find an *a/e* that would withstand 100% time in the Sun without freezing in the shadow, so the early Explorer satellites were designed to avoid orbits that placed them in 100% sunlight during their operational lifetimes.

Dan Gates along with Gene Zerlaut in the Materials Lab had started a program with the Illinois Institute of Technology Research Institute (IITRI) to develop thermal control coatings. Later supported by Ed Miller and others in Bill Snoddy's Group, they set out to develop surface coatings that had a wide range of *a/e* values. This effort led to a series of zinc oxide pigmented silicone-based and silicate-based paints such as S-13G and Z-93 that were to be used in many future satellites. Most paints deteriorate rapidly under exposure to vacuum ultraviolet light but, as it turned out, these coatings (especially Z-93, which is used on the International Space Station radiators) remained remarkably stable in the harsh radiation and atomic oxygen environment of low earth orbits (LEO), although the combination of these effects was not understood or appreciated at the time.

Electric Propulsion

Stuhlinger's dream of electric propulsion for spacecraft would allow interplanetary vehicles to carry much greater payloads to Mars and beyond than conventional chemical rockets. The famous rocket equations tell us that a rocket's final velocity (neglecting gravitational and drag losses) is given by its exhaust velocity times the natural logarithm of its mass ratio (ratio of its initial to its final weight, the difference being the amount of fuel expended). The exhaust velocity of a conventional H2-LOX engine is ~4,000 m/s, whereas an ion engine could have an exhaust velocity as high as 100,000 m/s. Unfortunately, the thrust of such engines is very small so they are not suited for lifting payloads into Earth orbit. However, once in Earth orbit they can slowly accelerate a spacecraft to the escape velocity needed to leave Earth's gravity and fly to other planets. The difference between orbital velocity and escape velocity is about 3.2 km/s. To achieve this with a chemical rocket would require a mass ratio of 2.22, meaning that 55% of the departure weight would have to be fuel. With an ion rocket, the mass ratio would have to be only 1.03, meaning only 3% of the departure weight would have to be fuel. However, an enormous power source in the form of a nuclear reactor with its shielding or large solar collectors would be required to accelerate the ions. Also, very large radiators would also be required to reject the waste heat, so the advantage of electric propulsion is not as large as it might seem. Even so, various forms of electric propulsion are presently being considered for future deep space missions.

To meet the huge power demands of electric propulsion schemes, the Rover program was started in 1955, in which the Los Alamos Scientific Laboratory (now the Los Alamos National Laboratory or LANL) and the Air Force set out to develop a nuclear rocket engine for missile applications. This program was transferred to the newly formed NASA in 1958 and would become the Nuclear Engine for Rocket Vehicle Applications (NERVA) for which MSFC would be given the responsibility for the Reactor In Flight Test (RIFT) upper stage for the Saturn vehicle. Russ Shelton had been hired to head the Nuclear and Ion Physics Branch to develop shielding techniques for dealing with the radiation from this powerful reactor. Although the NERVA program was cancelled in 1972, the shielding models developed by this team played a major role in all of NASA's manned spaceflight endeavors and are still used today.

Because the RPO had no in-house laboratory capabilities, all of the actual development and testing of ion and plasma engines were done by contractors such as Hughes, Electro Optical Systems (EOS), Plasmadyne, etc., while the theoretical aspects were handled by Bob Seitz and other in-house scientists. Jim Downey had the responsibility of coordinating all of these efforts until the program was transferred to the Lewis Research Center in August of 1961.

It should be remembered that the Army officially sanctioned none of this work because the Army had no official role in space after Project Orbiter was rejected. Everyone in the lab had secret clearances, and although very little of the work that was being done was actually classified (except for the actual launch times), the workers were encouraged not to talk about it on the outside. Col. John Nickerson released a classified memo detailing the capabilities of the ABMA team with the hopes of changing Secretary Wilson's restrictions on the range of Army missiles, and was court marshaled for his efforts.

Explorer I

James Van Allen (State University of Iowa) had worked closely with WvB and Stuhlinger to fly his Geiger counter cosmic ray detectors on the V-2s at White Sands. (Incidentally, Stuhlinger did his thesis work under Hans Geiger.) Lundquist recalls that as a student he heard Van Allen give a paper on his rocket research in cosmic rays at an American Physical Society meeting chaired by none other than J. Robert Oppenheimer. In May 1954, Stuhlinger went to meet with Van Allen at his home and revealed the Army's plan to launch a satellite. In November 1956, Stuhlinger phoned Van Allen of the success of the Jupiter-C and suggested he fly his cosmic ray experiment on the Jupiter-C in the event the Army was directed to launch a satellite. Van Allen's cosmic ray experiment had already been given top priority for inclusion in the IGY satellite plans by the Working Group on International Instrumentation (WGII). Van Allen agreed and designed his experiment package to be compatible with both the Vanguard and the Jupiter-C spacecraft. Originally it was planned to have the payload built by ABMA, but on November 9, 1957, William Pickering, the Director of JPL, persuaded Army officials to give the responsibility for the payload to JPL.

When the Army was given the go-ahead to launch their satellite, Van Allen approved his experiment to be transferred from the Vanguard to the Jupiter-C and his graduate student, George Ludwig, loaded his experiment equipment and his family in his car and headed to Pasadena to integrate Van Allen's experiment into what was to become Explorer I.

The Explorer I flight produced two unexpected results. The most important result was the behavior of Van Allen's Geiger counters. They behaved normally near perigee and showed an increasing count rate as the altitude increased, but then the counters exhibited no counts. Van Allen's graduate student, Carl McIlwain, pointed out that the counters would jam when the count rate exceeded their recovery times and that the unexpected behavior could be explained by the presence of intense radiation. Van Allen later interpreted these results as the presence of charged particles trapped on the Earth's magnetic lines of force, which became known as the Van Allen radiation belt.

The other unexpected result was rotational behavior of the Explorer I satellite (although it shouldn't have been unexpected). The fourth stage rocket casing stayed attached to the payload, giving it a cylindrical shape with a high aspect (length much greater than the diameter). The entire upper stage assembly is spun to stabilize it during its coast to apogee and the firing of the various stages, leaving the final payload spinning along its long skinny axis. It was expected that the satellite would remain spinning about this axis and would eventually slow down due to magnetic damping. Because he was counting on the spin to even out the heat input from the Sun, Heller had commissioned a study by Prof. Smythe at Cal. Tech. to calculate the magnetic damping rate to make sure the payload maintained sufficient spin during its expected lifetime.

Shortly after orbit insertion it was apparent from the antenna patterns that instead of maintaining its axial spin, the satellite was tumbling propeller-like. Lundquist quickly realized that this was the result of mechanical energy loss, probably due to the flexing of the stiff wires that formed the turnstile antenna. An object spinning about its minimum moment of inertia (the long axis) is in its maximum energy state. To conserve angular momentum, any energy loss due to mechanical flexing requires lowering the rotation rate while increasing the effective moment of inertia, the end result being a slower rotation about the maximum moment of inertia or a propeller-like spin.

Explorer II was essentially identical to Explorer I except its Geiger counters were adjusted to accept a much higher count rate without jamming. It was launched on March 5, 1958, but unfortunately its fourth stage did not ignite and it failed to reach orbit. This was followed by Explorer III, which was successfully launched on March 25, 1958. Explorer III carried a tape recorder to provide data throughout the entire orbit and confirmed the existence of the intense radiation that was suspected from the Explorer 1 data.

Cow Pasture, operated by ABMA, was one of the primary tracking stations for these early Explorer satellites (our antenna test facility and tracking station was located literally in a cow pasture – cows graze pretty much over most of the Arsenal, which had been farms before the Army appropriated it at the beginning of WWII). Every ninety minutes or so when a pass was due, someone would go to the tracking station to get the strip charts from latest pass and bring it to the lab for initial analysis before time-tagging it with the recorded signal from WWV, matching it with the position of the satellite from the ephemeris provided by the Comp Lab, and sending it to Van Allen. It was from the recorded signal strength on these strip charts that Lundquist noticed the tumbling motion of the spacecraft.

Explorer IV

Bob Naumann joined ABMA in February 1957 and spent his first year in the Structures and Mechanics Lab researching what was known about the potential hazard that micrometeoroids presented to spacecraft. He had just returned from a semester at the University of Alabama, finishing his MS course work, and started to work analyzing the rotational motions of Explorers I and III in Lundquist's Branch. He became curious about the green badges signifying Top Secret clearances worn by the senior members of the RPO. Obviously something exciting was in the making, perhaps involving the launch of Explorer IV scheduled for some time in July. The turnstile antennas were removed from Explorer IV with the thought that the transition the axial spin to a flat spin could be avoided or at least delayed. In addition, one of the Geiger counters that had been flown on the earlier Explorers had been replaced with a directional scintillation counter.

Naumann was joined by Bill Snoddy, a summer student with a fresh BS in Physics from the University of Alabama, who was assigned to Heller's branch. Billy Jones also came over from the Technical and Engineering Division in Ordnance Missile Lab to work with Heller. Heller had been concerned that the temperatures on the Explorers were quite different from his predictions and assigned Bill Snoddy to find out why. In order to develop Bill's skills in satellite temperature control (and to check his own model), Heller had him start from scratch and develop the theory of satellite thermal control from first principles. Bill found that the temperature of the satellite was very much dependent on the angle the long axis makes with the sun and showed that the measured temperatures could be reconciled with his model by assuming the satellite was in the tumbling mode rather than spinning about its long axis as had originally been assumed.

Meanwhile preparations were underway for the launch of Explorer IV in late July. Snoddy had been spending mornings over in the Guidance and Control Laboratory overseeing the thermal testing of Explorer IV. It was emphasized that it was essential that the spacecraft last at least 60 days, which seemed to have something to do with the green badges worn by the higher level employees in the Lab. This meant that the launch time had to be selected so that the orbit would not go into 100% sunlight for 60 days. The launch hour sets the angle between the Sun relative to the orbit plane, but this angle changes as the Earth moves around the Sun. Also, the oblateness of the Earth causes the orbit plane to precess around the equator and the line of apsodies (line between the perigee and apogee) to slowly rotate within the orbit plane, all of which must be considered in selecting the launch hour. Don Cochran, working with "Snuffy" Smith in the Comp Lab, had plotted the time to reach 100% sunlight against the launch hour in order to determine the launch window. Bill Snoddy had been looking over these plots and was struck by the abrupt changes that would supposedly occur in rather short times of only a few hours of the launch time. As he tells it,

> "It just didn't seem to make sense. I pointed out to Bob the fact that I simply couldn't understand how the shadow times could be changing in such a strange manner. The calculations had been made by another group, and so we had no convenient way of checking their equations. However, Bob suggested that we simply make ourselves a little model and see if it agreed with the calculations. This we did using a standard globe Earth and cutting a model of the orbit out of cardboard and placing it around the globe.
>
> "Bob and I then simply set the orbit up as it would be for a particular launch time and then slowly rotated it to simulate the 5 degree rotation it would have each day. One of us would pretend he was the sun and would stand as far back as we could get and see how much of the orbit we could see. The part we couldn't see would represent the part that would be in the earth's shadow. Sure enough, the numbers we came up with did not agree with the calculations. Turns out the calculations were wrong due to a miss-punched card in the computational deck, as we found later. This created a bit of a panic because the time of the upcoming launch, which was only a few days away, had been based on these incorrect calculations.
>
> "Matters then became even more interesting because our boss, Mr. Heller, was on vacation hiking in the Smoky Mountains with his family. Efforts to locate him were unsuccessful. Something had to be done and before we knew it, Bob and I found ourselves in a meeting with Dr. von Braun's Deputy, Dr. Eberhard Rees. Bob and I explained to Dr. Rees our

findings, and he gave us the job of coming up with the correct launch time for the Explorer IV launch.

"So we set out to verify our little model with computer computations of our own. Bob worked up a simplified set of equations, which we programmed into our office's brand new Burroughs E101 Computer. This desk size computer was the first one that our office had ever seen and was limited to a mathematical formula requiring no more than about a dozen calculations. Unfortunately our formulas had more than that. There was no time to find a more capable computer, and so we personally became part of the process. Three of us (Bob, me, and Mary Jo Smith – another summer employee) each took on a role in the calculations. I believe I was the square root routine, Bob was the sine and cosine routine, and Mary Jo was the arcsine and arccosine routine. The computer would start grinding away and when it needed a square root, for example, it would give me the number and stop, I would look up the square root in a book of tables and type in the correct number and then the computer would start up again. Same for the sines and cosines, etc.

"By working all night we were able to confirm that, indeed, the new launch times as determined by our little globe were correct and were able to come up with a new launch window."

Since Heller had signed off on the launch time, it was necessary for him to approve any change. Finally, the Park Rangers were able to locate him and his family and he was told to call his office. The situation was explained to him and he verbally approved the change in launch time based on the information the kids who worked for him had provided. The Explorer IV launch was successful, and it provided the primary data for the Argus experiment. Its role became even more crucial when its backup, Explorer V, failed to achieve orbit a few weeks later. There is a moral to this story: *never believe what the computer is telling you without doing a sanity check on the indicated results.*

There were a number of summer students in the RPO who had worked on Explorer IV and they were determined to see it launched. There was no way that they could have gotten into the actual Atlantic Missile Test Range (now the Kennedy Space Center) to view the launch up close, so Naumann and Snoddy decided to lead an expedition to Cocoa Beach and watch the launch from the beach. The group arranged to be "sick" the day before the scheduled launch, because no one had any annual leave to take. As Naumann recalls it:

"We drove down in two cars, leaving before daybreak and arriving in late afternoon. The first thing we did was to scout out an area on the beach where we could get the best view of the launch tower. Then we found two rooms in a motel, one for the girls and one for the boys, and had a delicious supper of boiled shrimp and beer. We knew the launch window opened at 0200 and closed at 0600, but the day of the launch was classified secret.

Therefore, we didn't dare ask the hotel manager for a wakeup call at that strange hour for fear of revealing that a launch was scheduled at that time. None of us had an alarm clock, but we thought at least one of us would wake up.

"The effect of the long drive plus the shrimp and beer took its toll, and soon we were all fast asleep. I believe I was the first to wake up and found to my horror, it was 3:00 AM. We got the girls up, and hurriedly dressed and drove out to the beach spot we had selected. Much to our relief, we could see that the big white Jupiter-C was still on the pad, bathed in the glare of the powerful lights trained on it. Unfortunately, there was no way we could hear the countdown, so we had no choice but to wait out the launch window on the cold wet sand on the beach, while swatting mosquitoes.

"Suddenly we saw this bright light on the horizon in the east and thought it might be the spotlight on a ship clearing the range prior to the launch. But we soon realized it was the planet Venus rising. I've never seen it so bright. As dawn approached along with the end of the launch window, and seeing the vehicle was still on the pad, we figured the launch had been scrubbed and our efforts had been in vain. The group of students I had been sharing the old Monte Sano Lodge with that summer (apartments were hard to find in those days) had been experimenting with model rockets. Determined that we would see a launch in one way or another, I had brought a two-stage model rocket with me and decided to set it off. Unfortunately, I had not perfected the art of staging, and the second stage took off first and went one way while the first stage went the other, similar to a Polaris launch that occurred a few months earlier.

"There was nothing to do but to pack up our belongings and head back to Huntsville, since none of us could afford to stay another day. When we got back, we found the launch had been scrubbed earlier that evening for technical reasons. However, we all did get to sit in the conference room to listen to the countdown and launch on the next day."

Bill Snoddy had originally planned to return to the University that fall to begin graduate school, but was offered a full-time job and he accepted. Graduate school could wait—this kind of an opportunity may never present itself again.

Despite the fact that the wire turnstile antennas had been removed from Explorer IV, it also transitioned into a flat spin just like Explorer I, although not as rapidly since the energy dissipation was much less. The payload shell was insulated from the rocket casing, and the two elements formed an antenna that had a distinctive skewed four-leaf pattern. As the spacecraft passed over a tracking station, it was possible to deduce the orientation of the plane of rotation relative to the station location. Taking advantage of the fact that the angular momentum vector remains more or less fixed in space, its spatial orientation could be determined quite accurately from multiple passes. Extending this analysis to include the modulation of the Van Allen radiation, with the help of Bob Holland and Stan Fields, who developed a method for solving the over-determined set of nonlinear equations needed to find the angular momentum vector, Naumann was able to deduce the angular distribution of the Van Allen radiation relative to the magnetic field lines and thus find the mirror points where the trapped radiation was reflected. Stan had begun his 38-year career in the Lab as a student aide, with the understanding that he would finish his undergraduate degree in night school at the University of Alabama extension in Huntsville (now University of Alabama, Huntsville).

The Argus Project

Nicholas Christofilos had been working at the Lawrence Radiation Laboratory (now the Lawrence Berkley National Laboratory) on the Astron machine designed to contain a controlled thermonuclear reaction. The Astron was a magnetic mirror machine in which charged particles are trapped on magnetic lines of force and are reflected back and forth by converging field lines at each end. Shortly after the launch of Sputnik in October 1957, he wondered if it would be possible to artificially seed electrons into the Earth's magnetic field to form a dense shell of electrons circling the Earth. And if this could be done, what would be the military implications? Could the electron density be made high enough to damage the electronics of ICBMs that would have to pierce the shell on the way to their target? Or could the synchrotron radiation from such a shell of electrons blind enemy radar and disrupt communications to mask a missile strike? The beta decay from the fission fragments from an atomic bomb would be a perfect source of electrons.

He discussed this idea with his supervisor, Herbert York, and Hans Mark, a colleague. When the Van Allen belts of charged particles trapped on the magnetic field became known, it was clear that his idea was feasible. On March 17, 1958, York was named Chief Scientist of the newly formed Advanced Research Projects Agency (ARPA). ARPA's mission was to manage projects that were either ignored by the military or were hung up because of interservice rivalry. Christofilos's idea seemed a perfect mission for ARPA to undertake. It was decided that a dense shell of electrons could have significant defense implications and justified a test. Thus, the idea went forward as Project Argus and was classified Top Secret. The Armed Forces Special Weapons Project (AFSWP) organization was given fiscal authority. The Air Force would provide three X-17A rockets to launch the nuclear devices and would also launch 19 sounding rockets into the Argus shell. The Navy would provide the USS *Norton Sound*, a guided missile frigate, and attendant vessels to make up Task Force 88. ABMA would launch Explorers IV and V to survey the Argus shell and assist in the data analysis. The Smithsonian Astrophysical Observatory (SAO) would track and refine the orbital data and would provide the magnetic field data. Finally, the State University of Iowa (SUI) would provide the detectors and have the primary responsibility for the data analysis. Time was of the essence, because a nuclear test ban treaty was being negotiated with the Russians, and everyone wanted to get these tests done before the ban took effect.

The Argus plan was presented to the White House on April 21, 1958, and was approved by President Eisenhower in early May. Task Force 88 was formed in

April and was proceeding to the South Atlantic toward the south Atlantic anomaly near Gough Island, where the geomagnetic field dips closest to the Earth. Meanwhile, the USS *Albermarle* was dispatched to the magnetic conjugate point in the North Atlantic to observe the aurora resulting from the explosions.

Explorer IV was launched on July 26 and the first X-17A carrying the Argus 1 shot was launched from the deck of the Norton Sound on August 27. Those with Top Secret clearances sat in front of the teletypewriter anxiously waiting for word of the launch. Suddenly the Teletype clattered, "buzzer away" and then "flashlight". The W-25 warhead was exploded at an altitude of 200 km and produced a yield of ~1.4 kt. Argus 2 was exploded at 240 km on August 30 and Argus 3 was exploded at 540 km on September 6, with similar yields. On August 21, the U.S. and the Soviet Union negotiated a moratorium on nuclear testing to be effective in October 1958.

Charged particles such as electrons can move along magnetic field lines but are deflected by the Lorentz force when they try to move perpendicularly to the field, resulting in a spiraling motion along the magnetic field line. The pitch angle of the spiral increases as the field lines converge until it reaches 90°, and the particle is reflected back along the field line. If the reflection point is above the Earth's atmosphere, the particle remains trapped, bouncing back and forth between the Northern and Southern hemispheres. If the reflection point is too low, the particles enter the atmosphere and cause an aurora. The radial gradient of the Earth' magnetic field causes electrons to drift eastward as they bounce back and forth between the hemispheres, causing the artificially injected particles to circulate around the Earth, forming the Argus shell of charged particles as predicted by Christofilis.

To get ready to track Explorer IV, John Hendricks and Frasier Williams, two summer students from the University of Alabama who worked in the Guidance and Control Laboratory, took a surplus SCR 584 radar mount and used it to make a steerable antenna to improve the signal quality from the satellite. The Comp Lab would provide the predicted azimuth and elevation of the satellite for each pass, but someone at the station had to there, even in the middle of the night, to track the satellite as it moved across the sky from radio horizon to radio horizon. Bill Snoddy would frequently go to the tracking station to get his thermal data in near real-time. He recalls the receiver's beat frequency oscillators would switch from a low frequency to a higher frequency and back when the radiation count reached a predetermined number. He recalls one pass that lasted almost 20 minutes when the sound went from "beeeeeee-boooooo0-beeeeeee-boooooo", when the satellite was below the Van Allen belt, to "be-bo-be-bo" as it entered the radiation belt. All of the data was recorded magnetic tape, as well as on 8-Channel Brush recorders in the form of strip charts that could be as much as 20-30 feet long.

Needless to say, those with green badges waited anxiously for the passes just after the Argus events. Lundquist, Shelton, and Al Weber from St. Louis University, who was visiting us that summer, would spread the long strip charts side by side in the hall of Building 4488, get down on their knees and match up the places where bursts of radiation could be seen as the spacecraft passed through the Argus shell. They then mapped the positions of the spacecraft on a magnetic map to locate the Argus shell in magnetic coordinates. Meanwhile,

Bill Snoddy was analyzing the thermal data and Bob Naumann was working with the recorded signal strength to obtain the satellite orientation in preparation for the total data package to be sent to Van Allen at SUI.

Later that summer, the tracking station was relocated to the top of Madkin Mountain, a large hill in the middle of the Arsenal, to broaden its radio horizon. The dirt road up to the station was bad enough when it was dry, but virtually impassible when wet. The Army was kind enough to loan us an old tank whose turret had been removed so we could have access to nearly real time data. One morning Frasier Williams arrived with the latest pass, but was soaking wet and covered with wet leaves. It had rained the previous night and on the way down the mountain, he decided to make his own road by driving through a grove of saplings, with predictable results.

In February 1959, a Top Secret meeting of all the participants was held at the Lawrence Radiation Laboratory. Edward Teller, the Director, personally participated in the meeting. It was determined that the Earth's magnetic field was too weak to contain sufficient Argus-type radiation to have military significance. However, the scientific value of having probed the Van Allen belts in a controlled manner in order to measure the trapping efficiency and lifetime of particles in the belt far outweighed the need for continuing the classification of the data and the group recommended declassification of the project. In fact, it was stated in the minutes, "The FLORAL (code name for series of nuclear tests that included Argus) series of tests in the South Atlantic undoubtedly constitute the greatest geophysical experiment ever conducted." The declassification debate continued until March 18, when it was learned the *New York Times* had known about Argus since December 1958 and was threatening to publish it, which it did on March 19. On the same day, the National Science Foundation and the National Academy of Science issued a joint statement that the results from Argus would be presented at the annual meeting of the National Academy of Science on April 27-29. This meeting represented the centerpiece of the U.S. contribution to the International Geophysical Year.

Perhaps the most remarkable thing about the Argus project, which involved all of the armed services, the Atomic Energy Commission, several government agencies, and a university, was the speed in which it was accomplished, less than a year from concept to approval to completion!

Pioneer III

After the first two Air Force attempts to launch a payload past the Moon and into a heliocentric orbit using the Thor-Able vehicle, ABMA was given permission to launch Pioneer III on a lunar fly-by mission using a Juno II vehicle on December 6, 1958. The Juno II (the Roman goddess Juno was the wife of Jupiter) is a modified Jupiter-A first stage with the JPL spin stabilized upper stages. (The Jupiter-C or Redstone booster with the spin stabilized upper stages was also referred to as the Juno I.) The payload was assembled at JPL, but ABMA was responsible for the thermal design. Bill Snoddy was sent to JPL to oversee the thermal tests. RPO personnel participated in the apex predictions and early tracking as in all Juno launches.

Like the first two Thor-Able launched Pioneer flights, Pioneer III also failed to pass the Moon and go into a heliocentric orbit because of a premature first stage burnout. However, it did reach an altitude of 102,000 km (~16 Earth radii), far enough to reveal the presence of a second or outer belt of trapped radiation known as the second Van Allen radiation belt.

Pioneer IV

Pioneer IV, similar to Pioneer III, was launched on March 3, 1959, and is the first vehicle to attain escape velocity from the Earth, placing it in a heliocentric orbit where it remains today. The mission provided additional detail about the outer Van Allen radiation belt as well as a valuable exercise in tracking deep space probes. Although JPL had the responsibility for the primary data acquisition, ABMA, Ballistic Research Labs, G.E. Schenectady, and the University of Illinois all participated using equipment that was available to them in order to compare different communication techniques. The University of Illinois attempted to use a Dicke-type radiometer, similar to the method used by radio astronomers, but ran into equipment problems. The ABMA system lost the signal on the third day when its 14-ft dish antenna had to look close to the sun. The G.E. 18 ft. dish was able to track the 180-miliwatt transmitter operating at 960 MHz until the vehicle's batteries were exhausted at a range of 655,000 km.

Able and Baker

To answer questions about the effects of low gravity and other aspects of the space environment before launching and recovering primates, Lundquist decided a life science expertise was needed in RPO and preceded to hire Richard (Dick) Young, a biology instructor at Athens State College. Don Cochran and Jerry Johnson assisted Dick Young in developing a series of animal experiments to be flown and recovered in test Jupiter nose cones. They flew sea urchin eggs and sperm, frogs, mice, DNA, drosophila, human blood cells, yeast and onion skin cells, corn seeds, mustard seeds, mold spores and monkeys, "literally a Noah's Ark so to speak," as Don described it. In their experiment with sea urchin eggs, fertilization was achieved and the offspring developed normally. This was probably the first demonstration that fertilization between sperm and egg could take place in the virtual absence of gravity. (The June 15, 1959 issue of *Life Magazine* carried an article describing these experiments.)

A short time later, two rhesus monkeys, Able and Baker, were the first primates to survive space flight. They were carried on a Jupiter IRBM AM-18, launched on May 28, 1959. An earlier Jupiter flight, AM-13, carried Gordo, a squirrel monkey, who survived the launch and recovery but was lost when the nose cone sank before it could be recovered. Able and Baker were in good condition after the 38g reentry, but Able died four days later from a reaction from the anesthesia administered when the medical electrodes were being removed. Baker lived in the Huntsville Space and Rocket Center until 1984. Dick Young and Jerry Johnson transferred to the Ames Research Center (ARC), where Dick established the life science activity at ARC in 1961 and headed the Exobiology Division. Don left later to form his own company, ENERTEC, in South Carolina.

The Juno II Satellites

The higher throw weight provided by the Juno II launch vehicle allowed satellites with useful masses up to 100 lbs. to be placed in low earth orbit (LEO). All of the satellites launched by the Juno II vehicles were funded by the newly created NASA, but built in-house by the ABMA Guidance and Control Laboratory with RPO being responsible for the thermal design and testing. Billy Jones was lead thermal designer for Explorer VII, and Bill Snoddy took the lead on the others. Stuhlinger initiated a policy of assigning one of his junior scientists to work with the various Principal Investigators on the satellites for which we had some responsibility. The purpose of these Project Scientists was to have someone close by who understood the objectives of the experiment and could represent the interests of the PI during the development and testing of the spacecraft. Also it gave our promising young scientists valuable experience by working with highly recognized space scientists on cutting edge projects.

Explorer VII (S-1 Payload)

After the successful launch of Explorer VI by the Air Force using a Thor-Delta, Explorer VII was successfully launched on October 13, 1959. This was the second launch attempt; a guidance system failure just after liftoff caused the vehicle to tilt over and point toward some of the observers of the launch just as range safety blew up the booster. Several RPO employees were at the Cape viewing the launch. "It was pretty scary," one of them related later. "When we saw that big Juno-II pointed right at us, we scrambled to get under anything we could."

The greater payload capability of the Juno-II launch vehicle allowed the satellite to carry a multitude of instruments, as well as solar cells for power. After separating from the fourth stage rocket, the payload was left spinning about its maximum moment of inertia, which prevented it from tumbling like the earlier Explorers. Like the pervious Explorers, Explorer VII carried radiation detectors from the State University of Iowa (SUI) that, because of its longer lifetime, gave information on the stability of the inner Van Allen belt and the temporal variations in the lower extremities of the outer belt. An ionization chamber also counted heavy primary cosmic ray particles and determined their energy spectrum.

As a side note, Van Allen was once asked by a reporter, "What good are the Van Allen belts?", to which Van Allen replied, "Well, I make a living from them." In another incident, Ernst Stuhlinger introduced Van Allen at a conference by saying, "Our speaker has become famous for his work on the Van Allen Belts—in fact he has become so associated with them that he has taken their name—Ladies and Gentlemen—Dr. James Van Allen."

Perhaps the most significant experiment carried by Explorer VII was Verner Suomi's radiometer designed to measure the radiation balance of the Earth, which is critical to the current question of global warming. The five radiometers carried by Explorer VII, each tuned to respond to a different spectral region, were able to determine the solar radiation incident on the Earth, the amount reflected back, and the infrared reradiated by the Earth at the position of the satellite. Large-scale fluctuations in outward radiation, apparently coupled to clouds and weather events gave the first indication of the complexity of determining the radiation balance of the Earth, a problem still confronting the climatologists of today as they attempt to understand the role of CO_2 and other greenhouse gases in global climate change.

After ABMA was incorporated into NASA and became the George C. Marshall Space Flight Center in July 1960, the agency continued to launch the remaining Juno II satellites that had been previously approved. These included Explorer VIII (S-30 Payload), the first in situ measurement of the ionosphere, the S-45 ionosphere beacon (failed to orbit), the S-46 radiation package (also failed to orbit), and the S-15 gamma ray telescope, which became Explorer XI. Again the RPO was responsible for the thermal design, with Bill Snoddy taking the lead for all of these payloads.

The Explorer XI gamma ray telescope deserves special attention, because it was the first attempt to locate gamma ray sources in the universe and in a sense was a forerunner to the AXAF of Chandra X-ray Observatory, which was managed by MSFC. The PIs were Bill Kraushaar and George Clark at MIT, who provided the gamma ray detectors. Since Explorer IV also transitioned from an axial spin to a flat spin even though the wire turnstile antennas were eliminated, it was decided to take advantage of this transition so that the telescope could scan the entire hemisphere. The orientation of the spacecraft was deduced using the antenna patterns as was done for Explorer IV. Since the spacecraft had to operate in the Van Allen belts, anticoincidence counters had to be used to separate true gamma ray events from charged particle events. During its four-month lifetime, it detected 22 gamma events with an angular resolution of 17° half-angle, thus giving us the first glimpse of the gamma ray sky.

In addition Explorer XI carried two carefully prepared discs with different thermal coatings that were isolated thermally from the rest of the vehicle. By monitoring the temperatures over the lifetime of the spacecraft, Snoddy was able to measure the combined effects of vacuum, ultraviolet, micrometeorite dust, and high-speed atoms, ions, and molecules that constitute the space environment on thermal control surfaces for the first time. Such measurements are crucial, since it was not possible to simulate the combined effects of all of these factors. It was found much later that the combined effects of ultraviolet and atomic oxygen were extremely important under vacuum conditions.

Bob Naumann had been tracking the charged particle detectors on Explorer *XI*, when he noticed an anomalous burst of Argus-like radiation when the spacecraft was over the Soviet Union in a region where they had been carrying out antiballistic missile tests, apparently the result of a clandestine small yield, high altitude nuclear explosion. This observation was reported to the CIA and was presented at the Beta-Beta and Delta Gamma Symposium sponsored by Defense Atomic Support Agency.

When it became clear that coming absorption of ABMA into NASA would give the future development and operation of small satellites to GSFC, the talented group of satellite engineers in the ABMA Guidance and Control Lab, led by Olin King and Bill Greaver, left to form Spacecraft Incorporated, which later became SCI and later merged with Sanmina to become Sanmina-SCI. Meanwhile the RPO was preparing for its new role under MSFC in the Apollo program.

Lunar Thermal Studies

In keeping with its mission to provide planning and research for future missions, much attention in the Research Projects Office was turned toward the Moon. In anticipation of a possible lunar mission, Heller also became intensely interested in the optical and thermal properties of lunar soil. Recall that the lunar disc at full Moon appears to have the same brightness near the limbs as near its center, quite unlike a diffuse reflector like a tennis ball that shows pronounced limb darkening. What kind of particles in the lunar soil could be responsible for such behavior? What are the thermal properties of this soil? What temperature extremes would be seen on the lunar surface? The Moon, after all, is a satellite that orbits the earth once a month, and as it goes around it keeps the same side facing the earth so the lunar "day" is four weeks long and its temperature can be calculated in the same manner as an artificial satellite, provided the optical properties of its soil are known. These properties, of course, were not yet known, but using properties of Earth soil, the temperature extremes were estimated to be approximately 200°C during the lunar daytime and -200°C during the lunar night, thus presenting a real challenge to the design a thermal control system for vehicles and astronauts operating on the lunar surfaces. To gain a better understanding of the optical properties of the soil, plans for an infrared observing program were begun using the newly developed cryogenically cooled infrared detectors. This work ultimately produced a series of some of the most definitive studies of lunar soil properties published in the journal, The Moon. (Reidel Publishing Company, Dordrecht-Holland.)

Selenography

Debates about lunar geology, the origin of the Moon's surface features and the possible origins of the Moon had gone on for many years, but few geologists had considered the matter beyond the speculations of pioneers like G.K. Gilbert (The Moon's face: a study of the origin of its features, Bull. Phil. Soc. Wash. 12, 241. 1893). Even R.B. Baldwin's The Face of the Moon in 1944 or J.E. Spurr's Geology Applied to Selenography in 1949 did not seem to excite many in the geological community. While many geologists in the oil industry used aerial surveys as a tool, the mining industry depended primarily on surface reconnaissance, drilling, and ground-based electronic surveys. For any who were interested in the development and use of instrumentation for geological studies from orbit and the chance to help verify the theories of geological pioneers in lunar geology, employment at a space agency seemed the only answer. WvB's popularization of lunar exploration helped to change the employment prospects for earth scientists; by the 1960s the U.S. Geological Survey received contracts for lunar-related work from NASA, and many geologists had been hired at Manned Spacecraft Center (MSC), Goddard, and NASA Headquarters.

When Sputnik was launched in October of 1957, John Bensko was in Boston, taking part in a conference on high velocity impact. The materials group had a dual interest in conducting high velocity impact investigations: (1) the effect of micrometeorite and space debris on spacecraft components and structures and (2) the origin and structure of lunar craters. Most of the meetings he attended on high velocity or hypervelocity impact, especially the equipment and procedures, were classified. There was enough unclassified work going on, however, so that it was possible to find equipment designs and researchers willing to give a lot of information that would be useful to the materials group in setting up their own experiments through a contractor. By the spring of 1958, he was able to select surplus materials from the surplus and scrap yard at North American Aviation to assemble lab equipment for impact studies on rock, plastic, and metal targets. Some of the results of the first six months effort were presented to the Lunar and Planetary Colloquium in January of 1959.

The study demonstrated the influence of existing target internal structure on the geometry of the craters produced by impact. Harold Urey (Urey was a physical chemist who won the Nobel Prize for the discovery of deuterium and became a self-appointed expert on the Moon) looked at photos of one of the targets and remarked that it looked precisely like two small craters in Mare Foecunditatis with two rays extending to the west.

Soft Lunar Lander Studies

Working with Robert Jastrow in NASA Headquarters (who later became the first director of the Goddard Institute of Space Studies at Columbia University), Bensko set up guidelines for the lunar landing research effort and sought out individuals who would lend their expertise to form a Research Projects Lunar Working Group with the goal of putting together a proposal for a soft lunar lander.

As it turned out, ABMA was requested by NASA in June 1959 to coordinate a comprehensive study on lunar exploration and experiment objectives. The study was to cover soft landings on the Moon for a stationary payload package, a package with roving capability, and a manned circumlunar flight with subsequent recovery. The study was originally scheduled for completion on January 1, 1960, but was informally extended to February 11, 1960. The working groups were expanded. Some of the original participants had moved to different projects. Stuhlinger thought Bensko's vehicle concepts could be improved in several ways and suggested he work with Josef Blumrich and Erich Engler (Structures & Mechanics Division) and with Georg von Tiesenhausen and Jean Olivier who translated the engineering drawings and artists' views of the lunar instrument packages into more capable vehicles. Bill Robinson (Research Projects Office) joined the working group and worked out the operation sequence and communication requirements for the packages.

The massive study, completed in February 1960, was classified Secret because it contained information about projected U.S. launch capabilities. Volume 9 of the study was compiled by Georg von Tiesenhausen and Jean Oliver, who worked in the Test Lab under Kurt Debus and focused on the mobility of vehicles on granular surfaces. This work ultimately led to MSFC's later mission to develop the

Lunar Roving Vehicle (LRV) as well as the work on lunar soil mechanics done in the Space Sciences Laboratory (SSL) by Nick Costes during the Apollo Program. Volume 10 was done by George Bucher, Jim Downey, Al Weber (a visiting Professor from St. Louis University), John Bensko and Herman Gerow. This chapter included the work on the lunar drill, the laser corner reflector, and the lunar seismometer that eventually led to the Apollo Lunar Surface Experiments Package that was left on the Moon by the Apollo Astronauts. Despite the expertise and effort demonstrated by the ABMA team, the Jet Propulsion Laboratory was given the mission to develop a vehicle to land a payload softly on the Moon, which became the Surveyor program.

The Research Projects Office did receive considerable publicity because of its researcher's involvement in the NASA lunar program and their interest in satellite instrumentation. A paper by Bensko read at the American Geophysical Union meeting in 1960 was picked up by Clifford Simac, a syndicated science reporter in the October 24, 1960 edition of the *Minneapolis Star* (Moon Landing? It's Possible!), and quoted in his book, The Solar System: Our New Front Yard. Conrad Swanson participated as a panel member in the Lunar Exploration Vehicles Panel at The Lunar and Planetary Exploration Colloquium in November 1960. In 1961, Chuck Lundquist gave a paper entitled "Instrumentation for Space Projects" at the 7th Aerospace Instrumentation Symposium that was later cited in *Space Daily* (June 2, 1965, page 166).

Despite the loss of a formal mission for lunar studies, these efforts placed the ABMA personnel, after their transfer to MSFC, in a good position to be major players in the Apollo lunar science endeavor.

Fab Lab

By Jay Foster

Fabrication and Assembly Engineering Laboratory was also known as Manufacturing Engineering Laboratory, ME Laboratory, and universally as Fab Lab for short. Fab Lab had two primary functions: vehicle production and engineering planning.

Fab Lab fabricated and assembled the vehicles that were assigned to MSFC. In the early days, Fab Lab did it all from the ground-up. They designed and fabricated the tools, built the airframe and assembled all of the components into the airframe. This was true for the Redstone and early Jupiter vehicles. Fab Lab also built the initial Saturn I test vehicle in-house, designated SA-T. The Saturn Program was initially funded by ARPA, the Advanced Research Projects Agency, the research arm of the Department of Defense (DOD). Saturn I was initiated well before President Kennedy's Lunar Program announcement. Initially, the task was to build the Saturn I, SA-T, 1st Stage, not to fly, but to undergo an all up static firing to establish the fact that eight engines and nine fuel tanks could be successfully clustered and perform successfully together. In fact, SA-T was known in the vernacular as "Clusters Last Stand".

The fabrication and assembly of the SA-T was unique, in that to satisfy schedule pressure, the stage was moved to the Static Test Stand when the stage was only about 80 percent complete. This required close coordination between Test Lab and Fab Lab. A daily meeting was set up every morning between Gordon Artley of Test Lab and Jay Foster of Fab Lab to coordinate tasks on the vehicle. Test Lab was endeavoring to install static test measuring equipment in parallel to Fab Lab endeavoring to complete vehicle assembly. The purpose of the meeting was to minimize interference between the two groups crawling around on the vehicle. This approach worked well and proved a satisfactory way to shorten the schedule. There was much interest throughout MSFC during this process as much of the center's activity was focused on the SA-T at that time.

The Fab Lab organization was built around the Redstone and Jupiter vehicle elements, and consisted of five divisions. Three of the divisions directly produced the vehicles and consisted of: a) The Fabrication Engineering Division, with Otto Eisenhardt, chief; b) The Assembly Engineering Division, with Max Novak, chief; c) The Plant Engineering Division, with Bill Potter, chief. The Plant Engineering Division was responsible for keeping all of the myriad tools, facilities, and equipment in first class working order.

The other two divisions were the Electro-Mechanical Engineering Division, with Robert Paetz, chief, and the Methods Research Division, with Bob Schwinghamer, chief. The Methods Research Division, as its name implies, was to develop and test new and improved ways to fabricate and assemble complex structures, components, and sub-assemblies.

The second major function of Engineering Planning was performed primarily by the Electro-Mechanical Division of Robert Paetz and its branches. A target launch date would be established by management. Then, the planning branch of Robert Paetz's division would go into action. Initially, Production Planning developed the "Fabrication and Assembly Plan". Then planning would establish required delivery dates for all design drawings. These dates would then be negotiated and finalized with the vehicle designers, including Airframe, and the many mechanical and electrical and electronic components. The engines were the longest lead time items and normally were started into design and production several years prior to management establishing a specific flight schedule. The scheduling of designs included obtaining flight component design completion in time for Production planning to establish requirements for tooling, with time for tool design and fabrication, before elements were required on the assembly floor. Beside the engines, the longest lead time was for the structural design, since time must be provided for determining tool requirements, tool design, and tool fabrication. Sometimes, the negotiations with the vehicle design elements were tough, but everyone knew the launch dates and all elements had to shoehorn into the time available. In this process Planning worked the "make or buy" decisions and developed requirements for tools and sent requests to the Tool Engineering Branch, with the criteria required for the various tools. The Fab Lab Engineering Planning Section also negotiated with design elements for ease of fabrication. Suggested design changes were accepted if Planning could convince design elements that manufacturing-suggested changes could enhance fabrication and reduce cost of production without impacting the function of the component. Engineering Planning Section also worked closely with other Fab Lab elements to develop cost estimates that were used to develop inputs to the overall MSFC budget process. The Planning elements laid out the fabrication and assembly schedules for the vehicles. Robert Paetz also had two Process Engineering Branches, Structure and Assembly. These branches would issue work orders to the shops to keep the hardware flowing to the production floor in accordance with the established Fabrication and Assembly Plan. The fourth branch in Electro-Mechanical Division was the Tool Engineering Branch with Jack Franklin in charge of all tool and fixture design.

One Fab Lab staff organization was very important in the Fabrication and Assembly Plan. This was the Technical Liaison Office. This office developed procurement documents and specifications for the many items, large and small, that engineering planning determined to be "buy" candidates. Individuals from the Technical Liaison Office worked closely with the Procurement Office and subsequently with the selected contractors to assure that items received fully met design, quality, and reliability requirements. This organization was headed by Jack Trott, an American engineer from New York.

It should be noted that of the several hundred individuals working in Fab Lab, less than 1% were original members of the WvB Team. The American team members were essential to the team and performed by far the majority of the tasks. The Fab Lab civil service work force was augmented by industry; initially Hayes Aircraft Corp. supported Fab Lab. Some of the key Americans working in Fab Lab included: Bob Schwinghamer, Bill Wilson, Walt Crumpton, Bill Potter, Gene Davis, Jack Franklin, Jack Trott, Jack Haire, and your author Jay Foster.

The MSFC nine laboratories all functioned together as a team that could produce and launch a complete vehicle, and did in the early days. These complete vehicles included Redstone, Jupiter, and early Saturn I's. The laboratories' primary missions were:

1) The Aeroballistics Laboratory carefully developed the environment that the vehicle traveled through. This information became the criteria that the mechanical, electrical, and electronic designers had to satisfy with their designs.
2) The Computation Laboratory provided the compute power for the entire organization. In the early days the MSFC computation capability was significantly ahead of most of the country. The Computation Laboratory also had a sizable group (~300) programmers to produce the specific programs (Apps) required by space flight. This capability

was significantly reduced after Bill Gate's Microsoft came along and you could buy apps off the shelf.

3) Guidance and Control Laboratory developed the brains of the vehicle. This capability provided for a preplanned trajectory and the ability to correct for wind shear and any other external force the vehicle encountered. G&C was also responsible for all of the electrical and electronic component design.
4) Research Projects Laboratory developed ideas for experiments, and developed new technologies for space flight. One idea developed by Dr. Ernst Stuhlinger, the director, was Ion propulsion. Later Ion propulsion research was transferred by Headquarters to Lewis Research Center to better balance the work load across centers. Dr. Stuhlinger was disappointed to see his project transferred to Lewis, but chose to stay with the WvB team.
5) Structures and Mechanics Laboratory was responsible for the design of all mechanical items, including the airframe and all the internal components.
6) Quality Laboratory was responsible for assuring that all components and assemblies were assembled, functioned, and installed in accordance with the design and specifications.
7) Test Laboratory developed test procedures and facilities to fully test all major assemblies and static fire complete stages to assure that the vehicle would perform in accordance with design requirements.
8) Launching and Handling Laboratory (Later Launch Operations Directorate). This organization was established at Cape Kennedy and was responsible for launching NASA vehicles. This organization initially reported to WvB, and later became the nucleus of the Kennedy Space Center and reported to the Office of Manned Flight in NASA Headquarters.
9) Fabrication and Assembly Engineering Laboratory. Fab Lab was responsible for fabrication and assembling the MSFC family of vehicles.

With this summary view of the laboratory functions, it is easy to see how the Fab Lab was integrated into the WvB team and played an essential and key role in the total organization.

I will provide one personal example of the power of the laboratories. In the late '50s, it was decided to set up a series of working groups to oversee the disciplines across all the contractors. One of the working groups was the Manufacturing Working Group. Hans Maus, the director of Fab Lab, asked me to chair this group. Oswald Lange was then the head of the Program/Project Offices. Dr. Lange objected; he believed that I was too young and in-experienced to head a working group. Hans Maus responded with a short memo, to Dr. Lange: "You are informed that Jay Foster is the chairman of the Manufacturing Working Group."

With the rapid buildup in the total activity under the Saturn Apollo Lunar Program, it became necessary for MSFC to significantly increase emphasis on utilizing industry's capabilities. Consequently, all of the major elements were contracted with industry. MSFC did retain the ability to build anything, but not everything. With the emphasis on industrial operations, more engineers were required by the other laboratories to exercise their automatic discipline responsibility. In an effort to resolve this impasse, WvB decided to combine Test Laboratory and Fab Lab into a new Test Lab.

Guidance and Control Division/Astrionics Laboratory

The Early Years

By Brooks Moore

When I arrived in Huntsville in March 1952 to accept a job with the U.S. Army's Guided Missile Development Division, I was assigned to work in the Control Section of the Guidance and Control Laboratory. Professor Theodore Buchhold headed up the Laboratory and Dr. Ernst Stuhlinger was his deputy. I learned right away that in the German professional hierarchy, "Professor" was a notch above that of "Doctor". In addition to being Professor Buchhold's deputy, Dr. Stuhlinger also headed up the Research Section and Dr. Manteuffel was his deputy. The Control Section to which I was assigned was lead by Dr. Walter Haeussermann, whose deputy was Mr. Hubert Kroh. Dr. Schlitt was head of the Guidance Section with Mr. Kuerschner as his deputy. A Gyro Section was led by Mr. Fritz Mueller with Henry Rothe as deputy. Mr. Otto Hoberg was leader of the RF and Telemetry Section and Dr. Schwedetsky lead the Measuring System Section. Mr. Dieter Grau was chief of the Electrical and Power Section, Mr. Josef Boehm led the Design Section and Mr. Wilhelm Angele lead the Precision Mechanics Section.

The Control Section was focused on the design and development of the control system for the Army's 200-mile range guided missile, the Redstone. The control system concept involved controlling the Redstone's direction of flight by four air vanes on the aft exterior of the vehicle, coupled with four carbon vanes that projected into the fixed engine's jet stream. The vanes were driven by servomotors.

Dr. Haeussermann apparently had a plan to train me in the development of all elements of the system, as I was assigned and periodically reassigned to his lead engineers for each element. Kurt Lindner had the responsibility for the servomotors and I was assigned to work for him for a few weeks, followed by a period with Gerhard Drawe working on the design of the Relay Box, which contained a multistage relay system that drove the servomotors. The overall design included a Control Computer to receive information from control sensors, and distribute the controlling signals to the servomotors. Ernst Lange was the lead engineer on the Control Computer. Portions of the Control Computer were still in the early development stage.

Vacuum tubes, the state of the art electronic devices at the time, were considered too fragile and unreliable for missile on-board use and transistors were not yet available. The plan was to use magnetic amplifiers for amplification and distribution of electrical signals. These devices had been used in some industrial application in earlier years and were being modified for missile use. This led to my being assigned temporarily to Dr. Manteuffel in the Research Section, since he was the technical lead for the development of the magnetic amplifiers.

Although bypassed by several generations of advanced technologies, magnetic amplifiers still maintained a niche in the development of the Huntsville rocket team's control systems two decades later. Because of their extreme ruggedness and unquestioned reliability, they were used in special critical applications in the Saturn V Launch Vehicle analog Flight Control Computer. Men were sent to land on the Moon using 1900s technology magnetic amplifiers to gimbal the control engines on the Saturn V Launch Vehicle.

During my training, I had one other informal mentor who was not a part of the Control Section. That was Mr. Dieter Grau, chief of the Electrical and Power Section. The two sections were located adjacent to each other in the old Hospital Building on Squirrel Hill on the northern edge of the 50,000 acre Redstone Arsenal. My office was two doors down from Mr. Grau's, and I had the opportunity to interface with him frequently. I thereby obtained a very good insight into the overall Redstone electrical system and established a lifelong friendship.

After my extensive training period, I was assigned to a long-term tour in the Simulation Group, headed by Mr. Hans Hosenthien. When I joined this group, there were already two other U.S. born engineers, Jack Lucas and Fred Digesu, working there. We worked together performing flight control studies to define the requirements of the Redstone control system to maintain a stable flight while being guided to the end target with a specified accuracy. We used an old 400-cycle AC analog computer built by the team to simulate the Redstone equations of motion in flight. The computer was very large and had very limited capability by current standards. It consisted of many racks of vacuum tube circuits, which covered one wall of what we called the "Sim Lab" in the old hospital building located on Squirrel Hill. This simulator was used successfully to support the design of the guidance and control system of the Redstone.

Jack Lucas and Fred Digusu are two of the many examples of how the young engineers of the early 1950s later served in key supervisory positions as their continuity of experience contributed significantly to the design and development of the Saturn V a decade later.

As I settled in to my first long-term assignment and began to have broader base of contacts, it became very obvious that although our official mission was the development of surface-to-surface guided missiles for the Army, it was no secret that WvB and the team were also studying how the developments could support the U.S. entry into space. Even though there was much informal discussion about space travel in those early years, those of us working on the Redstone guidance and control system had no thoughts at that time that before the end of the decade the little rocket that we were working on would play a major role in our nation's infant space program. By that time it had become known as the "Old Reliable Redstone" because of its demonstrated high reliability compared to other evolving missile developments. In January 1958 the first U.S. satellite, Explorer I, was successfully launched into Earth orbit propelled by the Redstone rocket.

Transfer of Rocket Team to NASA

When Marshall Space Flight Center was created in July 1960, it was designated the launch vehicle center for NASA. The Huntsville Rocket Team was essentially reassigned to NASA. There was a small contingent who stayed with the Army to organize and staff the new Army Ballistic Missile Agency, since the Army intended to remain very much involved in the development of military missiles in Huntsville.

The Guidance and Control Laboratory organization that was in existence in ABMA in early 1960 was transferred essentially intact to NASA. The staffing was also essentially the same except for personnel shifts to replace those who remained with the Army. Under the NASA/MSFC designation, there was initially a change in the organizational element nomenclature in that the ABMA laboratories became NASA "divisions", ABMA divisions became branches, etc. In 1962, the names of the major MSFC technical elements were changed back to "laboratory" with the internal organizations being divisions and branches. For simplification, the latter is the nomenclature used herein.

The new MSFC Guidance and Control Laboratory responsibilities included the electrical, electronic, guidance, control, instrumentation, communication, and related sub-systems. The director and deputy director, after the transfer, were Dr. Walter Haeussermann and Hermann Kroeger. Technical and staff Assistants were John Chase, Hubert Kroh, Frederick Brandner and Fritz Weber. Fred Digesu headed up the Advanced Studies Office reporting to the director. The Laboratory was organized into eight divisions: Instrumentation, Electrical System Integration, Gyro and Stabilizer, Navigation, Flight Dynamics, Electromechanical Engineering, Applied Research and Pilot Manufacturing.

Otto Hoberg and Lucian Bell were the chief and deputy of the Instrumentation and Communication Division, with staff assistants Heinz Kampmeier and Charley Chambers. The division was organized into four branches with the following technical leaders: Measuring Systems–J.T. Powell, Ted Paludan, Sanford Downs, Jim Derrington and Harlan Burke; Telemetry System–Jim Rorex, Walt Frost, Frank Emens, Leo Arsement, Bill Threlkill and Bob Eichelberger; RF Systems–Tom Barr, Rudy Becher, Warren Harper, Jack Duggan, Lee Malone and Carl Huggins; System Engineering–John Cox, John Price, Bob Ferguson, A. L. Bracker and Nathan Ginn.

Hans Fichtner and Phil Youngblood were chief and deputy of the Electrical Systems Division. The division was organized into four branches with the following technical leaders: Airborne Systems–Dick Smith, John Stroud, Bill Shields, Jim Stulting, Al Woosley, and Etheridge Pascal; Component Design–Bob Aden, Ellis Baggs, Bill Glass, Walt Goodhue, Bob Milner, W. Body and Roger Cole; Power Supply Systems–Josef Boehme, Gene Cagle, Wayne Shockley, and Bill Striplin; Ground Support Equipment–Werner Rosinski, Don Knott, Fred Roe, W. W. Guillebeau, Ray Flack, and Ben Swords.

Carl Mandel was the chief of the Gyro and Stabilizer Division with no deputy at the time of transfer to MSFC. The division was organized into five branches with the following technical leaders: Servo Systems–Herman Thomason, Lelous Wood, Clyde Jones, Lewis Cook, and Charles Lee; Stabilizer System–Fred Kelley, Tom Morgan, Clifton Sims, Tom Bolton, and Paul White; Accelerometers–Gene Fikes, George Ferrell, H.C. Powers, and John Harper; Component Design–Richard Tuggle, Erick Zeismer, Helmut Hoffman, and James Patterson; Research and Evaluation–Rudy Beyer, Walter Kasparek, and Al Amalavage.

Brooks Moore and Ludie Richards were the chief and deputy of the Navigation Division, with technical assistants Norman Gilino and Jerry Mack. The division was organized into seven branches with the following technical leaders: Guidance Theory–Gilbert Gassaway, Charles McMahen, and Melvin Brooks; Guidance System Design–Hugh Taylor, Lewis Wood, and Bob Asquith; Guidance Components Design–Charles Swearingen, Charles Glass, Cliff Kirby, and Richard Richards; Control System–Jack Bridges, Richard Tutt, John Caudle and Jim Blanton; Flight Controls–Mike Kalange, Russ Alcott, Bill Howard and Howard Hovis; Control Preliminary Design–Gerhard Drawe, Bernie Wiesenmaier, Bill Chubb, and Hans Kennel; Ground Support Systems–Fred Wojtalik, Gerald Turner, Jim Lewis, Bill Burdine, and John Noel.

Hans Hosenthien and Jack Lucas were chief and deputy of the Flight Dynamics Division. This relatively small division was organized into two branches, with the following technical leaders: Flight Simulation–John Blackstone, Harold Mink, and Jack George; Simulation Facilities–Ernest Orem, Hugh Coppock, Gene Lee and Charles Casey.

Josef Boehm and Dale Lamb were the chief and deputy of the Electromechanical Division. The division was organized into three branches with the following technical leaders: Design and Development–Adolph Hermann, Walter Kasparek and Marvin Macuch; Mechanical Systems–Helmut Pfaff, Manfred Kuebler, and Herman Wagner; Production–Steve Dobbs, Frank Matthews, and Herbert Williams.

Jim Taylor and Jim Reinbolt were the chief and deputy of the Applied Research Division. The division was organized into four branches, with the following technical leaders: Magnetics and Semiconductors–Dorrance Anderson, Carl Winkler, Al Willis, and Bill White; Control Systems–Dieter Teuber, Oscar Hovik, and John Gould; Digital and Analog Timing Systems–Bill Kreider and Jim Snellgrove; Test and Evaluation–Jim Reinbolt, Bob Simpson, and Ernst Lange.

Wilhelm Angele and Hans Martineck were the chief and deputy of the Pilot Manufacturing Division. The division was organized into four branches, with the following technical leaders: Mechanical Development–Hans Martineck, Ignatius Wagner, Theo Starkey, and John Lemay; Electrical Development–Ray Van Orden, Frazier Bright, and Bill Kelley; Operations–Harvey Chapman, James Word, and Bert Orton; Manufacturing–Joe Johnson, Bill Linstead, Andy Crutcher, and Charles Jennings.

The above named individuals made up the leadership of the Guidance and Control Laboratory Director's Office and the nine divisions that were transferred en masse from the U.S. Army's ABMA to NASA's Marshall Space flight Center in July 1960. The above were accompanied in that transfer by many other highly skilled and dedicated engineers, technicians and administrative personnel.

The Transfer and Lessons Learned

As the move was made from ABMA to NASA, the resolve was to continue to apply the lessons we had learned in the past as we looked forward to future challenges. Some of the design principles that had evolved in the Guidance and Control Laboratory were probably typical of those applicable in other laboratories as well. Some of those principles were:

1. Design conservatively. Incorporate additional safety factors wherever practical.
2. Keep it as simple as possible. This is ultra important in flight hardware, since complexity can decrease reliability. Any component left on the ground can't cause a flight failure.
3. Build on past experience. Don't be tempted to "improve" on a component or subsystem design that has worked well in past similar applications.
4. Use proven techniques and flight-tested hardware wherever possible. Resist the temptation to "re-engineer" unless different requirements justify.
5. "Avoid like the plague," the urge to incorporate new advanced technology just because it is now available.
6. Test, Test, and Test. Conduct extensive testing at all levels any item that is to be a part of flight hardware. Test at the component, assembly, sub-system and system levels. The ultimate in test desirability is to flight test a critical piece of flight hardware as a "passenger" before incorporating it in an active critical flight application.

Our main focus at the Marshall Center was continuing the development of the avionics systems for the Saturn I. The goal of the Saturn I program was to develop our nation's first "heavy-lift" space launch vehicle, a rocket designed specifically to place large payloads into low Earth orbit. Following past practices, the intent of the Guidance and Control Laboratory personnel was to use the Jupiter hardware to the maximum extent possible. However, there was the challenge of the added complexity of a multi-engine booster with a much more complex mission.

One of the areas significantly affected was the guidance and control system. Two of the components in the Saturn I G&C system that were to be redesigned were the Stabilized Platform and the Control Computer. Because of the extensive in-house experience in developing platforms and the excellent performance on Jupiter, it was decided to proceed with development a smaller version of the Jupiter with the added flexibility of an additional gimbal (designated the ST-124) to accommodate the Saturn I missions

For the Saturn I Control Computer, a decision was made to use the same technology and, where applicable, the same hardware that had been used on Jupiter. Of course, some major changes were required, one being the modification to accommodate the eight-engine Saturn I.

A third major element was the Guidance Computer. Due to the Saturn I mission complexity, the Jupiter Analog Guidance Computer was not applicable. A digital computer was required. Since our experience with developing flight digital equipment was minimal, a survey was made of industry to locate the best available candidate for Saturn I. The most desirable candidate located was a flight computer developed by International Business Machines' (IBM's) Federal Systems Division, designated the ASC-15. This computer was developed for the Air force's Titan II ballistic missile. ASC-15's were acquired and an extensive in-house evaluation was initiated.

Rapid progress was made on the two in house designed items and integration of the ASC-15 and the ST-124/Jupiter-type Control Computer proved to be a viable combination.

WvB Engages with NASA Headquarters

Although the initial transition from ABMA to NASA went relatively smoothly considering the magnitude of the change, a major disagreement developed in early 1961. That disagreement involved the mission of MSFC's G&C Laboratory. While the laboratory had proceeded with design and development of not only the guidance and control system for the Saturn I, but also all of the avionics subsystems, a totally unexpected directive was received at MSFC from NASA Headquarters.

The author of the directive was the director of Launch Vehicle Systems, an Air Force Maj. Gen. Donald R. Ostrander located in NASA Headquarters. The document directed that MSFC discontinue it's development of the Saturn I guidance and control system and by inference that of follow-on space launch system, and incorporate the Centaur guidance and control system consisting of a Minneapolis-Honeywell platform and a Librascope Digital computer.

As might be expected, this directive caused great consternation in the management and personnel of the Guidance and Control Laboratory. Since the development of the Saturn I system was already under way before we became a part of NASA, considerable progress was already being made toward the system design using the ST-124/ASC-15/Jupiter analog computer approach. The implications were much wider, however, concerning the technical responsibilities of MSFC vs. Headquarters.

In my position as chief of the Navigation Division, I wrote a painfully long "white paper" protesting the apparent new policy whereby major launch vehicle subsystems would be "selected" at the Headquarters level, rather than designed and developed using the extensive technical experience and expertise at MSFC. A quote from the MSFC Charter that was included in the letter was from a NASA report to Congress establishing the center, which stated that MSFC would be "responsible for the development of large space launch vehicle guidance and control systems."

The tone of the letter in general was very caustic. It inferred, without actually making the specific accusation, that conflict of interest might be a factor in this and other similar situations. The rationale stated was since a decision made by one or a few managers at Headquarters might be unduly influenced by a close relationship with the benefiting contractors, while decisions made by a much larger group of engineers in a field center would more likely be based on true technical merit.

The white paper closed with this dangerous statement: "I have no objection to the above comments being made known to MSFC and NASA Headquarters management, even if the revelation may result in my reprimand, demotion, or dismissal."

Of course, Dr. Haeussermann immediately took the letter to WvB, who decided to give Dr. Haeussermann a chance to appeal his case to have the order rescinded directly to the general. WvB made the arrangements for Dr. Haeussermann to visit the general. During Dr. Haeussermann's appeal visit requesting that we not be required to use the Centaur System, he chose to give the general the white paper. He read it and commented: "This is completely uncalled for!" The result of the appeal was that the general stated that the order would not be modified or rescinded.

Dr. Haeussermann reported back to WvB, and told him that due to the circumstances he, Dr. Haeussermann, was submitting his resignation. WvB chastised Dr. Haeussermann for threatening such drastic action and decided that he would visit the general and take Dr. Haeussermann with him. This is Dr. Haeussermann's written account of that meeting:

> "Wernher suggested that he see the general first and that I wait in the outer office to be available in case he needed me. I waited outside to be called in, when after barely 15 minutes Wernher came out smiling and obviously in a good mood. As we departed he said: 'The directive will be canceled.' When I asked how did he accomplish this so quickly, he answered still smiling: 'I told him unless the order is rescinded, I will resign!'"

The order was rescinded and the G&C Laboratory went forward with the development of the Saturn I avionics systems using the ST-124 platform, the ASC-15 computer, and some flight proven hardware from the Jupiter as applicable. The general left NASA a few months after the fateful meeting with WvB to resume his Air Force duties.

New Focus-New Name

Soon after the new focus of the goals of the NASA manned space flight program based on the expected expansion to meet the president's new challenge of landing an American on the Moon in the decade of the 1960s, there was some re-structuring and re-naming of some of the major technical discipline areas within MSFC.

The Guidance and Control Laboratory was one of those major elements that took on a new name. The name was changed to Astrionics Laboratory to more accurately match the much wider diversity of the many electrical/electronic/electromechanical subsystems that were included in the responsibilities of the laboratory, not just the guidance and control discipline.

In the same time frame as the renaming of the Astrionics Laboratory, the name of one of the laboratory divisions was changed. The Navigation Division became the Guidance and Control Division to better identify the major subsystems, which were the responsibility of the division. A new organizational element, the Guidance and Control Systems Branch, was added within the new Guidance and Control Division. This group was assigned the responsibility to do the analyses and simulations required to define in detail the requirements for the flight control and guidance hardware elements that would be developed within the Guidance and Control Division for the lunar landing mission. They also had the responsibility for the system engineering and integration of these flight systems.

Jerry Mack was assigned to be the branch chief, with Lewis Wood as his deputy. Other supervisors in the new G&C Systems Branch were Jim Blanton, P.D. Nicaise, Forrest Winkler, Melvin Brooks, Bill Chubb, Hermon Hight, Gerhard Drawe, Bernie Wiesenmaier, Hans Kennel, Dave Schultz, Harold Brown and Charles Jones. In addition, key contributors in this group included Paul Golley, Harold Brown, Gerald Nurre, Fred Applegate, E.C. Smith, Pat Vallely, Doug Hendrix, Harvey Shelton, Henry Waites, Glen Ritter, John Farmer, Jack Priest, Tom Fox and Don Cook.

Saturn V Moon Rocket Design Challenges

One new very critical consideration for the design of the astrionics subsystems was the reliability requirements of manned space flight. The Huntsville Rocket Team's only experience with manned space flight was Alan Sheppard's flight on the Redstone-Mercury vehicle on May 5, 1961, three weeks before President Kennedy's announcement of the lunar landing program.

For the Saturn V Program, specific reliability and flight safety criteria had been established. The Saturn V reliability requirements far exceeded those of unmanned programs. All components and subsystems had to be evaluated to determine if they were a "Safety Critical Item." A Safety Critical Item was defined as "a part, assembly, or sub-system the malfunction of which would cause a catastrophic failure, loss or serious damage to the space vehicle, personnel injury, or loss of life." Additionally, the stated requirement was that no Safety Critical Item could contain a "Single Point Failure", defined as "an independent element of a system (hardware or software), the failure of which would cause loss of objectives, the space vehicle, or crew." All astrionics subsystems were evaluated to determine if they were Safety Critical. Essentially, all elements of the guidance and control subsystem were identified as being Safety Critical, as were some items in other subsystems. The design of all elements proceeded in accordance with the stated criteria.

One overall approach to significantly improving reliability and meeting the no single point failure criteria was the use of the technique designated "redundancy." This means simply using multiple elements instead of a single element along with a means of determining which of the multiple elements is performing properly. The concept can be applied at the component, subsystem of system level.

The items of the guidance and control system identified as being Safety Critical were the Flight Control Computer, the Launch Vehicle Digital Computer, the Launch Vehicle Data Adapter, the Stabilized Platform, the Rate Gyro System, and the Control Actuators. All of the conservative design criteria described earlier were applied in the design of each of these elements. In addition, all incorporated duplex or triple redundancy at the component or system level so that the no single point failure criteria was met. There were no (even single point) failures in the guidance and control system in any of the 12 Saturn V flights.

Evolution of the Instrument Unit (IU) Concept

As we faced the reality of the significant increase in number and complexity of components in the astrionics systems of the emerging launch vehicles, a decision was made in the Astrionics Laboratory to consider a new physical grouping of the components. The practice used by the Army and other agencies developing guided missiles in the 1950s was to mount avionics components at various places in the missile structure where space was available.

The first step toward grouping all astrionics components in one physical structure was made on the early versions of the Saturn I (Block I) vehicle. The first four flights of the Saturn I were tests of the 8-engine first stage only. On these, the astrionics items were packaged into a canister mounted on top of the booster stage. On the fifth test (Block II) of the Saturn I test flights in January 1964, is when the concept of having a "stage" to house all astrionics components (except those which had to be located elsewhere because of their function) was tested. The concept was to have the astrionics stage integrated into the Saturn I (Block II) structure just forward of the second propelled stage.

The astrionics stage for this first test of the concept was built in-house at MSFC. The structure consisted of four pressurized containers attached to a central hub and also supported by braces extending to the outside wall. Although the concept worked very well functionally, the total stage structure was unnecessarily heavy. However, the concept proved to be very advantageous from the Astrionics Laboratory standpoints of design, integration, environmental control, subsystems integration, and pre-flight functional testing. Since weight was not a serious concern on the Saturn I test flights, the pressurized container astrionics stage concept continued to be used throughout the Saturn I program.

In the meantime, a much more weight-efficient concept was evolving in the Astrionics Laboratory. This concept had all astrionics equipment mounted on the inside walls of the stage structure. This proved to be a much more weight-efficient approach while maintaining the advantages of a separate stage. The separate stage concept also had the advantage of the flexibility of being used on multiple vehicle configurations. Specifically, as the Saturn IB two-propelled stage and the Saturn V three -propelled stage vehicle concepts evolved the astrionics stage could be easily fit into both with the stage being integrated into the structure just forward of the top propelled stage. The final design consisted of a thirty-three foot diameter (the same as the S-IVB), three-foot high stage that was mounted on the forward end of the S-IVB. That is the approach that was chosen for implementation on the Saturn IB and V space launch vehicles. The name that was chosen for this separate stage was the "Instrument Unit (IU)."

In keeping with the typical MSFC approach, a prototype of the Instrument Unit was built in-house at MSFC and a contractor was chosen to build the follow-on flight units. The contract was competed and IBM Federal Systems Division was chosen. Having IBM under contract to do the fabrication, system integration, and test of the IU's was a very excellent arrangement since they were already designing two of the more complicated and system critical components, the Launch Vehicle Digital Computer and the Launch Vehicle Data Adapter. As a part of the contractual arrangement IBM was required to perform the IU work in Huntsville. This arrangement was a requirement because of the desire to have daily oversight of this complex and critical activity by MSFC technical and management personnel. IBM built a facility for this purpose located only a short drive from MSFC.

Automated Checkout

In the 1950s the Huntsville Rocket Team and other groups developing guided missiles used manual checkout procedures at the individual system level, the assembled vehicle level and for pre-launch checkout. With the advent of the more complex designs of multi-engine and multi-stage space launch vehicles it became obvious that manual checkout was taking an excessive amount of time.

In the early days of the Saturn Program, Ludie Richards realized that there was a dire need for some means of assisting test engineers in the checkout of flight equipment at various stages, but in particular the overall assembled vehicle and pre-launch at the launch site. He initiated an in-house study to determine the feasibility of using a ground digital computer to do certain routine repetitive monitoring to assist the test engineers and speed up the process. In the same time frame as the early design of the Saturn I vehicle and clarification of its missions were evolving, it became obvious that an on-board digital computer would be required to accomplish the more complex navigation, guidance and control functions. The projected addition of an on-board digital computer in the flight critical guidance and control system further substantiated the desirability, if not the necessity, of a ground digital computer to communicate with the on-board system.

While the process of testing and integration of the on-board digital computer (ASC-15) was being lead by Charles Swearingen, Hugh Taylor, and Lewis Wood, the search for a ground digital computer suitable to assist in launch vehicle checkout was initiated. The Ground Support System Branch under the leadership of Fred Wojtalik, Gerald Turner, and Jim Lewis performed this search. In his position as deputy director of the G&C Division, Ludie Richards, who led the conceptualization and feasibility studies of using a digital computer as a part of the ground support equipment, continued to be heavily involved in this computer search and the overall automated checkout effort.

Locating a digital computer candidate for use in an automated checkout application was not an easy task. While the digital computer industry was expanding rapidly in the early 1960s, most computers were being developed to satisfy business applications. The typical input/output technologies in that time frame were punched cards and similar approaches, which satisfied business applications. What was needed for checkout of on-board equipment was interfacing with lights, meters, strip charts, etc. There was not an extensive list of companies that produced a computer that was viable for adaptation to the automated checkout application.

As the team investigated the possibilities, they found that RCA in Natick, MA had developed a computer targeted toward manufacturing companies for possible use in industrial equipment checkout. It had fairly extensive capabilities that could be adapted to the performance of space vehicle automated checkout, much more than any other candidate on the market at that time. The team selected the RCA-110 as the most promising candidate to adapt for the automated checkout application. A RCA-110 was purchased and the in-house evaluation and development of the required adaptations was begun.

The Electrical Systems Division lead by Hans Fichtner had responsibility for some elements of the Astrionics Laboratory ground support equipment design so this automation effort became a joint endeavor involving both divisions. Bob Aden was the leader of the Electrical System Division role in the automated checkout system development assisted by Ellis Baggs, Bill Glass, and Bob Milner.

The automation team developed interfaces with flight hardware and tests were run in the laboratory to demonstrate feasibility of the automated checkout process. Even though there was some minimal introduction of automation into the Saturn I Program, it was insignificant. The first five Saturn I test flights were flights of the first stage (the Jupiter/Redstone "cluster") only and all checkout at the subsystem level, the assembled stage, and pre-launch was done manually.

The last five test flights (designated Saturn I, Block 2) were the first of the multi-stage vehicles of the Saturn Program. The S-IV was added as the second stage. Although the addition of the upper stage increased the complexity of the vehicle and complicated the checkout process, the test procedures employed at the stage level and integrated vehicle level were still essentially all manual.

The transition to the Saturn IB program was where automated checkout began to have an impact. The Saturn IB vehicle consisted of an S-1 Stage, a S-IVB stage, with a single J-2 engine and an Instrument Unit. By this phase of the Saturn Program, Chrysler was under contract to build the S-I stage, the first few had been built in-house at MSFC. The Instrument Unit had evolved while being designed and built in-house to its final version and IBM was under contract to build the IU's for the Saturn IB and Saturn V vehicles.

As the test planning evolved for the multi-stage launch vehicles, it was decided that each stage would be comprehensively tested at the contractor's plant, with MSFC oversight and shipped to Kennedy Space Center where the vehicle would undergo a comprehensive full stack test. In this and all other aspects, the Instrument Unit was considered to be a "stage" the same as the propulsion stages, except that it reported to the Astrionics Lab. Fritz Weber served on the lab staff as the manager of the IBM IU effort and Luther Powell was the on-site MSFC representative in the IBM Huntsville IU facility.

As the Saturn program progressed, the Astrionics Lab automated checkout team continued with the improvements on the RCA-110 system. It had become obvious from the experience gained on the multi-stage Saturn I, Block 2 that total manual checkout was far too cumbersome and time consuming. A decision was made to introduce a hybrid manual/automated system for the Saturn IB. NASA had decided that it was necessary to phase in automated checkout at the launch site with the development of the system to remain with the MSFC Astrionics Automated Checkout Team, in close coordination with KSC test personnel.

Although several KSC personnel participated very closely with MSFC in refining the launch site requirements and system implementation, Richard Jenke was heavily involved in a leadership role and in that regard spent extensive periods of time in Huntsville. Other KSC personnel who assisted in automation incorporation at KSC included Ike Rigell, Frank Penovitch, Carl Whiteside, Bob Register, Joe Medlock and Bill Jefferies. Other Astrionics Lab personnel who made major contributions to this effort, in addition to those already named, were Charles Swearingen, Sherman Jobe, Jim Christy, Chris Hauff, Ralph Smith, Claude Boykin, Eddie Albright, Johnny Brothers, John Wolfsberger and Richard Beckham.

RCA was awarded the contract to provide the checkout computers for this considerably expanded application. The experience gained by the MSFC team had revealed a number of desirable modifications to expand the capability of the RCA-110. Most of the upgrades that were incorporated were to expand the input/output capability. This upgraded computer version was designated the RCA-110A.

The transition from manual checkout to partial automated checkout was a gradual process during the Saturn IB program. Also the ultimate goal was to make the stage checkout at the contractor's plant as compatible as possible with the overall assembled vehicle checkout at the launch site. Due to MSFC's cutting edge involvement in the automated system development and the fact that the earlier S-I and all the IU stages were built in Huntsville in close coordination with the MSFC subsystem designers, there was a somewhat easier transition to automation on the S-I and IU than on Douglas' S-IVB stage. In fact, since Douglas used all manual checkouts on the S-IV, the initial intent of their test engineers was to continue the same mode with the S-IVB. As the program progressed, some progress was made in incorporating partial automation in the checkout of all stages at the contractor's plants and at the launch site. It is estimated that by the end of the Saturn IB program, automated equipment controlled over 50 percent of the tests.

Although partial automated checkout proved to be an advantage on the Saturn IB vehicle, with the arrival of the Saturn V with its complex three propelled stages plus the IU, a higher degree of automated checkout was absolutely necessary from a time standpoint. All of the stage contractors, Boeing (S-IC), North American (S-II), Douglas (S-IVB), and IBM (IU) eventually accepted the necessity of converting to primarily automated in lieu of primarily manual checkout. At the stage level Boeing and IBM used the RCA-110A computer in their checkout systems, whereas North American and Douglas chose to use Control Data Corporation's CDC-924A computer. It was a later generation computer with some added test capabilities compared to the RCA-110A. Although the use of a different checkout computer for some of the stages at the contractor's plant versus the launch site was a matter of some concern, the coordination problems were overcome.

In spite of the obvious necessity for a high level of automation at the launch site because of time and accuracy considerations, there remained some serious concern on the part of test personnel to be totally dependent on the computers to independently monitor critical items and make critical "decisions" concerning launch readiness. To alleviate these concerns auto/manual switches were added on the consoles, which allowed the test engineers to switch to manual if they had any reason to doubt the computer information. This feature significantly increased the acceptance of automation on the part of the test personnel and the responsible subsystem engineers. As the Saturn V program progressed, the acceptance of and confidence in the automated checkout systems continued to increase. It is estimated that by the end of the program, all test activities were 90% automated.

It is not an overstatement to observe that among the many factors which led to the phenomenal success of the Saturn IB and Saturn V programs with no in-flight failures was the thoroughness, timeliness, and accuracy that was provided particularly at the launch site by the MSFC Automated Checkout System.

Astrionics Laboratory Staffing Changes

In late 1968, a decision was made to create a Central Systems Engineering Laboratory as a key part of the MSFC organization. There was a desire on the part of management to have the organization headed up by a senior leader from within the Science and Engineering Directorate. Dr. Walter Haeussermann, who at that time was continuing to serve in his role as director of the Astrionics Laboratory, was selected to fill the newly created management position.

Since the staffing and organization of the new Central Systems Engineering Laboratory was being created out of the existing managers and personnel of the Science and Engineering Directorate, Dr. Haeussermann naturally chose to take some key personnel out of the Astrionics Laboratory (as well as other laboratories) to assist him in implementing his new responsibilities.

The Astrionics Laboratory's responsibilities were not affected by the creation of Central Systems Engineering. However, a few changes were made in the organizational structure compared to that of the Astrionics Laboratory of the early 1960s. Brooks Moore was selected to be the new Astrionics Laboratory director and Bill Horton to be deputy director. Bob Kruidenier was the assistant director and there were three staff offices: Hans Hosenthien became director of the Research and Analysis Office, Heinz Kampmeier-Projects Office and Billy Payne-Operations Office. The laboratory was organized into seven divisions: Systems, Instrumentation and Communications, Electrical, Guidance and Control, Computers, Electromechanical, and Technology.

The Systems Division was a new division created within the Astrionics Laboratory. It was an expansion of the old Systems Engineering Office with responsibility for the guidance and control and vehicle dynamic analyses, generation of the requirements for all astrionics subsystems, and overall systems integration. Fred Wojtalik and Gilbert Gassaway were chief and deputy of the System Division. Norm Gilino was head of the Engineering Office and Charles Riley was head of the Systems Development Office. The division was organized into four branches with the following technical leaders: Electrical Systems—John Stroud and Don Woodruff; Communications and Data Systems—John Cox and Leo Arsement; Guidance and Control Systems—Melvin Brooks and Jim Blanton; Vehicle Dynamics—Harold Scofield and Harold Mink.

The Instrumentation and Communication Division responsibilities remained essentially the same as they had been throughout the Saturn program with new leadership: J.T. Powell and Jack Duggan were chief and deputy of the division with staff assistant Jim Atherton. Grady Saunders headed the Theoretical Studies Office. The division was organized into four branches with the following technical leaders: Communication and Tracking—Tom Barr and Warren Harper; Telemetry and Data Technology—Walt Frost and Bob Eichelberger; Measuring Instrumentation—Ted Paludan and Tom Escue; Imaging—Jim Derrington and Carl Huggins.

The Electrical Division also retained essentially the same responsibilities with new leadership. Bob Aden and Phil Youngblood were chief and deputy of the Division with Staff Assistant Werner Rosinski and an Advanced R&D Office lead by Bill White. The Division was organized into three Branches with the following technical leaders: Design and Development—Jim Stulting and Bill Glass; Power—Charley

Graff and Lelous Wood; Advanced Development and Technology—Ellis Baggs and Jim Felch.

The Guidance and Control Division did undergo some change of responsibilities. Essentially the previous Gyro and Stabilizer Division was combined with Guidance and Control, and the responsibility for computers was separated to create a new division. The new leaders of the Guidance and Control Division were Carl Mandel and George Doane as chief and deputy. The new organization had two staff offices: Engineering Planning lead by Bryce Alsup and Robert Stacy and Theoretical Studies led by Richard Tuggle and Joyce Neighbors. The division was organized into four Branches with the following technical leaders: Sensors—Pete Broussard and Tom Morgan; Control Mechanisms—Mike Kalange and John Caudle; Advanced Technology—Clyde Jones and Bobby Walls; Flight Qualification—Fred Kelley and Charles Lee.

The Computer Division was a new division created by transferring out of the Guidance and Control Division the responsibility for all flight and ground digital and analog computers. Charles Swearingen and Jack Lucas were chief and deputy of the of the Computers Division, with Gerald Turner heading up the Advance Computer Studies Office and Ernest Orem heading up the Engineering Planning Office. The division was organized into three branches with the following technical leaders: Flight Computers—Jack Bridges and Harrison Garrett; Ground Computers—Jim Lewis and Sherman Jobe; Flight Systems Simulation—Huey Coppock and Gene Lee.

The Electromechnical Division retained the same responsibilities that it had throughout the Saturn Program. Josef Boehm and Glen Barr were chief and deputy with Staff Assistants Ernst Lange and Walter Kasparek. Duane Counter and Joe Rackley lead the Engineering Analysis Office. The division was organized into three Branches with the following technical leaders: Mechanical Engineering—Helmut Pfaff and Herman Wagner; Electronic Engineering—Jim Blanche and Ralph Kaufmann; Environmental Test—Cliff Kirby and William Allen.

The Technology Division also retained the same responsibilities as with the previous organizational arrangement. Jim Taylor and Earl Reinbolt were the chief and deputy with Technology Staff Assistants Carl Winkler and Alvis Holladay. Bill Kreider headed the Planning Office. The division was organized into three branches with the following technical leaders: Microelectronics—Dorrance Anderson and G. Bailey; Optics—Joe Randall and Chuck Wyman; Digital Applications—John Gould and Klause Jurgensen.

The above named individuals made up the leadership of the Astrionics Laboratory after the reorganization, which became effective in early 1969. These leaders were supported in carrying out their duties by the several hundred highly skilled and dedicated engineers, technicians, and administrative personnel who made up the Astrionics team for the last four years of the Saturn V Lunar Launch Vehicle Program. This was the period during which the flights with the mission of landing astronauts on the Moon and bringing them safely back to earth was fulfilled. There were six successful lunar landing missions for a total of twelve American astronauts to walk on the Moon and three additional successful manned flights, which orbited the Moon.

The Saturn V Flight Test Program

The Saturn V Flight Program was an amazing success, considering the complexity of the system. There were of course many instances of anomalies, "glitches," component malfunctions and failures that occurred during the comprehensive test programs leading up to the assembly of the composite launch vehicle at the launch site. Even during pre-launch testing at the launch site, anomalous conditions continued to be detected and corrected, often with launch delays. The launch-site automated checkout system used extensively on the Saturn V launches helped immeasurably in the detecting, analyzing, and correcting problem areas in a reliable and timely manner.

In spite of all precautions taken in designing, testing, and extensive pre-launch re-testing, there were still a few in-flight anomalies that occurred. However, no in-flight anomaly or failure in any Astrionics sub-system ever endangered the vehicle or the safety of the on-board crew during the twelve flights of the Saturn/Apollo Program. There were some mission compromises where all missions goals of each flight were not entirely met, but all were either partially successful or totally successful.

The mishap on Apollo 13, the explosion in the Service Module on the way to the Moon, was the most dramatic and publicized example of the abandonment of the primary mission of landing two crewmembers on the Moon. However, the revised primary mission of saving the crew and "returning them safely to Earth" was 100 percent successful, thanks to the ingenuity of the Johnson Space Center team. Even on this fateful flight, it should be noted that the Saturn V performance was normal with no anomalies during the launch phase or in the injection of the spacecraft on the lunar trajectory traveling at the prescribed velocity of 25,000 mph.

In analyzing the performance of the Saturn Astrionics Lab flight subsystems during the twelve Saturn flights, the conclusion is obvious that the functional performance and the reliability of all were outstanding. This was due at least in part to the practice of ultra conservatism in the design and development process. The incorporation of previously flight-tested techniques and components was also a factor as was the emphasis on simplicity to the maximum extent possible. The extensive yet judicious application of redundancy techniques in flight critical subsystems added immeasurably to the overall reliability of those elements.

Some of the same comments made above about the importance of the overall vehicle test program on the phenomenal success also apply to the Astrionics Lab subsystems. In particular, the comprehensive test programs at all levels in the laboratory detected defects, as the tests were intended to do, and corrective actions were taken. This philosophy of thorough testing at all levels of assembly did not eliminate but minimized anomalies in the Astrionics subsystems. However, no in-flight anomaly or failure in any Astrionics sub-system ever endangered the vehicle or the safety of the on-board crew during the 12 flights of the Saturn/Apollo Program.

There were examples of situations on some flights where an Astrionics subsystem performance exceeded expectations and the specified requirements. One example of that was Apollo 6, SA-502. This was the second unmanned Earth orbit test. The first similar test, Apollo 4, SA-501 launched on Nov. 9, 1967 had gone very well with no significant anomalies. However, the Apollo 6 launch on April 4, 1968 had several propulsion anomalies. The first stage propelled the vehicle normally, even though there was an unanticipated high level of propulsion-induced vibrations during the first stage burn. However, soon after ignition of the second stage, two of the five engines malfunctioned resulting in thrust termination on both. Both failed engines were outboard engines and were two of the four normally gimbaled to control the orientation of the vehicle.

To meet the no single point failure criteria, the system had been designed so that the thrust of one engine could be lost and the second stage could continue with the thrust of only four engines. Likewise, the control system had been designed so that the vehicle could remain under control with only three gimbaled engines and this control had been verified with extensive analyses and computer simulations.

There was no requirement or expectation that the system could continue its mission to reach earth orbit propelled by three engines and controlled by two. However, the automated system adjusted the two remaining gimbaled engines to maintain control and the vehicle continued to move forward at a reduced rate of speed with 60 percent of normal thrust. Since the vehicle was still moving forward in a controlled manner, a decision was made not to abort. The automated system delayed sending the engine cut-off signals and thereby extended the flight time sufficiently for the vehicle to reach Earth orbit. Remarkably, most of the original mission objectives were met.

Another example of where the astrionics subsystem performed beyond expectations and the specific criteria for which it was designed and tested was on Apollo 12, SA-507. The previous flight, Apollo 11, SA-506 launched on July 16, 1969, had been a remarkable success. Neil Armstrong and Buzz Aldrin had successfully landed and walked on the Moon and been returned safely to Earth.

Apollo 12, with Pete Conrad, Alan Bean and Dick Gordon on board, was launched on Nov. l4, 1969. The countdown proceeded to liftoff even though there were rainy conditions in the area. At 36 seconds into the flight, the vehicle was struck by a tremendous bolt of lightening. Most electronic systems in the command module were blacked out, and the lights and meters being monitored by the astronauts, to use their terminology, "went crazy" and they were "flying blind."

The redeeming feature to the story is that a decision had been made quite early in the development of the Saturn V—at the great displeasure of many of the astronauts—that the guidance and control of the vehicle during the launch phase would be totally automated. For reasons not completely understood the sensors, Stabilized Platform, computers, engine actuators and all other elements of the Astrionics Lab guidance and control system continued to guide and control the vehicle without the slightest deviation and the injection into Earth orbit was completely normal. Post-flight analysis showed not even the slightest "glitch" in any of the critical systems at the time of the lightning strike.

Fortunately, the spacecraft electrical systems were not permanently damaged and the astronauts were able to reset the power systems and computers and bring them back on line before they were needed for the translunar flight. The rest of the mission was relatively uneventful, and Pete Conrad and Alan Bean became the third and fourth men to walk on the Moon.

Afterwards, the crew thanked us profusely for our design of a lightning-proof system, which disregarded completely the massive lightening bolt that might otherwise have aborted our second manned mission to the Moon, or resulted in even more dire consequences.

In Retrospect

I consider myself most fortunate to have had the privilege of participating as a member of the Huntsville Rocket Team in our great nation's conquest of space. The culmination of the first major phase of that conquest was the safe landing over a period of three years, beginning in July 1969, of twelve astronauts on the Moon and returning them safely to earth. The feat has been lauded by many as being the greatest technological accomplishment of the 20th Century.

It was also gratifying to have had the opportunity to participate with the team during the decade of the 1950s in the development of U.S. Army rockets. The progression of developments and the experience gained on those guided missile programs during the 1950s contributed immeasurably to the success of the lunar landing program of the 1960s.

I am convinced that my very rewarding career has been due in great part to "being at the right place at the right time." I have taken great pride in being a part of the WvB team. WvB was uniquely qualified to lead our space program and we are deeply indebted to him as a nation for his service, and I am also deeply indebted to him personally for the honor of having been associated with him over a period of 18 of his 20 years in Huntsville.

I had many mentors among the 120 "original" WvB team members, but the one to whom I owe most is Dr. Walter Haeussermann. He initiated my training program when I was first assigned to him in 1952. Other than the mentors to whom he temporarily assigned me that first year, I reported directly to him for the next 16 years in the following positions: deputy chief, Simulation Section; chief, Control Section; chief, Navigation Branch; and chief, Guidance and Control Division. When Dr. Haeussermann left the Director's position to take over the new Central System Engineering, he insisted that I be named the director of Astrionics Laboratory, the highest honor that I could ever expect to receive. In the latter position, I became a member of the MSFC senior staff, reporting to Mr. Herman Weidner and WvB.

In addition to my mentors, I also want to recognize and thank the many dozens of highly skilled and dedicated friends and associates who worked with me during the early and middle years of my career and for the hundreds of highly skilled friends, fellow workers and supporters who worked with me on the Astrionics team. I consider it a great privilege to have served with each of you in my "team leader" positions. You deserve full credit for the amazing once-in-a-lifetime accomplishments of our Astrionics team.

Test Lab

By Ed Buckbee

Huntsville has always been something of a mystery. Why Huntsville? Why is NASA here? What attracted politicians, going back to the days of U.S. Sen. John Sparkman (D-AL) and Congressman Bob Jones (D-AL)? The arsenal or what some referred to as the "Bullet Factory" was hopefully going to develop into something more than a sister to the Watercress Capital of the World. During the missile days, not many politicians came to visit what were mostly classified programs. As soon as NASA, the civilian space agency was formed, President Dwight D. Eisenhower arrived to dedicate the new NASA center in honor of his wartime friend, Gen. George C. Marshall. When President John F. Kennedy visited to witness a Saturn booster ground test firing, that was the coming out party for Huntsville. Suddenly, Huntsville was overwhelmed with visitors from Washington. Houston (Manned Spacecraft Center) had astronauts. Huntsville had WvB and the fire and smoke show.

Marshall's Test Laboratory was the most highly visible laboratory within the complex. Flying into Huntsville, you couldn't miss seeing those tall test stands, some of which were the tallest structures in the state. As you arrived, you couldn't fail to hear the sounds, see the fire and smoke and feel the vibrations. It was known as the "dangerous zone" at Marshall Center. Because of the test firings conducted there, it was the attraction and showplace for VIP visitors such as Kennedy and many congressional leaders and members of the press.

Testing was important. Test the small components first and assemble them into the big components and finally put them all together into the system. Within the Test Lab, there existed the capability to test fire a range of rocket engines from 500 pounds to 7.5 million pounds thrust.

Jack Conner who worked in the Test Lab most of his career shared his thoughts: "It is quite evident from those associated with space launch vehicle development that extensive ground testing simulating actual operational environments of components, sub-systems and complete systems contributed immeasurably to the 100 percent successful flight records of the Saturn I,

IB and V. During the Saturn program, the Lab was divided into the East and West test areas. Major facilities included component, single engine, and booster test stands where the hot firings were conducted. The blockhouse served as the control center. The dynamic test stand enabled the entire Saturn V launch vehicle to be assembled vertically and tested. Before a Saturn vehicle was shipped to the Cape for launch, we had static fired the engines, assembled the vehicle vertically in the dynamic test stand and tested the launch tower swing arms. I was proud to be a member of the Test Lab team."

WvB loved to orchestrate events, especially when VIPs were coming to visit. Ground test firings of the Saturn rocket were incredibly popular. In fact, Marshall was the only show in NASA in the early '60s. In Huntsville we called it the "fire and smoke" show. Initially, we put everyone in the dark and drab blockhouse where they could observe the firing, protected behind portals of several layers of glass and thick concrete walls. Although quite close to the test stand, they couldn't hear the sound nor feel the heat or vibration of the firing. WvB didn't like it. He felt the whole point of the demonstration had been lost. WvB's trusted secretary, Bonnie Holmes, related what occurred: "If they were in the blockhouse, they didn't get the full feel of the test firing. WvB suggested they put them on the roof of the blockhouse so they could get real close to the test stand, feel the vibrations, the heat, and all of that. So when this was done, they were very impressed by getting to witness some of those Saturn firings right there in the Test Lab."

This worked for awhile, until the safety people declared it unsafe for visiting dignitaries. It was suggested visitors observe from an open bunker on what became known as " Heimburg Hill," named after the director of the Test Laboratory, Karl Heimburg. The bunker had a roof, and a concrete wall with a slit for viewing the test stand. It was open in the back and affectionately referred to as the "cow shed." Upon completion of the next firing, the VIP's were not only impressed, they were startled and some scared out of their wits. Suddenly, we had a flood of Washington politicians coming to Huntsville to witness one of "WvB's fire and smoke shows." Standing 2,000 yards away in the "cow shed," first timers watched a giant rocket ignite generating 32 million horsepower, shaking the ground, creating shock waves, pressure against their chest like a hammer, and heat flashing directly upon their faces. It turned out to be a crowd-pleaser that WvB took great pride in orchestrating.

The visit of President John F. Kennedy was important to the center. It was a huge deal. JFK had challenged the U.S. to beat the Russians to the Moon within the decade. He was counting on us in Huntsville to deliver the rocket that would get our astronauts there. White House and the Secret Service staffers came to visit. We were informed the president's call sign during the visit would be "Lancer" and WvB's, "Rocket Man." We presented our tour schedule for the president and described our plans to conduct a test firing of the Saturn I booster for two-plus minutes. The staff looked concerned, thinking we were talking about launching a rocket. Once they saw the Saturn I test stand and viewing area from Heimburg Hill, they asked about a potential failure and possible blast zone that might harm the president. Finally, after several meetings and lots of hand wringing, the firing time was reduced to 30 seconds requiring less fuel and reducing the blast distance in the event of a failure.

Heimburg Hill became the site for the president to observe the test. Standing 2,000 yards away, the president was tense as he watched the Saturn booster ignite, shooting fire and smoke skyward, vibrating the ground, generating 32 million horsepower. After the firing, President Kennedy was excited. Displaying that famous Kennedy smile, he shook WvB's hand, congratulated him and said, "That was fantastic, impressive, it feels so powerful." They climbed in the limousine and rode to the Blockhouse, stopped and asked who was in charge of the firing. The Blockhouse crew answered, "Saidla." Bob Saidla made his way to the limo and was congratulated and shook hands with President Kennedy and WvB.

During the 1960s, the S-IC Saturn V booster test stand, became the focal point in Test Lab. This mammoth structure was connected to four huge concrete pedestals anchored to bedrock, 40 feet deep. Initially it was designed to test a rocket stage generating in the range of 6-7 million pounds thrust but was modified to accommodate a Nova-class booster of 10-12 million pounds thrust.

In March 1965, the first Saturn V booster was delivered to the Test lab to be erected in the S-IC Test Stand. The Public Affairs Office (PAO) had invited the press to witness this milestone event. When we arrived at the Test Stand, Bart Slattery, the public affairs officer said, "The rocket has no engines. Where are the engines?" No one seemed to know and Slattery said, "Where's Heimburg?" The Manufacturing Engineering Lab had shipped the stage without engines, according to Heimburg, to maintain the delivery schedule. Plans were to mount the engines after the stage was erected by Fritz Vandersee in the Test Stand. This was not well received by the PAO people who had planned an elaborate press event featuring WvB viewing the first Saturn V booster being prepared for test. This S-IC-T stage was known around the center as the "T-Bird," short for test bird.

Bob Saidla, test conductor, was there that day and describes the event: "When we were finally told to proceed with the lift into the test stand, the wind had increased to a real stormy force. The S-IC-T had to be raised up and over the load platform to set down on the hold down arms. There was very little clearance. The stage was so large that the wind had a big sail to work with. We were not properly prepared. The stage was banging from side to side in the load platform as it was lowered. We had crewmembers all around it at every platform level to guide it in with our hands and feet as was our custom. We could not control it and we were literally using our bodies as fenders to keep from damaging the stage. It was a very tense hour, but we did get it in and set down with no major damage."

To observe a static firing of a S-IC stage close-up was much more impressive then watching a launch. At a launch you were three miles way from the rocket on the pad. As ignition occurred, the rocket lifted off skyward and within seconds was gone as the rocket engine sounds decreased.

Static test firings were most often held in the afternoon. In the mid-'60s, a full duration test for 2 1/2 minutes with outboard engines being gimbaled or swiveled was the featured attraction in the Test Lab. For test firing viewers, the process begin with alarms blaring and lights flashing indicating a firing was about to be held. Announcements were made: "All personnel, evacuate the area immediately." Water at the rate of 320,000 gallons per minute was released to prevent the flame bucket from melting. Standing in an open field, observing the firing from about 2,000 yards, you listened as the countdown went to single digits and you heard "Two, One, Zero." For a brief moment, there is no sound, just a huge ball of fire coming from the engine compartment. Suddenly, the shattering sound is upon you. The shockwave and force hits your chest like a hammer. The ground is moving and vibrating under your feet. You feel the heat on your face. The flame climbs 1500 feet into the sky. You look at your buddy as he leans toward the rocket and his clothes are tight against his body as if he's in the midst of a severe windstorm. As the engines are swiveled or gimbaled, the thundering sound increases and reaches decibel readings of 130-plus (threshold of pain). As the fuel is consumed by the engines, the sound increases and reverberates off nearby structures. Suddenly, shutdown occurs and no sound. There is dead silence, just a giant smoke cloud climbing into the sky. It truly was, "Thunder Over Alabama."

During the Saturn program, I observed 40 test firings of single engines or boosters, mostly with first-time viewers who were VIP's or members of the press. One of those members of the press was a *Life* magazine photographer invited to witness the first full-duration firing of the S-IC booster stage. I was instructed by my boss to, "get him as close to the test stand as you can." I drove a NASA pickup truck, entering the Test Lab through a back gate used by the technicians who always seem to have a better viewing site then anyone else. As the photographer was setting up his camera, I explained what was about to occur: Prepare yourself—for severe noise level, vibration, shock waves, heat—even though we are in the safety zone for this test. After he set up his camera on a tri-pod and sighted toward the test stand, he said to me, "Look son, I have covered wars, floods, hurricanes, tornados and any other destructive event you can imagine, I think I can handle this little rocket firing."

The Saturn booster produced 7.5 million pounds of thrust. That's equivalent to 160 million horsepower. Someone compared that to a string of automobiles, bumper-to-bumper from Seattle, Wash., to New York City with the accelerators to the floor. All horses were running that day. As the booster was ignited and the sound and vibration came upon us, the *Life* magazine photographer knocked over his camera and ran across the field, seeking shelter. Although he failed to get one picture, I convinced a Marshall Center photographer to give up one of his 4x5 negatives from his big box camera to share with our distinguished photographer from *Life* magazine.

SATURN V PROJECT OFFICE

By Bill Sneed

Editor's Note: Bill Sneed, a long-time member of the WvB team, was one of the young American engineers mentored by German scientists in the 1960's. He was head of the Saturn V Program Control Center proclaimed by NASA Administrator James Webb as, "one of the most sophisticated forms of organized human effort that I have ever seen anywhere." This was before computers. Sneed provides an overview of the organization structure and shares with us the critical operation of the Saturn V project office.

The Office of Manned Space Flight (OMSF) was established to oversee and direct all activities necessary to accomplish the lunar landing mission. Dr. Brainerd Holmes was initially selected to be the Associate Administrator for OMSF, reporting directly to the NASA Administrator. He was succeeded early on by Dr. George Mueller, who served in that capacity throughout the remaining duration of the lunar landing program. Under the previous NASA organizational structure, the three NASA field centers who were so instrumental in accomplishing the lunar landing mission reported directly to the NASA Administrator, Mr. James Webb. Under the new organizational structure, the field centers reported directly to the OMSF Associate Administrator so that the centers could direct their entire attention to the lunar landing mission. The three centers involved in this reorganization were: Marshall Space Flight Center (MSFC) under the direction of Dr. Wernher von Braun, Johnson Spaceflight Center (JSC) under the direction of Dr. Robert Gilruth and the Kennedy Space Center (KSC) under the direction of Dr. Kurt Debus.

Program offices were established within the OMSF to oversee and direct the three programs essential for preparing for and accomplishing the Lunar landing mission, including the Mercury, Gemini and Apollo Program Offices under the direction of Gen. Samuel Phillips. Program directors were selected for each of the above programs with the responsibility for overseeing and directing all activities relating to their respective programs. These program directors, having total programmatic responsibility, worked directly with the program managers at the field centers who had responsibility for implementing their assigned program element. The field center directors retained institutional responsibility for the successful accomplishment of their assigned program elements.

A number of "functionally oriented" offices were established within the OMSF program offices which were located in Washington D.C., as well as within the program offices at the three field centers. These "functionally oriented" offices had the primary responsibility for coordinating the day-to-day activities of the lunar landing program. The five functionally oriented offices established were Program Control, Systems Engineering, Test, Reliability and Quality and Flight Operations. This organizational arrangement greatly enhanced the response time for communicating program status and problems areas, problem resolution and timely decision making. Similar organizational structures were put in place for other Apollo program elements, e.g. Saturn I/IB, Engines and Apollo spacecraft.

Saturn V Space Vehicle

The Saturn V space vehicle was a vitally important element of the Apollo lunar landing program. It's two primary requirements were to inject (as a minimum) 90,000 pounds of payload into a lunar transfer trajectory, and support a lunar landing before the end of the decade.

Meeting these requirements was a most challenging assignment. It required the development of a launch vehicle an order of magnitude larger than any vehicle available at that time. After many years of effort, these requirements were met, and even exceeded for the lunar exploration program that followed the first lunar landing.

Much has been written about the successful technical achievement of the lunar landing program and its significance to the prestige of our nation. Not so much has been written about the management achievements and the thousands of people within NASA and the American industry that made it all possible. Hopefully, the following sections will provide some insight into how the Marshall Center team went about accomplishing the monumental task of developing the Saturn V launch vehicle, making it possible for NASA to land a man on the Moon by the end of the decade.

Marshall's Role for Developing the Saturn Family of Space Vehicles

The Marshall Space Flight Center, under the direction of WvB, was assigned responsibility by NASA for designing and developing the Saturn family of launch vehicles necessary to accomplish the lunar landing mission. MSFC was assigned this role because of (a) its extensive expertise and experience in the development of numerous rocket and launch vehicle systems (b) the fact that MSFC had an experienced and proven team in place (i.e. the WvB team) and (c) the center possessed many of the essential manufacturing and test facilities necessary for the development and testing of such a complex system for launching man into space.

In 1962, it became apparent to WvB that it would be necessary for MSFC to restructure its organizational structure in order to implement its expanded role for the design and development of the space transportation systems required for the lunar landing mission. MSFC's expanded role would require the involvement of extensive government and industrial capabilities throughout the United States. Tens of thousands of people and numerous manufacturing and test facilities would be required to design and develop the Saturn family of launch vehicles. The new MSFC organization was comprised of basically two main organizational entities:

–the Research and Development Directorate which contained all the laboratories and facilities needed for the design and development of the Saturn family of launch vehicles. Contained within these laboratories was the technical expertise so essential for MSFC's accomplishment of its expanded responsibilities. The directors of these laboratories were members of the "WvB" team who had come with him to NASA from the Army's Ballistic Missile Agency. These laboratory directors served as mentors to the laboratory division directors, many of which were staffed by up and coming young American engineers and scientists.

–The Industrial Operations Directorate which contained the management organizations that were responsibility for managing the various launch vehicle systems, i.e. Saturn I/IB, Saturn V and Engines. These program offices relied on the technical expertise of the Research and Development Directorate for the technical support of its respective programs.

Managing the Saturn V Space Vehicle

One of the key offices established under the Industrial Operations Directorate was the Saturn V Program Office, which was assigned the responsibility for managing and directing all the in-house and contracted activities necessary for the design and development of the Saturn V launch vehicle. Dr. Arthur Rudolph was selected by WvB to be the program manager for this office. He was selected because of his past experience in the management of large, complex programs such as the Pershing Weapons System and the Redstone Missile System while with the Army's Ballistic Missile Agency. Additionally, in Dr. Rudolph's earlier career he had extensive experience in the design, development, manufacturing, testing and quality assurance for various rocket and engine systems. Dr. Rudolph's appointment as manager of the Saturn V program was instrumental in the success of the Saturn V launch vehicle since many of his subordinate managers had very limited experience in the management of large, complex launch systems. He mentored, advised and directed these managers on the intricacies of their respective responsibilities based on his vast knowledge and past experiences.

The Saturn V Program Office was comprised of basically two levels of management. One level of the management structure contained five functionally oriented offices mentioned earlier, that were primarily responsible for defining launch vehicle level requirements; allocating these launch vehicle level requirements to subordinate elements of the launch vehicle system; and monitoring the implementation and successful implementation and accomplishment of these requirements. These five functionally oriented offices and the responsible managers were: Program Control—Bill Sneed; Systems Engineering and Integration—Lucian Bell; Test—Howard Burns; Reliability and Quality Assurance—Jewell Moody; Flight Operations—Arthur Rowan.

The second level of the organization structure contained five "hardware oriented" project offices for the launch vehicle stages and the launch vehicle ground support equipment. These offices were responsible for managing and directing both the in-house and contracted activities for their respective areas of responsibilities. The five hardware oriented project offices and the responsible project managers were: S-IC Stage—Matthew Urlaub/Jim Bramlet; S-II Stage—Elmer Field/Col Sam Yarkin/Roy Godfrey; S-IV Stage—Roy Godfrey/James McCullogh; Instrument Manager—Frederick Duerr; Vehicle Ground Support Equipment—Spencer Smith.

Both the "functionally oriented" and the "hardware oriented" project offices relied extensively on the technical expertise that resided in the various Research and Development laboratories. These laboratories contained the corporate memories from many years of experience obtained from the design, development and testing of numerous rocket, launch vehicle and propulsion systems. This laboratory expertise was vital in providing the all-important technical support to the various Saturn V Offices. The support provided by the various laboratories was one of the most—if not the most—important contributing factors to the success of the Saturn V launch vehicle in placing a man on the Moon.

The design and development of the Saturn V launch vehicle could not have been accomplished without the participation of American industry. MSFC had a sizeable capability for manufacturing and testing of the Saturn V launch vehicle system, but not large enough to accomplish the total job in the time available to complete the lunar landing mission. MSFC designed, manufactured and tested the ground test articles and the early flight articles for the S-1C Stage and the Instrument Unit. MSFC had to reach out to highly qualified and experienced companies to assist with its in-house efforts as well as to assume responsibility for designing, developing and manufacturing other major elements of the Saturn V launch vehicle system. The companies selected and the responsible executives for this critical effort were: S-1C Stage-Boeing—Richard Nelson; S-11 Stage-North American Aviation—Bill Parker/Robert Greer; S-1VB Stage-McDonnell Douglas—Jack Bromberg/Ted Smith; Instrument Unit-IBM—Clinton Grace; Ground Support Equipment-General Electric/Systems Engineering & Integration-Boeing—Lionel Alford.

These companies worked in close collaboration with the MSFC managers and technical personnel in the laboratories in accomplishing their respective responsibilities. Early in the program, the relationship between MSFC and its contractors was a little contentious. But, as the working relationship between the two parties evolved, it became more harmonious and subsequently developed into an outstanding team effort that was so essential for the successful and timely completion of the lunar landing mission.

Program Control Office

A program of the magnitude and complexity and having the national significance of the Saturn V launch vehicle, required that an effective management information system be developed to ensure that all technical requirements and schedules be met within budget commitments. Knowing this need, Dr Arthur Rudolph, the Saturn V Program Manager, asked his Manager of Program Control, Bill Sneed, to take on the challenging task of developing a management information system for accomplishing this pressing need. It was imperative that a Saturn V management system be designed that would integrate all elements of the Saturn V program (hardware, software, facilities, etc.) and up-to-date information in sufficient depth necessary to ensure that all program objectives would be met in accordance with program commitments.

The Boeing Company was contracted to assist in the development of the Saturn V management information system because of its prior experience in developing a similar system for the Air Forces' Minuteman program. Extensive effort was put into determining what information should be contained and displayed in the management information system and who would be responsible for developing and maintaining the information on a real time basis. The resulting product of this effort was the development of the Saturn V Program Control Center.

The Saturn V Program Control Center displayed a cross-section of information determined to be essential for assessing program status and problem areas for all elements of the Saturn V program. Following are examples of some of the types of data displayed in the Program Control Center:

- Summary PERT Logic network identifying and interconnecting critical activities and events for the entire Saturn V program.
- Projected payload capability for the Saturn V launch vehicle compared to the payload capability commitment.
- Ground test program plans for each of the five (5) ground test vehicles.
- Qualification and reliability status for critical components of the Saturn V launch vehicle.
- Configuration Managements data reflecting outstanding change orders, installation and verification of modification kits to be installed in each launch vehicle, etc.
- Comprehensive development schedules for each of the propulsion stages, (i.e. S-IC, S-II, and S-IVB), Instrument Unit and Ground Support Equipment.
- Detailed integrated schedules for each flight vehicle (i.e. AS-501, AS 502, etc.)
- Identification of top 10 program problem areas, including a plan of action to resolve each identified problem area.

The Saturn V Program Control Center was an invaluable management information system for assessing program progress and problem areas. The data displayed in the Saturn V Program Control Center provided the Saturn V program and office managers with essential information and visibility needed for assessing program progress, identifying problem areas and their impact on the program, and in problem resolution. The Program Control Center was used for weekly internal Saturn V Program Office staff meetings, monthly program reviews with other MSFC supporting organizational elements, briefing key NASA officials and others on the status of the Saturn V program.

In 1965, the NASA administrator, Mr. James Webb, came to the center for a briefing on the Saturn V program. Following the briefing, he spoke to a select number of MSFC employees at which time he made the statement that he had just seen," one of the most sophisticated forms of human effort that I have ever seen anywhere." As a result of Mr. Webb's visit to the Program Control Center, he invited numerous dignitaries from other government agencies, industry, and academia to visit the center.

Saturn V Flight Program

There were a total of fifteen Saturn V launch vehicles built for the lunar landing program. Six (6) of these were needed to accomplish the first lunar landing, six (6) used for additional lunar missions, and one (1) used to place Skylab, the world's first space station, into Earth orbit. The two (2) remaining vehicles are on display in space museums. Of the thirteen (13) vehicles flown, twelve (12) were declared to be a total success, a truly remarkable success story when considering the size and complexity of the Saturn V launch vehicle. The other one (AS-502) did not accomplish all of its mission objectives.

Three (3) of the six (6) Saturn V launch vehicles used for the lunar landing were of significant importance to the success of the lunar landing program and thus warrant some special attention.

First launch—AS-501

First manned flight—AS-503

First manned lunar landing—AS-506

First flight of AS-501

The preparation for and the launch of the first Saturn V launch vehicle, AS-501, was one of the most important milestones for the timely accomplishment of the manned lunar landing mission. The fact that the first launch of the Saturn V launch vehicle encompassed the use of all live propulsive flight stages added to its significance, a configuration referred to as the "all up" concept. The all up concept was a new approach for the Marshall Space Flight Center, a concept that was reluctantly agreed to by MSFC officials. MSFC's preferred approach for the first flight of the Saturn V launch vehicle was to use only a live S-1C first stage with dummy upper stages. Additional live upper stages would have been added to subsequent fights. Had the MSFC's preferred approach been followed, the use of all live propulsive stages would not have occurred until the fourth flight of the Saturn V launch vehicle.

Dr. George Mueller, the Associate Administrator for Manned Space Flight, made the decision to go with the "all up" concept for the first flight as a means for conserving time to ensure that the first manned lunar landing could be accomplished within the decade. After much deliberation, WvB and his team members agreed to support the "all up" concept for AS-501.

The launch of AS-501 on Nov. 9, 1967, was the culmination of many years of hard work, long hours, and personal sacrifices by many NASA and contractor people involved with the Apollo program. The days, weeks and months leading up to the first Saturn V launch vehicle were extremely intense and pressure packed for all the Apollo team members at MSFC, JSC, KSC and their contractors. The decision to launch AS-501 was a long and arduous process necessary to determine if the launch vehicle had been adequately qualified for flight. Complicating this decision was the fact that numerous failures had occurred during the ground testing of the various Saturn V stages. These failures brought into question the readiness to launch the first flight. But, after completing a comprehensive and thorough Flight Readiness Review, an unanimous decision was made by all senior OMSF and prime contractor officials to proceed with plans for the first launch. Coincidently, the launch of AS-501 occurred on the birthday of Dr. Rudolph, the Saturn V Program Manager.

The task of preparing AS-501 for launch fell to the launch operations team and its supporting contractors at KSC under the direction of Dr. Rocco Petrone. To complete this task required an around the clock, seven day work week by the launch operations team. The magnitude of preparing AS-501 for launch can be put into perspective when considering that the initial flight of the Saturn V launch vehicle brought together for the first time:

- All live Saturn V flight stages (S-1C, S-11 and S-1VB), Instrument Unit, Ground Support Equipment and flight software.
- Launch operations facilities (i.e. Vertical Assembly Building, Crawler Transporter and Launch Complex) and supporting systems.
- Apollo spacecraft flight hardware (i.e. Command Module and Service Module)
- Mission Control Center at JSC.

The successful flight of AS-501 was essential for NASA maintaining its schedule for landing a man on the Moon by the end of the decade. The confidence in achieving this objective was greatly enhanced with the success of AS-501. There was great excitement and celebration by the NASA and contractor team members with the successful launch of AS-501, but there was much work left to be done before landing a man on the Moon and safely returning him to earth.

To All Civil Service and Contractors: "My heartfelt congratulations to everyone who contributed to the stunning performance of the Apollo/Saturn V (AS-501) on Nov. 1967. This was a critical test, and the Saturn V looked like a thoroughbred..." Von Braun (Buckbee 2010)

First Manned Flight of AS-503

The excitement and enthusiasm following the first flight of AS-501 was short lived. The next launch, AS-502, launched on April 4, 1968, encountered some very serious flight anomalies. Excessive vibration problems on the S-1C stage caused some major failures on the S-11 and S-1VB stages, which did not allow the launch vehicle to meet all of its flight objectives. The source of all the problems encountered on AS-502 was immediately identified from a review and analysis of all flight data, and subsequently replicated through a series of ground tests. Design fixes were made for each of the anomalies. Necessary flight hardware modifications were made, qualified by ground testing, and installed on all remaining launch vehicles. The OMSF established a launch criteria that required the first manned flight be preceded with two successive, successful unmanned flights. Clearly, the problems encountered on AS-502 did not meet this criterion. This presented OMSF officials with a major dilemma on whether or not to proceed with plans to man the next launch. After many reviews and hours of deliberation, the OMSF officials made a monumental, gut-wrenching decision to proceed with plans to launch AS-503 with astronauts on board. The launch of AS-503 occurred on December 21, 1968, and was declared to be a total success, with all mission objectives being met. With the successful flight of AS-503, NASA was back on track to meet the goal of landing a man on the Moon within the decade.

First Manned Lunar Landing - AS-506

Following the successful launch of AS-503, two additional flights (AS-504 and AS-505) were completed for the purpose of qualifying the Apollo spacecraft and accomplishing other mission operation objectives for the ultimate lunar landing mission. On July 16,1969, AS-506 lifted off the launch pad at KSC with three astronauts on board the Apollo 11 Spacecraft on the first manned Lunar landing mission. On July 20, 1969, astronaut Neil Armstrong set foot on the Moon, followed shortly by astronaut Edwin "Buzz" Aldrin. After spending four days on the Moon, these two astronauts rejoined Michael Collins, who was orbiting the Moon in the Command/Service Module. The three astronauts safely returned to Earth on July 24, 1969, thus achieving President Kennedy's goal of landing a man on the Moon and returning him safely to Earth. The rest is history.

LUNAR ROVING VEHICLE

By Ed Buckbee

One of the most unique projects undertaken by Marshall was the Lunar Roving Vehicle, a manned vehicle to be deployed on the Moon. Normally this would have been a Houston project, because anything an astronaut sat in belonged to the Johnson Space Center. For this to be designed, managed and built by another center was rare. Astronauts who drove the rover on the Moon expressed surprise to its success. Word was that Houston made the design requirements so difficult it would be impossible to deliver a vehicle that would meet the strenuous requirements of weighing no more then 500 lbs, being stored in the lunar module exterior compartment and having automatic deployment when the astronauts activated the system.

Marshall was responsible for the design, development and testing of the rover, which was nick-named the "Moon buggy."

It was a fragile looking, open-spaced vehicle about 10-feet-long with large mesh wheels, antennas, tool caddies and cameras. It weighed 460 pounds on Earth, but on the Moon in 1/6th gravity only 77 pounds. Powered by 36-volt batteries, it had four one-fourth horsepower drive motors, one for each wheel. The weird looking machine was collapsible for compact storage until needed, when it could be unfolded by hand. Marshall engineers contributed substantially to the design and testing of the navigation and deployment system. The rover was designed to travel in forward or reverse, negotiate obstacles about a foot high, cross crevasses about 2-feet wide, and climb or descend moderate slopes. Its speed limit was about nine miles per hour. A rover was used on each of the last three Apollo missions.

The spacecraft on wheels performed faultless during three Apollo missions on the Moon. It extended the astronaut's exploration distance by being driven more then 50 miles. It was given high marks from all who drove it on the Moon.

Saverio "Sonny" Morea served as Marshall's project manager. The Boeing Company was selected as the prime contractor conducting program management, engineering, final assembly and testing. General Motors Delco Electronics Division was the major subcontractor. The first flight article was delivered 17 months from the formal signing of the contract. This is believed to be the shortest development, design, qualification, and manufacturing cycle of any major item of equipment for the Apollo program. The management of this project impressed NASA Headquarters and led to Marshall being considered for other astronaut-related projects such as Skylab.

Morea made this observation after retiring: "A final issue deals with not the Lunar Roving Vehicle story, but rather it's legacy. High school and college students from the U.S. and other countries are invited to Huntsville to participate in a "Moon buggy" design and race competition. The race is conducted at the U.S. Space & Rocket Center. Competition is fierce. Designs are judged by some of the original rover team members. It is a marvelous site to behold." The "Moon buggy competition" celebrated its 18th anniversary in 2016.

Neutral Buoyancy Simulator

By Jim Splawn

In the early 1960s, ground test firings made Marshall the NASA center with spectacular experiences and the place to visit. The "Tank", as the Neutral Buoyancy Simulator was referred to around the center, became a huge attraction to the press and visiting dignitaries in the late '60s. During that era it was so typical of the Marshall team to invent something not in existence, challenging to design, built in-house that enabled astronauts to save the Skylab Space Station. It established the Marshall center's connection to the astronaut corps.

Introduction

Throughout our U.S. history, man has responded to challenges and opportunities to expand his current surroundings. Our space program offered yet another opportunity to follow this trait and push our boundaries to new levels. This new frontier would offer unlimited opportunities for all existing skills and generate demands for yet to be defined dreams.

A nationwide interest in the mechanics and promises of this evolving new frontier was contagious. Some were simply curious and intrigued by the unknowns of space travel; some were spellbound by the "heroes" of the risk-taking astronauts; some were curious of the scientific and technological advancements; some desired the exciting fracture of and advancement in educational boundaries; some saw significant business opportunities; some were simply 'caught up in pursuing the unknown'; while others were simply enamored in witnessing the mystery and excitement of accomplishing something man had never before done.

These intriguing dreams would dictate that many new and innovative tools, both mental and physical, be conceived and implemented to meet the new parameters of space travel and temporary habitable homes.

The Neutral Buoyancy Simulator (NBS) became one of these visionary tools that enabled engineers, scientists, technicians, and numerous other skills and trades to become active contributors to assist man in making space become a viable and technology-changing asset.

In the mid-to-late 1960's, it became very apparent that NASA would meet the national goal established by President John F. Kennedy to land man on the Moon and return him safely to Earth. In full belief the once-in-a-lifetime fantasy boundary would soon be a reality, it followed that "manned space" would definitely be established as the new frontier.

In natural form, NASA began looking for the "next step" following the lunar landing. NASA desired to define the capabilities of just what could man do in space, and establish the relative limits of that emerging performance.

WvB and his various laboratory directors began discussions to define a game plan. This new frontier would hold many mysteries and benefits not previously pursued for manned inhabitants. It would be a follow-on goal to capture the public's imagination for technology advancements, while maintaining the national goal for the United States.

Marshall's Manufacturing-Engineering Laboratory (ME Lab) formerly the Fab Lab, had the responsibility for manufacture and assembly of rocket hardware both for test and flight purposes. These tasks required many skills in machining, welding, metals treatment, electrical harnesses, and support tooling. With this mixture of machinists and technicians, both mechanical and electrical, coupled with the accompanying engineering support team, it was a natural organization to quickly grasp the vision of man's future mission of working in space.

Charlie Cooper and Charlie Stocks, co-workers with engineering and technical skills, respectfully, grasped the vision and began to compare ideas in mid-1966. Cooper and Stocks were co-located in a small building that contained many pieces of shop equipment from lathes to a new sewing machine. Cooper grew up in a "tinkering" environment and, as a high school student, began machining and assembling numerous mechanical devices with a keen interest in water-displacement devices. Little did he know how this interest would play in his future. Cooper became the face of the tank as the primary test subject. Together, Cooper and Stocks were working the prelude ideas that would be applicable in an extraordinary future within NASA. Two additional technicians were soon added to the team, Don Neville and Charlie Torstenson.

At this point in my NASA career, this writer, Jim Splawn, an engineer, was also working in the ME Lab on the Apollo lunar landing program. Being young engineers and technicians in a work environment where all ideas were encouraged because no one yet had "the answer" for the myriad of problems the astronauts would face, it was a haven for dreaming and ingenuity.

Over the next few months, and as we evaluated many areas of interest, multiple items seem to always creep into the discussion: How do we simulate weightlessness? And believing our astronauts would be working in space soon, how could they maintain stability and a specific orientation at a work station while having both hands free for handling tools and conducting experiments? How could they function efficiently without burning precious energy? How could they be trained in our 1-G environment for weightless performance?

Cooper, Stocks, Neville, Torstenson and Splawn, all volunteering their time outside normal work hours, had arrived at their challenge! But, oh my goodness, where to start? There had been several attempts to create devices that might work (mechanical devices that "floated" on a column or pad of air, generated by an air bag with an orifice at the base of each leg on an epoxy hardened floor; flying in an aircraft with a padded cargo bay while flying a continuous series of parabolas, similar to a roller-coaster]; or several other mechanical devices of springs, counter-balances, etc.

During a discussion one summer day, one of the guys innocently asked: "Have you ever watched your wife swim under water?" We all responded, "Yeah!" while pumping our fists in the air. But the response quickly came, "Hey, I'm serious…did you notice her long hair in free-float?" Somewhat in shock, we stared at each other quizzically. And that was the added spark that really started us thinking that underwater simulation just might be the path to weightless training. Cooper was smiling.

Many questions followed in rapid succession. The key ones being: Where can we find a large body of water for efficient training? How can we place a crewman in a pressure suit in water and eliminate his floating on the surface? What support equipment is required to assure the safety of the crewman/test subject? What medical precautions are needed? How do we acquire duplicates of flight hardware on which to train the crewman? And the list grew.

A few years earlier, the ME Lab had developed an experimental 25-ft diameter, steel-lined pit in an open field to evaluate the possibility of explosive-forming curved segments of a dome for subsequent closure of propellant tanks for rocket engines (liquid fuel and liquid oxygen). The pit had been abandoned, so it became the first step of developing the idea.

It quickly became apparent that scuba divers would play a key role in this astronaut training media. So, each of us learned scuba.

Pressure Suits

After some early shirt-sleeve testing, the search for pressure suits began. To our disappointment, the astronaut office in Houston informed there were no pressure suits available; all were committed to current training operations. After evaluation of our alternatives, contact was made with the U.S. Navy in San Diego on the potential availability of high-altitude flying suits used by Navy pilots. "Available" came the reply. Torstenson and Splawn traveled to San Diego for training, both "dry" environment and "wet" environment, learning emergency procedures in a swimming pool for our specific application. On the return flight to Huntsville, four pressure suits were simply checked as personal baggage. These pressure suits were used satisfactorily for many months until, ultimately, we were successful in acquiring flight training suits from Houston in preparation for official Skylab mission training.

Early Facility and Operations

In the meantime, an inter stage section of the Saturn V rocket, previously used as a test article, was available for our use. The inter stage was installed in a field relatively close to an adjacent building with accommodations of restrooms and showers. A homemade makeshift tent was erected over the new 33-ft diameter, 25-ft deep pool to permit year-round training and protection for electronic equipment.

After evaluation of several options, the ability to make the pressure suit become "neutrally buoyant" was perfected. The eventual answer used a parachute harness, modified with small attached pockets in which pouches of lead shot could be placed. This harness is placed on the main torso of the pressure suit, making sure the umbilical that provides air, communications, and medical data is not encumbered. Similarly, individual strap-on weights would be strategically placed on wrists, thighs and ankles as needed.

The test subject would enter the water by walking down steps to a platform, where the weight harness was placed on the suit. The safety scuba divers would then turn the test subject in all body positions, head up, head down, left side, right side forward and backward, adjusting the weights from pocket to pocket as needed. When the pressure suit does not rotate to a bias position, the test subject is now suspended, basically helpless in the water, i.e: neutrally buoyant. The scuba divers then move the test subject/astronaut to the work station on stationary hardware to begin the defined test. For each test subject/astronaut, a matrix of the distribution of weights is recorded for pre-loading the harness for subsequent tests. Although the test subject/astronaut is fundamentally weightless as in space, the medical parameters of the individual are still 1-G, heart circulatory system, blood pressure, etc.

With the success now of being able to place a test subject in a pressure suit underwater with two scuba divers at his side for safety reasons, the tempo for exploratory testing defining what man could do was expanding quickly, while being very exciting and demanding, The part-time work for the small staff of volunteer test subjects and divers was moving toward overload. Accordingly, Cooper, Torstenson, Stocks, Neville and Splawn were assigned as full-time for these space simulation activities.

As simulation tasks progressed from the remedial to the more complex, the demand for task-specific hardware escalated. A design engineer was added to oversee the manufacturing of appropriately configured on-orbit hardware while meeting the requirements for underwater operations. For example, underwater operations required that open-mesh material be used in many situations in lieu of sheet metal so that lighting, photography, TV coverage, and scuba diver line-of-sight could always assure the safety of the test subjects.

Several MSFC organizations became involved in designing projects to seek answers to the myriad of questions of how zero-gravity would influence on-orbit performance.

The Next Step

In mid-1967, the NBS staff felt sufficient progress had been made to the level where senior MSFC management should become more aware of the capability. This included WvB, his laboratory directors, and reporting staff.

WvB visited the NBS to witness a test in-progress. Even with the rather crude appearance of the facility and related conditions, he offered a huge "thumbs up" and commented, "Good…keep going," in his wonderful English with a German touch. We did not learn until later that he was an accomplished scuba diver. Subsequently, we received a call from his executive assistant, Bonnie Holmes, that WvB wanted to personally be a test subject for a firsthand experience.

November 1967 was a big month for the Neutral Buoyancy Simulator (NBS). Early November was set as his test date. But WvB did not come alone; he invited Astronaut Gordon Cooper to be his guest. It was a smart move. If the MSFC director invites you to witness his becoming a test subject in a pressure suit in a new weightless simulation scheme, it must be something special and have strong potential. A lengthy umbilical was connected to WvB's pressure suit containing feed-and-return air lines, medical sensor cabling, and water lines for circulating water through the undergarment to keep the test-subject cool. The gloves were secured, the helmet was carefully positioned and snapped secure, air valves were opened, communication check was affirmative. The weight harness was installed and trimmed out for neutral buoyancy. WvB was a superb and gracious test subject. He talked constantly over his headset to the test conductor and safety divers, asking questions, and summarizing his feelings and sensations of performing tasks in a pressure suit. The de-brief after test completion exhibited confidence and encouragement in what he had just experienced. He encouraged future "thinking outside the box". His big question was: "What's needed to go forward?" Similarly, Astronaut Cooper was favorably impressed and commented that WvB would make a good astronaut.

From this point on, things began to happen fast. In fact, they exploded. WvB invited several astronauts, congressmen, and NASA executives to witness a test

Plans for a new facility were initiated, and a formal organization was established. A long-range plan was to train the flight crews for the Skylab program. In November, WvB made the decision to build the large 75-ft diameter x 40-ft deep tank as an in-house design-and-build project.

Organization and Staffing

In preparation for astronaut training for Skylab, the Neutral Buoyancy Simulation Branch was formalized and staffed: Branch Chief Jim Splawn, Secretary Gail Moss; Tech Writer Pete Nevins; Simulation Engineering Operations Supervisor Charlie Cooper; Elmer Bizarth Test Conductor; Sam McLendon Hardware; Dave Riggenbaugh, Engineer; Billy Edwards, Mechanical Tech; Clifford Troupe, Mechanical Tech; Clyde Graham, Mechanical Tech; Dave Javins; Mechanical Tech; Don Hammer, Mechanical Tech; Control Room Supervisor Bill Cruse; Charle Torstenson, Instrumentation; Clarence Cruse, Instrumentation Tech; Edward Shelton, Instrumentation Tech; Charlie Stocks, Electrical Tech; Pressure Suit Supervisor Don Neville; Glenn Dobbs Suit Tech; Jim Martin, Suit Tech; Bill Lee Suit Tech; Alton Puckett, Suit Tech; Medical Team Dr James Spraul and Dr Tulio Figurola and Photo Lab Ernie Hardin.

In the Fall of 1972, it became apparent the demand for scuba divers would increase significantly with the rapidly expanding scope of activity supporting the Skylab Program. The staff of the Space Simulation Branch was heavily involved in developing the many systems and subsystems required for the "manned operations in the future" and, while we all were divers, our skills needed to be focused elsewhere. Similarly, the many "volunteer divers" scattered throughout MSFC organizations were likewise experiencing an increase in workload in their home organizations. The decision was made to approach the U.S. Navy for a loan of divers to Marshall. After discussions between the two parties of how the Navy divers would be used, approval was ultimately received. Ten Navy Seals reported for duty in early 1973. Our diving operations of routine tasks in a comfortable and controlled environment were remedial for the breadth of training these Seals had received. While very individualistic, they were an amazing team and served us extremely well throughout the heavy astronaut training phase, the Skylab repair and preservation days. These Navy Seals were: Al Ashton, Michael Bennett, Joe Camp, Tom Earls, Frank Flynn, Roger Gant, Pat Gruber, Richard Pouliot, Charles Fellers, and Ken Wilkerson.

The Control Room provided all the amenities for a typical manned test operation: TV pan and tilt capability, TV monitors, audio, a large quantity of electrical racks to monitor medical data, test progress, data recorders, etc. All equipment was high quality. The TV and data recording equipment was media quality. The test director, control room supervisor, mechanical and electrical techs, staff doctors, hardware tech, etc. manned all critical stations. The complete control room was designed, equipped, wired, checked out, and maintained by Space Simulation Branch and ME Lab staff.

The Pressure Suit Office and Lab was located in an adjacent building. Special environmental conditioning equipment, elevated cleanliness standards, and strict security measures were required due to the nature of their work. The techs took great pride in their individual and collective assignment to assure all equipment was in mint condition for successful test operations. Their record of 100 percent success – no pressure suit malfunctions that scrubbed a training exercise – was a result of attention to details and personal pride. Obviously, this organization possessed great responsibility since they prepared personal equipment to be used by many astronauts and test subjects. They were highly successful in meeting their responsibilities.

A sister simulation organization Man/Systems Integration Branch, was established in the Propulsion and Vehicle Engineering Laboratory (P&VE) to concentrate on the same Skylab hardware configuration for evaluations in 1-G shirt-sleeve environment, and weightlessness evaluations for short periods of time, approximately 30-second intervals on parabolic flights in a specially equipped aircraft. J.R.Thompson was chief of this organization and key personnel included Dick Heckman, Charles Lewis, Ed Pruitt, Bob Shurney and others including contractors. There was very close coordination between our two organizations, sharing findings of our individual tests and how they might enhance the man / machine interfaces for the flight crew.

"Big Tank" Design and Construction

The design of the "big tank" was steady and meticulous. The steel floor and wall segments were procured from industry. Test Laboratory construction crews, including welders and riggers, handled the heavy construction work. There were numerous systems and sub-systems required for this unique indoor training facility with dimensions of 75-ft diameter, 40-ft depth, containing 1.3 million gallons of deep blue water. Underwater support systems included: pan & tilt TV, fixed and portable, with broadcast quality, audio for scuba divers to monitor all communications to and from the control room and test subject, lighting, electrical power for experiments, mounting stanchions for large test hardware, a safety diving bell at the 35-ft level as a rescue haven in case of an emergency, and an air-lock through the tank wall permitting a doctor and medical supplies to enter the diving bell. Three exterior platforms circled the tank for viewing the test activities via multiple 2-ft port holes. Clear translucent roof panels were installed to assure good natural lighting levels. A passenger and small equipment elevator served the top deck. A two-chamber recompression chamber was positioned on the top deck for the purpose of treating any test-subject or scuba diver emergency. A hot line to the medical center was located both on the top deck and in the control room. The building of this facility was a huge task. It contained many systems and subsystems, all of which had to be integrated, tested, and certified to meet man-rated training requirements. The construction workers even caught the spirit and vision of what this facility was committed to achieve and, consequently, gave their very best efforts.

In parallel with the tank being built, the Skylab Program Office was making significant progress in defining the interior of the S-IVB Stage known as Orbital Work Shop or OWS for meeting mission objectives. Similarly, a major effort was under way with defining hardware that must be fabricated for installation in the tank. This hardware must be accurate in dimensions and function configuration, free of any sharp edges or snag points that could rip or tear a pressure suit, provide as much natural daylight as possible for interior operations, light-weight for ease of installation in an underwater environment by scuba divers, etc. The overall dimensions of Skylab were 118-ft length, 22-ft diameter. Inside the tank, the 22-ft diameter was not a problem, but the length had to be shortened while retaining the primary astronaut work stations.

Skylab Project Office

With the curtailment of the last two Apollo missions, the S-IVB Stage became available. The large fuel tank would become "home and laboratory" for

NASA's first Space Station; the much smaller oxidizer tank would become the waste management area. The mission of Skylab was basically two-fold: 1) to determine how man could live and work in a confined weightless environment for increasing longer periods of time, and 2) to determine the psychological and medical impact of the weightless environment on the human body.

Each of the initial three Skylab missions would utilize a three-man crew for durations of 30 days, 60 days, and 90 days. The experience base of the three-man crew for each mission would be mixed with veterans that had flown previously and first-timers with no flights. This was NASA's first manned Space Station. Of utmost interest was how the human body would react to extended weightlessness, how the flight crew would respond to extended isolation, mentally and psychologically, and how the productivity of man/machine would compare with the 1-G baseline.

The Skylab Program Office continued its hard work to define the interior of the S-IVB habitat stage and the Multiple Docking Adapter (MDA). Similarly, the Apollo Telescope Mount (ATM) continued definition along with the development of the five telescopes that would study continuously the sun and space unknowns. This total configuration was nicknamed the "Skylab Stack".

Since the astronauts had only experienced weightlessness for short periods of time in the specially equipped aircraft, and the fact that our first space station was actually being built, there were many experiments planned that would baseline/expand the knowledge of human body response: diet limitations, ability to accomplish constructive work, work/rest ratio, on-set of fatigue during EVA, impact of extended isolation, person-to-person inter-relationships in a confined and closed environment for long periods of time, etc.

Experiments were being designed to acquire the impact of weightless operation for comparison to our 1-G knowledge. School kids were also involved in originating experiments. Each crewman would be continuously monitored on basic body functions and key bio data to determine if heart, lungs, muscles, etc. show signs of diminishing.

Big Tank Operations

In May 1968, the construction of the 75-ft tank was completed; hardware installation began the same month. To activate a man-rated facility with a hazardous environment and minimum previous performance experience was a demanding hill to climb. The development of this idea was in keeping with the ingenuity environment of the NASA team and was in line with the spirit, vision and challenge of the Marshall center. It is important to state that Fritz Vandersee and his dedicated and skilled construction crew from the Test Laboratory performed miraculously to complete construction within an unbelievable schedule deadline. In early Fall 1968 the Neutral Buoyancy Simulator became fully operational.

Hardware

With the successful completion of the Operational Readiness Inspection, emphasis now shifted to hardware readiness. All hardware was designed and fabricated to actual flight configuration where astronaut/test subject work-stations or translation paths were located. Other hardware configurations complied with "envelope" dimensions and functionality. That said, however, the materials used were somewhat different. Since the NBS is a manned-rated simulator, safety measures were absolute. There could be no sharp edges or corners (a major requirement) that might cut the pressure suits and place the astronaut /test subject in an emergency situation in this hazardous environment. Wire-mesh segments were used wherever possible for natural daylight penetration to the interior areas of the hardware, roof translucent panels were used over the tank area, and to accommodate underwater TV coverage for safety purposes and video recording. Multiple stanchions of varying heights were utilized to support the hardware approximately 8-ft above floor level. The fidelity, up-to-date design accuracy, and functionality of the hardware was a major and continuous task. The ME Lab manufacturing organization, led by Otto Eisenhardt, performed magnificently.

Skylab Tests

In Fall 1969, the NBS staff initiated simulations of the man/machine interfaces as defined by the Skylab Program Office. These man/machine devices would be contained within the shirtsleeve environment inside the Orbital Workshop (OWS) as well Extra Vehicular Activities (EVA) outside the OWS in a pressure suit configuration. As would be expected, the early-on devices for evaluation seem somewhat rudimentary to today's standards.

A stationary bicycle had been identified as an item to maintain body conditioning while collecting biomedical data for a body burning energy for an extended period of time in a weightless environment. So, a stationary ergometer bicycle was secured to an underwater platform for evaluation by a non-pressure suited scuba diver. Restraint systems for securing the test subject on the seat, feet on the pedals, and the biomedical package mounted on the crewman's waist area were required. These restraints were designed, tested, modified, and re-tested for flight applications. Similarly, a personalized sleep restraint sleeping bag or "cocoon" type device was tested for ease of operation, comfort, and easy access to closure devices. Also, a medical monitoring chamber of the lower body negative pressure (LBNP) of the blood circulatory system was tested for flight operations. While these tasks seemed somewhat remedial, it was essential tasks to assure efficient and accurate data be generated and recorded for the multiple astronauts over an accumulated period of 180-days.

In mid-1970, the design of Skylab Stack had matured to the point the EVA (Extra Vehicular Activity or space walk) requirements were being defined. Once the Orbital Workshop (OWS) was stabilized on-orbit, the automated deployment of the operational configuration was initiated: the protective shrouding around the Multiple Docking Adapter (MDA) and Apollo Telescope Mount (ATM) was discarded via explosive bolts, followed by the ATM structure deploying to a 90-degree position for the telescopes to be looking directly toward the sun for detailed film recording of the sun-surface activity. The hatch in the forward dome of the OWS, once opened outward and stowed, provided access to the MDA and the ATM for EVA operations. The MDA provided docking ports for the temporary stowage of the return-to-earth Command Module and a second port for docking another CM. Various equipment and experiments are stowed in the MDA. Similarly, the super structure surrounding the ATM permits pre-launch mounting of experiments and potential access to its sun-face workstation for camera film exchange via EVA. Throughout 1970 and 1971, many hundreds of hours in the NBS were spent in simulation tests: Skylab interior and exterior configuration upgrades, experiments maturing in hardware and operational procedures, translation techniques for flight crew and materials handling, and fine-tuning of the overall workstations. The ultimate objective was to have a workable, productive, efficient space station for the nine Astronauts to achieve maximum productivity from experiments and human relationships for future pursuits to unlock the mysteries and benefits of space.

Through 1972 and early 1973, 15 astronauts were trained, nine prime crewmen and six back-up crewmen. Continuous testing by both NBS test subjects followed by astronaut test subjects indicated a rotation scheme should be established for efficiency and thoroughness. Accordingly, the following routine was generally used:

> The NBS test subject would evaluate the test hardware and make changes dictated by the test results. The test hardware would be modified and retested by the same or alternate test subject. This iteration would continue until the best option was base lined. The astronauts would then evaluate the NBS recommended configuration. Based on their likes/dislikes, additional testing may or may not be required. Over time, multiple astronauts would provide their suggestions as they prepared for their flight. This routine of multiple evaluations by multiple test subjects builds integrity into the man/machine hardware for success on-orbit. This procedure can be spread over several weeks depending on the availability of the flight crews. One can see the cumulative time required to fine tune hardware design for varied tasks. In actuality, this sequencing is very exciting to witness as the attention to details seeks perfection.

SL-1 Skylab Launch

The Skylab mission, designated SL-1, was initiated with a beautiful launch on May 14, 1973. As the rocket rose through the bright sky, smiles abounded from the launch team and all observers. Unaware of anything other than another successful launch, the excitement was thick that NASA's first Space Station was on its way. Once orbit was achieved and automatic deployment of the Skylab hardware was moved to its permanent cluster configuration, instrumentation on the ground began receiving real-time data on the health and well-being of Skylab. An out-of-limits internal temperature was the first indication of a problem. Then, practically no power was being received from the solar panel system. Anxiety soared; checks and cross-checks were conducted. Rather quickly, it was confirmed we had a "stranded ship". Time and data would eventually reveal that a weak area in the external skin of the OWS had yielded to the forces of max Q, the area of maximum dynamic pressures for a launch vehicle.

A section of external skin (heat shield / micrometeoroid shield) had torn away, taking with it one of the solar panels that was neatly designed and placed on top of the failed skin area. On the opposite side of the vehicle, the solar panel survived flight okay, but a structural strap had wrapped around the edge of the solar panel, limiting its deployment beyond about 5-degrees (rather than the planned 90-degrees). A trickle charge did confirm, however, that one solar panel system was still available and trying to function.

Fortunately, the ATM solar power system was unaffected by this failure sequence. When launched, the ATM is protected by a shroud covering. Once on-orbit and during the automatic deployment operation, the ATM is indexed to a 90-degree position. Then the four ATM solar panels deploy. These panels look like a windmill and provide some power, but insufficient to meet the large total power demands of the OWS. This low power, however, was a lifesaver for Skylab. They provided the much needed minimum power for early-on critical systems and sensor communications with the ground.

SL-2 Mission

The launch of the SL-2 flight crew, Charles "Pete" Conrad Jr., Dr. Joe Kerwin, and Paul J. Weitz, was postponed from May 15 to an unknown period of time by consensus vote from MSFC, Huntsville; Johnson Spacecraft Center (JSC), Houston; KSC, FL; and NASA Headquarters, Washington. Little did we know the significant role the NBS would play in the days ahead. While monitoring the launch of SL-1 and its early on-orbit transmitted data, Splawn sensed a volume of simulations would be required to support the repair of the conditions on Skylab. Consequently, an all-systems check was issued to assure top working condition for all systems associated with NBS activities.

Rather quickly it was determined the Skylab basic structure was sound, but several systems required immediate attention. One, a scheme to protect the now-exposed exterior surface of Skylab from the scorching rays of the sun to decrease the internal temperature, making the environment habitable for the crew, preserving the on-board food, avoiding heat damage to experiments. Next, free the remaining solar panel to the deployed position for generation of power lights, air conditioning, power for communications, data transmission, and experiments. These two major tasks would ultimately be worked simultaneously in the NBS and would actually share some tools to correct both problems.

Tools

After many man-hours of brain-storming and evaluations at the 1-G and NBS simulators, a set of four tools offered the greatest potential for correcting the problem areas; a pry bar, a "shepherd's" hook, cable cutters, and sheet metal cutter.

Actual Simulation Tasks to Save Skylab

At MSFC, two separate organizations were mobilized to define an answer to solve the elevated interior temperature situation. The Materials Lab successfully defined a treated "mylar" material that showed promising results. Significant testing confirmed the promise. Simultaneously, JSC proposed an "umbrella/parasol" type deployment scheme. These two ideas eventually merged for deploying a sun shade through an existing scientific airlock port in the OWS. On-orbit, the 19-ft by 21-ft device was deployed by the SL-2 crew, and the interior temperature dropped. Skylab was habitable for an unknown period of unknown time. Since this option was developed in 1-G conditions and would be deployed from inside the OWS, there is no further discussion herein.

The ME Lab/NBS was working an alternate scheme. The following descriptions of successful tasks that ultimately saved the full mission of Skylab are the final operative solutions. Many, many ideas were pursued and tested, requiring hundreds of man hours, before these primary solutions were selected.

Twin Pole Sail

Two days after SL-1 launch, multiple astronauts had arrived in Huntsville and were jointly involved with ideas and potential solutions. As an idea matured, the shops fabricated the hardware with a 24-hour rotating staff. As quickly as hardware was ready to be tested, an NBS test subject would don a pressure suit and run an evaluation. If suggestions resulted from this test, hardware modifications would be made, and another test would follow, possibly with a crewman. This pattern of testing resulted in quick development via multiple test subjects with different ideas and variable skills. All hardware and tools to be used to correct the conditions found in Skylab had to be transported via the Command Module (CM). Thus, weight and dimensions were critical.

Options began to surface for a more permanent solution for the unacceptable temperature conditions in the OWS. A V-frame configuration for deploying a sun-shade/sail seemed most likely. The deployment of the dual (or twin) pole sail was an EVA task. Two crewmen were required for the assembly and deployment.

A base plate was fabricated with a "V"-shaped receptacle to accept the twin poles for supporting the sail. The twin pole structure was 1" diameter aluminum tubing with a male / female fitting on alternate ends of each 5-ft length. A locking mechanism, designed and fabricated by a shop machinist, secured each segment at its mating interface. Each support pole when assembled would be 55 feet long in 11 sections. A continuous rope loop would be pre-threaded through eyelets attached to the first and last section of each pole assembly. A permanent clip was pre-attached to the rope for mating with the sail.

One crewman would exit the hatch of the Airlock Module and translate to the workstation vicinity of the ATM (used for normal film exchange in the telescopes) and secure the "V" shape base plate at a pre-selected location on the super-structure for positioning the sail across the top of the previously deployed parasol. The second crewman would remain in the hatch area of the

Airlock Module, assemble the 55-foot pole and manually pass to the crewman at the ATM who would secure the pole in the "V" shaped anchor on the base plate. The second pole would be assembled in similar fashion, manually passed, and secured in the "V" shaped base.

The International Latex Corporation (ILC), located in Dover, N.J. and fabricator of the Apollo Pressure Suits, was requested to send seamstresses to Huntsville to assist in the fabrication of the sail. Two seamstresses arrived with their own sewing machines and special thread. For days they worked side-by-side with the Materials Lab staff, Bob Schwinghamer in charge, to fabricate several configurations of varied materials for testing. Once the final material and configuration was complete, the folded sail was processed through a vacuum chamber to condense the packaging profile and sealed for storage. The ILC ladies made the 14"x14"x8" storage bag for flight.

For the crew training exercises in the NBS, the sail material became an open-mesh with reinforced edges. The simulations were very good representations the flight crew would experience in managing the deployment.

On orbit, the storage bag containing the 30-foot by 40-foot sail would be transferred via a powered materials handling boom/transfer device from the airlock area to the crewman at the super structure. For deployment, the crewman would pull the corners of the sail from the storage bag, attach to the rope hooks, and gradually pull the ropes in an alternating fashion to fully deploy the protective sail. The last two corners pulled out of the storage bag had straps for pulling the sail taut and tying off at the base plate. After many tests, the man/machine transportation and deployment procedure was finalized, practiced, and perfected in the NBS. The final acceptance test was conducted by the SL-2 flight crew.

Two Navy safety divers were sent to the Cape to assure the hardware was packaged correctly prior to being place in the Command Module for launch. One diver was an expert in parachute rigging and had been instrumental in the folding of the sail and packaging in the storage bag. The other diver had intimate knowledge of how the pole segments needed to be packed for proper sequencing during the on-orbit deployment operation.

Meanwhile back at the NBS, local test subjects would be duplicating the deployment sequence of the on-orbit flight crew. The air-to-ground audio of the two astronauts would be piped to our control center and relayed to our suited test subject and scuba divers underwater. We would simply play "follow the leader" on identical hardware in a backup mode. Should they experience a problem, we would be prepared to understand that problem, duplicate the same problem, and offer an immediate real-time solution.

The temporary parasol sun-shade mentioned earlier worked satisfactory for the full mission of SL-2, so the deployment of this twin pole sail was deferred to the SL-3 crew.

Solar Panel

The failure of the solar panel to deploy was a critical concern. Data transmitted from Skylab to the ground indicated only a trickle of energy was being generated. And that trickle came from only one solar panel. If the Skylab mission was to succeed, it was compulsory to have power from at least one fully functioning Solar Array System (SAS)...and soon.

Engineers evaluated the likely progression and impact to neighboring hardware if one SAS had actually sheared away during launch. Knowing the configuration and strength of materials in the failed vicinity of the partially deployed solar panel allowed them to anticipate the likely material and configuration of the debris that entrapped the solar panel on the opposite side of the vehicle. Given this data, the manufacturing shops of ME Lab fabricated a short section of the solar panel and "wrapped" a correctly cross-sectioned piece of debris on its base. This defined not only what the flight crew should expect to see, but it gave some clues of the tools that night be used to free the panel. This is an area of the vehicle where there are no workstation handrails or foot restraints to allow a crewman to anchor himself and do constructive work.

To free the panel, it was decided the most logical solution would be to perform a "stand-up EVA" from an open hatch of the Command Module while flying in close proximity to the OWS solar panel. It would be "sporty but doable." JSC flew a Command Module to Huntsville while divers rigged a support structure for mounting the CM at a reasonable distance from OWS and the short section of the solar panel and strap mentioned above.

Astronauts Pete Conrad, Paul Weitz, and Joe Kerwin would be the flight crew. The planned procedure follows. Once the CM was in close proximity to the OWS, the suits would be pressurized; the hatch would be opened; the commander would assure station-keeping of the CM; one crewman would remain seated with seat belt fastened and manage the tools; the other crewman would basically stand on the thighs of the seated crewman who would bear-hug the legs. The standing crewman would use a section(s) of the twin-pole hardware with a selected tool attached and attempt to free the strap from the solar panel.

As the SL-2 crew approached the on-orbit Skylab, what they saw matched very closely with the engineering description/mental picture of that created on the ground. They performed the "fly around" and relayed their observations. The crew wanted to attempt the "stand-up EVA" before entering the OWS. Conrad would maintain station keeping of the CM, Kerwin would be the anchor for Weitz, and Weitz would attempt to free the solar panel. Weitz attached the shepherd's hook to sections of the 5-ft pole segments, hooked it around the pesky strap and yanked hard. Kerwin held tight to Weitz's legs. Something moved. But the solar panel remained firm. The jerking force simply pulled the CM toward the Skylab. Oh, the magic of zero gravity.

The crew decided to delay this task to another day. They moved to the MDA, docked, and entered the OWS. The stowed tools, and all candidate repair equipment were moved into the OWS. For the next 10-days, Astronauts Rusty Schweickart and Ed Gibson along with our own test subjects worked fervently in the NBS to determine the solution to free the solar panel. Of all the options they pursued, here's their ultimate recommendation for the EVA repair and how it actually happened on-orbit.

Using an assembly of five (5) each of the 5-ft aluminum poles and the modified cable cutter with increased mechanical advantage, Conrad and Kerwin began an EVA and worked their way down the superstructure to where they could see the strap/solar panel. Spare tethers and gray tape were attached to the front of their pressure suits. Just in case the cable cutter was unsuccessful, Kerwin remembered there was a dental kit on-board that contained a bone saw, a high-strength very small cable with integral cutting barbs and a circular key-chain type finger loop on each end. Kerwin tethered this cloth storage pack to his suit as a backup. Kerwin exited the airlock and secured his boots in one of several hard-mounted foot restraints at strategic locations throughout the hardware work area and began pole assembly operations. Weitz helped with managing the umbilical. Kerwin moved to an area underneath the ATM struts to the edge of the Fixed Airlock Shroud known as the "A-frame". From this vantage point, Kerwin could see the trapped solar panel. From here, there is no structure, handrails, or foot restraints—just OWS skin.

Conrad moved down the pole to provide some stability for Kerwin to keep inching toward the solar panel. Using a chest tether he had attached to the base of a radio antenna, Kerwin was able to hold the pole assembly with both hands and aim the cutter toward the strap. After rearranging his chest tether, he was able to get his feet on the OWS skin. Gaining this stability, he now had better control of the pole assembly. Ever so slowly, the jaws of the cutter were slipped around the strap. Kerwin eased the jaws tightly closed, then checked his body position and in coordination with and coaching from Conrad, pulled with significant tension. The strap was cut! The solar panel responded slightly.

Just as Schweickart, Gibson and engineers at MSFC had anticipated, the hinged control damper had frozen during the long cold-soak period of time on-orbit. Nominal force would open the hinge. Kerwin and Conrad had been briefed by Schweickart and Gibson of this potential. So, Conrad and Kerwin proceeded with the next step. Kerwin attached the Beam Erection Tether (BET) to the solar panel beam and secured the running end of the rope to the superstructure. Both crewman stabilized their feet beneath them, wiggled their bodies underneath the rope in a somewhat deep-knee-bend position, and with an upward motion like a jumping jack, the frozen hinge yielded. The solar panel immediately began to deploy. Ingenuity and solid performance won again. As Conrad floated around in space, tethered to Skylab, he said to Kerwin, "This is just like the big tank at Marshall, only deeper."

Meanwhile, back at the NBS, a follow-the-leader exercise was again active in case it were needed. Joyous celebrations erupted. From the power standpoint, Skylab was functional.

SL-2 Routine

Considering they were the first Skylab inhabitants, the SL-2 crew performed many tasks to make the OWS as tidy as possible. They conducted their assignments via the checklists: performed on-board experiments, conducted medical evaluations, exchanged the ATM film canisters with EVA's, solved unanticipated maintenance, evaluated food menu's, etc. All these data would help the follow-on crews. The collective crew paid tribute to the training they had received for all aspects of their mission, both planned and unplanned events. After securing all systems and subsystem to their temporary state, the crew undocked and headed home with good feelings about their performance over the 28 days.

SL-3 Mission

The second crew, Owen Garriott, Jack Lousma, and Alan Bean was launched on July 28. This crew benefited from the debriefs of the SL-2 crew while training hard to polish the basic understanding of the experiments and tasks they would be performing in just a few days.

Knowing it would be their task to deploy the twin pole sail, Garriott and Lousma were intent to understand every detail of that sequence. Many tests were conducted in the NBS to fine tune the process and their familiarity. This permanent sail would be placed over the existing temporary sail, which was still functioning but diminishing in effectiveness. The specifics of this critical EVA are detailed above in the SL-2 section since it was unknown if the task would have to be accomplished during their mission.

After a little over a week on-orbit, Garriott and Lousma suited up and exited the airlock. Installing the portable foot restraint was among the first things that had to be accomplished. By practicing the uncharted path across the truss structure in the NBS, Lousma was successful in latching the foot restraint to the truss; a critical step was accomplished. Meanwhile, Garriott was assembling the 5-foot pole segments with locking mechanisms and the accompanying lanyards. Garriott then passed each 55-foot pole to Lousma who had secured the V-shaped base. Each pole was now successfully in place and secured. The storage bag containing the sail was transferred to Lousma and secured for sail deployment. When pulling the sail from the storage bag, it became evident the adhesive used during the folding operation for packaging had not had sufficient time to completely cure before packing. Lousma had to manually free the folds, making deployment more difficult. But determination paid off. Lousma completed the deployment of the sail to the ends of the poles and the permanent sail was in place to control the interior temperature for the duration of Skylab. The temperature decreased from 130-degrees to 80-degrees.

During this critical EVA, the NBS was in a follow-the-leader testing mode with full crew to assist Garriott and Lousma if assistance was needed. The flight crew performed superbly, so there was celebration at NBS as well as on-orbit. Skylab was now functional for its full mission. The remaining EVA's, primarily for the retrieval and resupply of film for the ATM camera, were fully planned and relatively routine for the SL-3 crew. The SL-3 mission duration was 59 days.

SL-4 Mission

The third and final crew was launch on Nov. 16 with Astronauts Gerald (Jerry) Carr, Ed Gibson, and Bill Pogue. This crew received three months of "learning" from the two previous occupants of the first Space Station for the U.S. Ed Gibson was a well-known and respected crewman at the NBS since he had been a frequent test subject at NBS. Carr and Pogue were less known, but as specific training for SL-4 mission evolved, their skills and dedication became apparent and rounded out a competent crew. There were no unforeseen emergency tasks requiring an EVA, so the NBS staff had full confidence in the training procedures and practices as demonstrated by the SL-4 crew. They would be performing the same ATM camera film exchange as had the previous crews. Should there be any anomaly on-orbit the NBS staff remained anxious and available to assist.

All other experiments would be conducted internally in the OWS.

The three missions on Skylab set an endurance record of 171 days of manned occupancy, and provided significant data from the on-board experiments and an impressive baseline of medical data from weightless exposure. These data would be of great benefit to the designers of the next Space Station for the U.S. The SL-4 mission was on-orbit for 84-days.

The Neutral Buoyancy Simulator (NBS) was a fantastic facility. The founding of this capability to simulate weightless conditions for extended periods of time proved worthy of training astronauts to accomplish unforeseen and unimaginable tasks in the new frontier of space.

The staff of young engineers and technicians caught the vision of space exploration. They committed themselves to "push their personal envelope," function as a team, and dream the big dream. They can be, and are, proud of their contributions. As a reward for the Neutral Buoyancy Simulator staff, for what they accomplished, for their dedication and personal sacrifices, for the significant contributions to NASA / MSFC, Splawn requested of MSFC management that the entire NBS staff be in attendance at the launch of the SL-4 mission. The request was approved. It was a great launch, and a fitting reward.

Postlude

During the operational period of the NBS, it became a significant "show piece" for Marshall. Many significant dignitaries, both inside/outside NASA ranks, visited the facility. On different occasions WvB personally invited Walt Disney to witness a test operation, an internationally known diver Jacques Piccard, a Swiss Oceanographer – engineer known for developing underwater vehicles for studying ocean currents at depths of 30,000-ft, to dive in the NBS to understand the flight hardware used for astronaut training. Many Senators and Congressmen visited the facility, both for educational purpose and understanding how funding was being utilized, and WvB also convinced the NASA Deputy Administrator that he should "suit up" and experience weightless training to better understand and appreciate the importance of astronaut training—it was a great day! These are only a few examples.

The Space and Rocket Center was very pro-active for the Skylab Program and the many organizations at MSFC with lead-role responsibilities. Bus tours from the Center occurred every two hours and exposed vast numbers of the general public from all walks of life to Marshall's involvement. The NBS was a very popular drawing card for these tours, particularly when planned tests were underway.

The most significant VIP event occurred following the landing on the Moon. In 1970, Astronauts Neil Armstrong, Buzz Aldrin, and Michael Collins made their celebratory tour to countries around the world as a thank you and goodwill gesture. During their visit to Russia, they extended an invitation to the Russian Cosmonauts to, in turn, visit the U.S. They accepted. When they came to Huntsville and the Marshall Center, the astronauts invited a cosmonaut to participate in a "follow-the –leader" exercise in the NBS. Again, they accepted. In mid-October, 1970, Cosmonaut Sevastyanov, Cosmonaut Nikolayev, and Russian Interpreter Barsky visited the NBS. Cosmonaut Sevastyanov agreed to participate in a "follow-the-leader" underwater training exercise with Astronaut Schweickart. Schweickart gave a detailed hands-on briefing of our U.S. pressure suit to the Cosmonauts who, obviously, were very attentive and asked lots of questions. Sevastyanov was given a cursory medical exam by our attending doctor and given clearance to proceed. After suiting-up and many photos, Schweickart and Sevastyanov rode the elevator to the top level. Sevastyanov observed as Schweickart entered the water and was trimmed to the neutral buoyancy state by divers. Sevastyanov followed, while being continuously briefed by the cosmonaut and the interpreter in the control room. Safety divers then moved both test subjects to the interior of the Skylab hardware and the test began. After Sevastyanov's assurance that he was OK and the monitoring doctors reporting that his heart rate was acceptable, Schweickart initiated the test sequence. Throughout the approximate 50-minute follow-the-leader test, it was obvious by TV and audio monitoring that Sevastyanov was comfortable and enjoying the exposure. The cosmonauts were obviously pleased to inspect and wear a U.S. pressure suit and were complimentary of the underwater simulation of weightlessness. When asked if they trained in a similar fashion in Russia, they said "no, but as right now, we're thinking about it".

This was a very exciting, but pressure-packed event. All NBS systems performed perfectly; all personnel performed with excellence and professionalism. It was a day to remember.

The Apollo Telescope Mount (ATM) Project

By Rein Ise

Introduction

We are amazed by and love the sun. It is what gives us life and keeps us warm. Those are special moments when it rises in the east and sets in the west, particularly when it decorates clouds with brilliant colors. Yet it represents many puzzles that we do not understand. How was this thermonuclear bomb created and what keeps it from exploding? What causes its spectacular eruptions and how do those affect us on Earth? Mankind has worshiped the sun and studied its activities since ancient history and continues to do so today. So I accepted with great interest and enthusiasm the opportunity to participate in a project started in the late 1960s to launch an advanced solar observatory to perform advanced studies of the sun from space. This program called the Apollo Telescope Mount (ATM) not only involved a quantum step in understanding the Sun, but also provided the first opportunity ever to engage man in the hands-on operation and maintenance of the solar telescopes while outside the Earth's atmosphere. In addition, the ATM was the last major project that enjoyed a unique arrangement, which was to accomplish the work in-house at one of the NASA centers — fortunately the Marshall Space Flight Center (MSFC) where I was employed. This story provides some of my recollections of the significant and unique experiences and events as the program was conceived and accomplished by MSFC. I will not go into detail on the technical details of the ATM, the NASA team members, principal investigators, and participating contractors, as such information is adequately covered in other publications referred to at the end of this story. However, I am compelled to mention Eugene Cagle (ATM project chief engineer), William Horton (Astrionics Laboratory deputy director and ATM lead), and William Keathley (ATM experiments lead manager), as key leaders within an outstanding MSFC team responsible for the project's great success.

Defining the ATM

The ATM was originally proposed in the early 1960's as a potential passenger in one of the Apollo Spacecraft Service Module's empty bays. It was conceived as a battery of a couple relatively small solar telescopes, which would be extended out of the spacecraft's bay after reaching Earth orbit and then operated by the crew. However, the limited space and brief observing time available in this configuration proved that such an effort would not be much of an advancement over what had already been accomplished with unmanned satellites. Various ideas for accommodating a larger observatory and longer observing times were considered, including a free flyer using the Apollo Lunar Excursion Module's (LEM) cabin and return stage for the observatory's manned control room. As NASA was struggling with this challenge, a manned Earth orbiting workshop concept was being considered using the third stage of the Saturn V launch vehicle, other available Apollo Program hardware, plus some special components to accomplish long duration experimentation in the zero gravity condition of Earth orbit. This configuration was later named Skylab. An early concept drawn by NASA's manned space flight director on a flip chart while visiting MSFC in 1966 (which confirmed ATM as a major payload to be launched as part of Skylab) was to launch the ATM on top of the workshop, deploy it in orbit tethered to the workshop with umbilical conduits, and utilize these conduits to permit power and communication exchange with the ATM while the observatory controls remained inside the workshop. In this concept, the purpose of the flexible tether was to avoid transmitting crew induced and other disturbances to the ATM telescopes, which required precision pointing. In the meantime, the selected principal investigators had refined the ATM concept to include eight major telescopes, some requiring camera replacement with extra vehicular activity (EVA) by the crew from the Workshop. Also, Skylab preliminary design indicated that the ATM would be called on to provide overall station attitude control and supplement some of its power requirements. As a result, it was decided that a permanently attached observatory deployed to the side of the workshop would be a better concept. However, this required that a two-degree-of-freedom pointing system with fine pointing and roll capability would have to be added to structurally support the solar telescopes package within the ATM and precision point it on the sun. The attached configuration turned out to be a fortuitous decision as a serious mishap on the workshop side during its launch phase would have resulted in a failed mission would it not have been for the ATM's capability to provide electrical power support and overall attitude control for the entire Skylab to control critical thermal overheating during its first few days in orbit.

An In-House Project

The Skylab was proposed as a follow-on to the Apollo project as that project was winding down, making MSFC's facilities and manpower readily available. Also, program funding for the Skylab was limited, making it imperative to find the lowest-cost approach for accomplishing the ATM as well as the Skylab efforts. At that time, NASA program funds were separated from the civil service salaries, allowing the MSFC manpower costs to be "free" for accomplishing the required ATM work. So, in late 1966, the manned space flight director, while flying over Huntsville on his way back to Washington, phoned the MSFC center director and asked him if MSFC could accomplish the ATM project for less than $100 million. And without hesitation (based on some preliminary cost studies) the center director agreed that it could. As a result, the project responsibility was assigned to MSFC and detailed design was started.

The ATM became a project within the Skylab program office and I was asked to manage the project, which I gladly agreed to. It turned out to be a marvelous opportunity as I was associated with the program from the beginning to its very successful end. Such an experience does not occur very often for major projects as they usually require a time span, which encourages the pursuit of career advancement before the project is completed. The hardware design and build phase for ATM fortunately took less than seven years — a remarkable accomplishment for such a complex project. This was largely due to the project's good fortune in being able to employ some outstanding people to make exceptional and timely contributions both in the project office and in the technical work within S&E. As an example, I recall when a couple of times the work in the shops was stopped because of some missing paperwork, but the chief engineer accosted the assembly team by telling them that they knew how to do the work without the missing paper, and threatened to get his screwdriver and pliers and do the work done himself if the problem was not resolved in a day. And the problem was indeed resolved in a day — probably to prevent having to face him again.

The ATM project office was responsible for the successful accomplishment of all the project-related activities, including the development of the solar telescopes.

The ATM in-house work was to be accomplished by what was then called the Science and Engineering (S&E) Directorate and its various laboratories. Each laboratory had the responsibility for a key discipline, such as structure, guidance and control, or test. Since the biggest challenge of ATM was to meet the demanding experiment pointing and control requirements, the Astrionics laboratory responsible for the electrical and pointing and control systems was also assigned the responsibility for integrating all the disciplines and design. To assure this, the S&E activities were set up as a virtual subcontractor effort to the project, with a chief engineer, acting essentially as a sub-manager, responsible for all the in-house effort. The Structures and Mechanics Laboratory within S&E established the overall ATM structural concept, which had to accommodate the interface to the Saturn V launch vehicle and the Skylab, while providing the maximum available space for the selected solar experiments. The structure was established as an octagonal framework supported by outriggers on four sides during launch and accommodating four large deployable solar arrays between the outriggers. Various components of the Skylab and ATM control systems and the power system were also mounted to this framework. The experiments package was a large thermally controlled cylinder mounted inside a structural ring system, which provided roll capability and flexible supports with actuators to provide two degree of freedom fine pointing.

The ATM and its Skylab support functions required the design and development of several state-of the-art systems, including very large Control Moment Gyroscopes for maintaining Skylab orientation in space, the system for fine pointing of the experiments package, and the largest deployable solar arrays, launched into space as of that time. Some of these developments, such as the solar arrays were done entirely in-house, but other systems were contracted out on a sole source basis to recognized companies with established expertise in the required fields.

For the solar instrument efforts, formal contracts were established with each of the principal investigator (PI) organizations, with the PI fully responsible for the instrument design and development activity. Where necessary, the PIs in turn awarded design and development contracts for their instruments to contractors with extensive working experience with science investigators and their delicate instruments. Within the ATM project, experiment managers were assigned to provide the oversight and liaison to each of the experiment development activities, and an overall experiments manager was appointed to manage and coordinate the integrated experiments requirements and efforts.

The project utilized existing capabilities and facilities within MSFC to the maximum extent. The center had extensive structural and electrical fabrication and test capabilities at that time. However, because of its size and complexity, a number of specialized facilities had to be designed and constructed. These included:

a. A Skylab attitude control simulator where the control moment gyros, the control computer, and associated software were thoroughly tested in a configuration that duplicated the Skylab control setup.

b. A full scale fine-pointing test bed to test the sun sensor, pointing actuators, and computer system for developing the control algorithms.
c. A solar array deployment test fixture where full configuration arrays and their release and deployment mechanisms functions were exercised in a sidewise suspended orientation in order to minimize the effects of gravity.
d. An ATM full-scale mockup/trainer that supported the design of the astronaut translation and positioning aids for the extra vehicular activity required for experiment film camera retrieval. The trainer was also used to develop in-orbit astronaut procedures.
e. An experiments control station with a complete functioning control console and simulated experiment functions to allow experiments procedures development and astronaut training. This simulator was later used also to support procedures modifications during the missions.
f. A large clean room to permit the flight ATM assembly and checkout under super clean conditions.

The expertise and momentum built up within MSFC with the Apollo program carried over to the Skylab and ATM efforts. The responsible individuals within each laboratory performed their efforts, including supervision of subcontractors, with great skill and efficiency. The project was conducted during the still-early days of NASA, where the work was done with a minimum of documentation, limited oversight and decisions were left to the experts, with only such coordination and management necessary to assure that no requirements or interfaces were violated. Also, trade studies and review panels were held to a minimum. Solutions were left mainly to accumulated experience of the experts. Quality oversight was primarily to assure no safety issues and to verify that the hardware had proven itself to be flight worthy as the result of extensive testing. Any safety related issues were identified and resolved by the technical experts with oversight by management. Large reviews were held to a minimum and issues raised during reviews or other activities were handled as a limited number of action items assigned to the responsible individuals for coordination and resolution. Flight readiness was assured with a Design Certification Review, which showed management that all required tests were successfully completed and technical issues satisfactorily closed. Final flight approval was obtained at a flight readiness review, which verified that not only the hardware but also the operational aspects were ready for the mission.

The ATM required and enjoyed great dedication from the project team. Fortunately the team was sufficiently young, and the work interesting and challenging, that there was little turnover of personnel. The dedication to the project was clearly proven when the technical support and test team was required to move to the Manned Spacecraft Center (MSC) at Houston for several months to accomplish the overall ATM thermal vacuum testing, and subsequently to the Kennedy Space Center for launch preparations. This required the personnel of the test and engineering support team to be in travel status for nearly a year, which resulted sometimes in family problems because of the extended absence. Yet, the work was successfully completed with a minimum of problems.

Working the Requirements

The initial ATM concept had to address mainly those requirements imposed by the experiments and the crew interfaces. This included the design of the optical bench for mounting the experiments, an enclosure for the entire experiments package, providing thermal control/aperture doors/camera access to the telescopes, a pointing system including rate gyros and sun sensors, an electrical power source with storage capability, the experiments control console, and EVA/crew access aids for film camera exchange. However, as Skylab matured the ATM was also required to provide overall attitude control for Skylab and some electrical power to support the planned long duration stay in space. Thus significant additional responsibilities were added to the ATM project, which included the development of large Control Moment Gyros, the computer for the attitude control system, long life battery packs for electrical energy storage, and a system to provide roll capability and two degrees of freedom gimbaling for fine pointing of the experiment package. This also required additional close coordination with designers on the Skylab side, numerous trade-offs to achieve the best design solutions, and additional contracts. As a result, the final cost for accomplishing the ATM project increased significantly. Again, it was fortunate that NASA in those days was able to handle such increases without major trade-off studies or other efforts and delays to the program. I recall traveling to NASA Headquarters early in the program to present a budgetary increase explanation to the responsible manager in a one-on-one environment, and had it approved without any further questions or complications.

One of the key interfaces was with the astronaut crew, which was destined to operate the ATM and perform camera exchange for the ATM instruments in space. As a result, many design reviews and work sessions, including hands on tests on trainers and neutral buoyancy simulators, were conducted with the assigned crew. This became a challenge when the astronauts sometimes differed with the designers and even among themselves in their opinions and inputs regarding the design or operating concept, requiring arbitration and/or decision by management. Yet all of this was resolved in a cooperative and constructive manner without major controversies or problems. A major issue occurred with respect to the overall layout of the experiments control panel, which required a major redesign. The redesign also was resolved by the MSFC team with only a minimum delay and proved to be a major improvement to the operational efficiency of the ATM.

Probably the most demanding interface was with the principal investigators. Their requirements were of course the basis for most of the ATM design. The early issues involved the accommodation of their instruments within the experiments package, where each PI hoped to maximize the space available to them. Other issues dealt with the quality of fine pointing performance, thermal control of the instruments, and astronaut access to the replaceable cameras. Later issues involved operational procedures for the telescopes and observing priorities. Such issues were typically handled in special working groups where all interests were equally represented, and solutions were mainly determined through a democratic process of compromise. These efforts were always constructive with good cooperation among MSFC, the PIs and the astronauts. I must say that working with the ATM scientists was always a pleasure as they had almost always solid justification for their requirements. The PIs, their teams, including subcontractors were all highly capable and well experienced in their fields. Communications and negotiations with them went smoothly, and fortunately project budgets had sufficient reserves, such that most of the solutions could be achieved with affordable cost while avoiding major trade-off studies or other delaying efforts.

The Result

The ATM successfully passed all its system environmental tests at JSC and system checkout and integration activities at the Kennedy Space Center (KSC). It was installed on top of the Orbital Workshop within the Saturn V payload shroud and successfully launched into the desired orbit and deployed into its operating position on May 14, 1973. The crew was to follow and dock with the workshop a week later. However, the OWS unfortunately lost one of its solar arrays and consequently thermal protection during launch, placing the entire mission into serious jeopardy. Fortunately, the ATM was fully functional, and was able to provide overall OWS attitude control and electrical power to maintain some control over the growing OWS thermal problem. This required intense 24/7 support by the ATM team at MSFC's mission control center during the early emergency phase, as well as the rest of the Skylab mission. MSFC, with support from JSC and other NASA centers was able to come up with a temporary fix for abating the thermal problem and the fix was launched with the first crew on May 25, 1973. The fix was successfully applied and the solar experiments were turned on a few days later. Everything worked as planned, including the crew that had to function at somewhat elevated temperatures, and the ATM was able to obtain solar data for nearly three weeks. The first Skylab mission lasted 28 days and was followed by a 59-day mission (including a more elaborate fix for the thermal problem), and that was followed by an 84-day mission. The ATM performed flawlessly throughout. So did the ATM mission support team. The team was instrumental in programming the ATM to provide the delicate attitude control required during the initial period when the thermal problems were threatening the entire mission. And whenever issues arose, the team was able to respond with rapid and practical solutions, such as fixing a stuck relay on one of the ATM battery packs by asking the astronauts during an EVA to hit the pack on the appropriate side with a few hammer taps to unstuck a balking power relay. This worked as expected. Toward the end of the mission one of the CMGs started losing spin speed, and the support team reconfigured the system and the algorithms as needed to allow completion of the mission as planned.

Skylab with the ATM remained in orbit for several years after the last crew returned on February 8, 1974. I remained associated with Skylab and the ATM for another year with responsibility for program closeout, the continuing efforts of ATM data analysis, and the publication of several NASA publications to record the spectacular results of the entire Skylab program. My final responsibility on the program was to keep MSFC management informed of the predicted re-entry and impact of the Skylab on Earth. There was of course concern regarding the probable survival of some of the dense and heavy masses, such as the ATM CMGs, through the re-entry heating, possibly causing damage on the ground. After various efforts by NASA to control the reentry trajectory, Skylab, with the ATM, re-entered the Earth's atmosphere on July 11, 1979 over the South Pacific. There was no resulting damage recorded anywhere, but quite a few articles (such as a scorched workshop fiberglass water tank) were later found in Western Australia — a very satisfactory end to the program.

As stated by Leo Goldberg, a leading solar scientist at that time, "the ATM telescopes and instruments exceeded the highest aspirations of astronomers" and "the study of ATM observations has led to many new discoveries about the nature of the sun and the fascinating events that occur there". Over 150,000 high resolution photographic exposures of the Sun in various wavelengths plus other quality telemetered information were obtained. The data analysis of this new information by the solar scientists lasted for many years and resulted in a large number of valuable scientific publications.

It is with the greatest satisfaction that I look back at my and the other ATM team members' contributions to achieving this most successful project and mission. I hope that others in today's NASA could enjoy a similar experience. For more information about the ATM mission and its results, refer to NASA publication SP-402, "A New Sun: The Solar Results From Skylab" and to NASA publication EP-107, "Skylab – A Guidebook".

Program Development

By William R. Lucas

Editor's Note: In the late '60s, competition for projects within NASA had become hotly contested as the Saturn Apollo program was coming to an end. WvB added program development as a major element to the organization. He moved some of his best and brightest people into this new organization for the sole purpose of finding new work. This was his way of diversifying the organization, and it paid off with new programs being captured like the space station and Hubble. Some lab directors and program managers objected to this effort, because they were losing good people to this new organization. It was another example of WvB anticipating the changes that were occurring in the manned space flight program. This development was typical of the WvB team. They did not wait for someone to tell them what their next mission might be. They created a 'sales and marketing' initiative that brought new missions and contributed to the Center's future success. WvB named Dr. William R. Lucas the first director of Program Development. Lucas began his career with the team in 1952. He was recognized for his work in materials research in the Propulsion and Vehicle Engineering laboratory. He became the laboratory director, and continued to advance through key positions at all levels. He was named the Marshall center director in 1974 and held that position longer than any other center director, retiring in 1986. Lucas describes how Program Development came into existence.

In the fall of 1968, the launch and flights of Apollo SA-501 and SA-502 had been thoroughly analyzed and the few hardware anomalies had been resolved. Preparations for the launch of SA–503 (the first manned lunar orbital mission) were under way. Confidence was growing that the national goal of landing a man on the Moon and returning him safely to Earth within the decade would be met. Yet, NASA in general and the Marshall Space Flight Center (MSFC) in particular also faced a different set of problems. With the maturing of the Apollo program, a developing surplus in the workforce was becoming available for reassignment or release. The firm assignments to MSFC beyond Apollo were not sufficient to justify its workforce or to maintain its unique capability for developing large rocket system. Already, MSFC was losing significant numbers of its team and was being pressured by NASA headquarters to reorganize for additional reductions in view of anticipated budget decreases. This pressure would continue along with the consideration of closing one or more NASA centers, especially MSFC. NASA's budget was being challenged by a national administration apparently less committed to space exploration than its predecessors in the early 1960's. There were too few new programs under consideration by NASA headquarters, and the centers were left to compete for the few potential new assignments.

In this environment, WvB led a small group of his key people, administrators, laboratory directors and project managers on a weekend retreat to Jekyll Island, Georgia, to consider how the center should prepare for the future. On the first afternoon of the retreat, the group met for a review of NASA's situation and its impact on the center's future.

At a break for dinner and a bit of relaxation before what would become a long and intense evening session, Ed Schorsten, the representative of the MSFC Public Affairs Office who was in charge of logistics for the retreat, came to me with a dinner menu for my selection and a request from WvB for me to have dinner with him in his room. This was far less than thrilling to me. I couldn't imagine what I had done or said to warrant this attention from the director. That meeting turned out to change the course of my professional life on a permanent basis. It was then that WvB revealed to me his assessment of our situation and what he thought should be done about it. In short, he thought that, due to the pressure of getting Saturn V ready to fly, we had not been aggressive enough in defining and seeking new mission assignments and that we could not depend on headquarters to do this for us. His proposal was to establish a new organizational element of MSFC made up of highly motivated people, devoted exclusively to identifying and selling new NASA missions in which MSFC would have a significant part. Then he discussed his view of my future and stated that he wanted me to head this new organization.

This was a shock and a release; a release in that I was not scheduled for discipline for wrongdoing, but a shock in that I had never visualized myself in that role. He asked for my commitment to undertake the task. I was enjoying my role as director of Propulsion and Vehicle Engineering Laboratory (P&VE) and thought that I was reasonably successful in it. I really didn't want to leave that role for a loosely defined new assignment with so many impediments to success. However, in the philosophy that one either does to the best of his ability what the boss wants him to do or gets out, I agreed to undertake the task if asked to do so. This was done with his assurance of the support of some of the best of MSFC whom I would have to recruit, the kind of people that the directors of the laboratories from which they must come would not want to release.

When the session convened after dinner, WvB began revealing his thoughts about what we should do. His plan was new to me (except for the dinner discussion) and to most of the people there, I believe, but I am confident that he had discussed it in considerable detail with a very few of his closest associates. He did not reveal a comprehensive plan, but presented his thoughts progressively in a way that evoked frank and intense discussion. The retreat ended with the conclusion that in anticipation of the completion of the Apollo program, the center must restructure its goals and objectives to facilitate maximum future benefit to the nation of the technological team and the facilities which had been developed at MSFC into a national treasure and to assist the NASA in the development of a progressive space exploration program. Several smaller organizational changes were discussed in the course of the retreat, but the immediate focus was to establish a new organizational element at MSFC to be devoted exclusively and aggressively to conceiving or acquiring, planning and implementing new programs. This organization, unnamed at the end of the retreat, would encompass the missions of some other planning activities at the center at that time and would involve some of the best personnel of the center on permanent assignments and/or on temporary assignment. WvB referred to the proposed organization as a sales office and to the director of that office as his vice president for sales. Advanced planning and studies were not new to MSFC. However, the effort must now demonstrate a much greater sense of urgency. The sense of the retreat was that a relatively small organization of well-balanced capability would lead the new center planning activity.

Implementation of the intent of the retreat began immediately upon our return to the center. Jay Foster, a long-time member of the MSFC team and a member of the MSFC executive staff at the time, was assigned to work with me on the administrative details of establishing the new organization. Jay was particularly valuable because he had participated in the retreat and had a good concept of the decisions reached. Jay and I selected the name, Program Development, for the new organization. The first recruits into the organization were my secretary of many years, Gertrude Conard, and Alexander Flynn, the head of the administrative staff of P&VE laboratory of which I was director going into the retreat. Al Flynn, with a background in a personnel office and in laboratory administration, was invaluable in the details of staffing the new organization.

Program Development began functioning as a center organization on December 16, 1968, with the transfer to Program Development of the entire Advanced Systems Office of MSFC, headed by Frank Williams, a long-time member of the MSFC team. The entire advanced systems office was transferred in mass subject to subsequent screening to identify those who could not be readily fitted into Program Development because of skills or inappropriate grade levels and would be made available for other assignments. Frank and the team brought with them approximately 25 small advanced study contracts.

On December 20, 1968, the initial documentation for the establishment of Program Development as a separate center organization reporting directly to the Director of MSFC was submitted and approved shortly thereafter. This document was in sufficient detail to identify the mission of the new organization and to justify its request for resources. Yet, ample flexibility was left for significant modifications after recruitment of key staff members and after additional consultation with center officials. The ultimate organization of Program Development also reflected input from attendees of the Jekyll Island retreat and others who were particularly helpful in effective organization, personnel relations and communications, including Harry Gorman, Dick Cook and David Newby.

The new organization, Program Development, would be unique. No other such organization existed in NASA, equipped with comprehensive system design capability and whose sole assignment was the generation of new business. This organization would continue to support advanced studies of long-range interest, but the greater emphasis would be on the development of hardened plans for new center assignments in the short term. A permanent staff of about 200 people, representing the broad technical capabilities of the center, was recruited from center organizations, primarily the Propulsion and Vehicle Engineering Laboratory, Aero-Astrodynamics Laboratory, Astrionics Laboratory and the Research Projects Laboratory. The priority was on system analysts and designers who had experienced the difficulty of transforming lines on paper into functional hardware. Bob Marshall, Charles Darwin and Carmine DeSanctis are examples of such recruits who had been through the ups and downs of Saturn design. Bob and Charles would each ultimately become director of Program Development. At the time the new organization was formed, it was apparent that potential transportation system assignments would require augmentation with other assignments such as the development of payloads for research and exploration. To meet that end, several individuals with primary backgrounds in science rather than engineering, such as Jim

Downey, Bill Snoddy, Jean Oliver and others were recruited from the Research Projects Laboratory. The permanent cadre of program development was supported by many specialists from throughout the center, either on part-time or on a temporary co-located basis.

The product of the new organization was to be complete definitions of new projects. But, these new projects had to be sold, first to the management of the center, and then to NASA headquarters and/or other sponsoring agencies. Thus, the ability to communicate across the center, with NASA headquarters, with other NASA centers and with the space contractor community dictated the selection of several leaders in the organization such as O.C. Jean, Bob Marshall, Bill Huber, Terry Sharp and Harry Craft. Because the leadership of the center, particularly those who participated in the retreat, understood the significance of the mission of Program Development, recruitment of staff from among the brightest and best of the center was successful.

In the early days, MSFC was recognized primarily for its engineering excellence. However, during the development of the Saturn family of vehicles in the 1960s, management techniques became much more a part of the center expertise. Therefore, from the beginning of development of new projects, the Program Development organization gave appropriate attention to the business and management aspects of the eventual project. To that end, two early recruits into Program Development were James Murphy, a retired Air Force colonel, with many years of management experience in Air Force rocket programs and more recently as deputy manager of the Saturn V program, and Bill Sneed, who had been responsible for the exemplary Saturn V Program Control Center. These two individuals exercised great influence on the consideration of management control of all new projects even during the definition phase of the project. They also indoctrinated new project managers in the fundamentals of their jobs as well as the expectations of MSFC management as to quality of product, schedule, and cost. This would ease the transfer of new projects at the end of definition from the supervision of Program Development to the supervision of the center's Program Management directorate. Murphy would eventually become Director of Program Development and Bill Sneed would advance in center management.

Advanced studies were underway continuously within the permanent cadre of Program Development with a goal of identifying potential new projects. When the studies and preliminary analyses showed sufficient promise, a more formal approach to project development began. An adaptation of the Phase Project Planning (PPP) Guidelines document, NHB 7121.2, dated August 1968, was used as the primary instrument for planning new projects. This document outlined four phases for the project: Phase A – preliminary analysis and feasibility; Phase B – definition and preliminary design; Phase C – design; Phase D – development. At the beginning of Phase A, a task team was selected, primarily from the permanent cadre of Program Development to guide the study. If the Phase A study confirmed the feasibility of the concept, a Phase B Team (pre-project office) was organized. A leader/manager was selected, preferably one with project management experience, to lead a small group that would grow as the study matured. Phase A and Phase B were both done under the guidance of Program Development. Upon successful conclusion of Phase B and the approval of the project, the team would transfer with the project to the center's Program Management Directorate for phase C/D. Having the core of a management team, which was knowledgeable of the background and status of the project, to transfer with the project was a distinct advantage in maintaining momentum through the transition from definition and preliminary design (Phase B) to hard design (Phase C).

The first project to go this route in development was the Lunar Rover Vehicle under the management of Saverio "Sonny" Morea, an experienced manager. That project was transferred from Phase B to Phase C earlier than anticipated due to schedule pressure. However, the project was very successful and was completed at a cost less than estimated initially during Phase B.

One of the earliest initiatives of the new Program Development organization was what might be called a market assessment, based on the assumption that the nation would want to continue a preeminent space program through NASA, a program to explore the new space frontier to develop new knowledge that might be used to improve the human environment and to do so by utilizing to the maximum extent the Apollo hardware and the knowledge already gained. Dr. Homer Newell, Associate Administrator of NASA, recently had led a working group that projected future NASA activity at various budget levels. Using this work as a baseline and ideas from many sources, Program Development organized a potential NASA plan extending several decades into the future. This plan was presented in four different areas: earth orbital, lunar, planetary and transportation systems. This plan was far too optimistic as to timescale. For example, one of its projections included a Space Base in low earth orbit to accommodate 100 astronauts doing science simultaneously, a lunar base and a Mars base within the century. While these projections were not realistic from budget or technical considerations, the composite projection included many things that have been accomplished subsequently. Furthermore, the writer believes that some spacefaring nation or conglomerate of spacefaring nations will eventually accomplish, to some degree, most of the other long range projections.

The dream of WvB and his colleagues at the MSFC had long been of an earth orbital space station and such a station had wide support throughout NASA, especially among the manned space flight centers. At that time, it was believed that the high cost of expendable launch vehicles would not be acceptable to support a space station and that a reusable launch vehicle would be required. The early studies of these two potential programs went along simultaneously. However it was soon apparent that the budget would not support developing the two programs at the same time and ultimately a decision was made at the national level to do the reusable launch vehicle, Space Shuttle, first. Actually, doing both programs simultaneously probably would not have been a good idea even if budget support had been available. Both the Space Shuttle and the space station went through the center's phase project planning system. Roy Godfrey was selected as the first manager of the MSFC Phase B Space Shuttle project and Bill Brooksbank, who had been involved in the development of Skylab, was recruited to head the space station Phase B.

MSFC committed substantial resources to defining each of these potential projects, and their eventual approval seemed highly likely. However, if either or both of the programs were stretched out in schedule, or if MSFC did not receive a substantial part of the space station project, there might not be adequate assignments to justify the continued existence of the center. Both programs stretched out and a gap did develop. To fill a potential gap in assignments, a decision was made for the center to pursue aggressively assignments of direct involvement in significant projects of space research and exploration in the recently accessible space environment. For a long time, many at MSFC had aspired to become intimately involved in large spacecraft for research and exploration. This decision, motivated to a significant extent by the threat of extinction of the center, became the tipping point for the diversification from its primarily large propulsion system development to include major space science projects.

MSFC had been involved to a limited extent in the development of payloads for Apollo and Skylab. Some of the earliest interests were in materials research and in biological separations. Dr. Robert Naumann was one of the earliest researchers in materials research in space. Roger Chassay, managed a small materials research program done on sounding rockets, and called Space Processing Applications Rocket Program (SPAR). Other projects in which Marshall scientists were involved, often with other scientists from outside the center, were Gravity Probe-A (GP-A), the Laser Geodynamic Satellite (LAGEOS) and several experiments on Skylab. These were good training experiences for ultimate collaboration on much larger space projects.

In the late 1960s, NASA developed the concept of outfitting the third stage (SIV-B) as a temporary space station, and MSFC was assigned the lead center for that development. This also involved the building of certain experiments and the integration of several others. This whetted the interest of MSFC in payloads and allowed the center to demonstrate competence for such involvement. In the 1970s, MSFC was assigned a NASA responsibility for the integration of shuttle payloads. O.C. Jean was one of the leaders in pushing MSFC toward seeking such assignments.

Early in 1972, a task team, headed by Fred Vruels, was established to study a simple pressurized container to accommodate experimenters in the cargo bay of the Shuttle and called the Sortie Can. Eventually, the Europeans Space Research Organization (ESRO) was allowed to develop and build the Sortie Lab, later named Spacelab. ESRO was made up of 10 European countries, each with a vested interest in the program. The European people were competent but lacked experience, and needed help from the U.S. However, providing that help required great tact. Jack Lee was appointed to manage the program, spread across Europe. Jack's job, which he did exceptionally well, required more diplomacy than engineering. John Thomas and Jerry Richardson were key members of Jack's team. Although there were cost and schedule problems along the way, the program was very successful and enabled years of utilization of the cargo bay of the shuttle for people to do hands-on research in space pending the availability of the space station. It also added markedly to the experience base of MSFC. In time, Jack Lee would become Director of MSFC.

At the end of the Apollo program, MSFC did not have the science capability to compete with the NASA science centers for the large science payloads. Similarly, these centers did not have the MSFC competence in systems engineering to integrate the large science packages into the spacecraft. Thus, the basis for collaboration existed. The few scientists of MSFC were unusually competent and the center was able to recruit a few others such as Dr. Bob Odell and Dr. Martin Weisskopf as time went on. These, along with the permanent cadre of Program Development such as Jim Downey, Jean Oliver, Herman Gierow, Bill Snoddy and others, were invaluable in bridging the communication between the MSFC systems engineers and outside scientists and making the ultimate collaborations on several projects much more efficient.

The first major step on the road to the diversification of center projects was the assignment of MSFC to manage the high-energy astronomy program with the Goddard Space Flight Center as the lead for science. The program consisted of three satellites, HEAO 1, HEAO 2, HEAO 3, designed to study sources of x-rays, gamma rays, and cosmic rays. Jim Downey was key to the definition of the program that had to be down-sized from initial intentions due to budget. Dr. Fred Speer was the project manager, Fred Wojtalik was the chief engineer and Dr. Tom Parnell of MSFC was project scientist for HEAO 3. The other two project scientists were from The Goddard Space Flight Center. The program was very successful and led to the MSFC becoming the lead center for the Advanced X-ray Astrophysics Facility (AXAF) a grossly more powerful facility than HEAO.

Fred Wojtalik was appointed the lead center manager for the development for AXAF, and Jean Oliver was the deputy manger. Noted astrophysicist, Dr. Martin Weisskopf, was employed specifically for AXAF and continues to be its chief scientist since the beginning. AXAF was launched on July 23, 1999 and renamed Chandra X-ray Laboratory. That observatory, one of NASA's greatest programs, continues to generate copious amounts of important data that scientists may be generations in evaluating.

Another deviation from the traditional MSFC role in the development of large propulsion systems was the Large Space Telescope (LST), later named the Hubble Space Telescope. A telescope in space, above the obstruction of the Earth's atmosphere, had been a dream of astronomers for many years. In fact, telescopes in space had been used as a prime justification for going into space. James Downey and his team, including Jean Oliver at MSFC, had studied a large space telescope as a potential project since the early 1970s. When the prospects of a large space telescope were encouraging, Bob Odell, former director of the Yerkes Observatory at the University of Chicago, was employed by MSFC. Bob and Jim Downey and his team were instrumental in getting the assignment of the LST project to MSFC. Bob was also very important in selling the program within NASA and the government as well as to the ordinary citizens of the country.

From the beginning, the Hubble project was a great challenge due not only to its complexity but also because of relationships within NASA, between MSFC and the contractors, and because of certain restraints placed upon the penetration of the contractors by MSFC. The telescope was planned first for launch on an expendable rocket but this had to be changed to launch on the Space Shuttle. After launch, a manufacturing a defect in the mirror was discovered that hampered the performance of the telescope. That defect was corrected and now 25 years after launch, the telescope continues to deliver exciting information. It is certainly one of NASA's greatest scientific missions to date. Overall, MSFC may have encountered more difficulty in this project than any other in its history, but it also may rank at the pinnacle of the center's performance, extending from the definition phase and the saving of the project from cancellation, to the technical management during development and to the correction of a problem after launch. All the MSFC heroes could not be listed but among those not yet listed are William Keithley, Jerry Richardson, Fred Speer, Fred Wojtalik, Jim Kingsbury, Gerald Nurre, and James Odom.

The search for new business was not restricted to space but included the application of space technology to the general economy. For example, the application of space technology to earth sensing for use of agriculture, forestry or other cases of surveying broad areas was demonstrated and promoted. When manpower was available, a program was undertaken to apply space technology to the operation of an automatic coal mining machine situated in a hostile mine deep in the earth. This demonstrated economic advantages as well as reducing the exposure of people. Significant work was done for the Department of Energy in demonstrating the heating and cooling of buildings by solar energy.

Program Development, a subordinate element of the MSFC formed late in 1968, was successful in securing new business for the center. It helped define new NASA projects, and it led the center in applying its extraordinary systems engineering capability, sharpened by years of developing large propulsion systems, to the development of a wide array of experiments and large science payloads. It led the center in gaining two of NASA's outstanding projects of all time, the Hubble Space Telescope and the Chandra X-ray Observatory. Without sacrificing its traditional role as a developer of large rocket propulsion systems, MSFC became the center of NASA with the broadest base for the development of space hardware. Some believe that the diversification of the assignments of the center was responsible in substantial part for the survival of the center as an institution through the 1970s. The judgment of the Jekyll Island retreat was validated.

Part III
Management

Management

By Ed Buckbee

What kind of person was WvB and how was his management style a reflection of his character? He was far more than a rocket scientist, but was also a leader adept at managing massive scale projects. He was also a pioneer for space and a man of great vision. He had a nobility about him that sat him apart from others in the space program. He believed in the decency of man. He believed man's reach could exceed his grasp.Those who worked closely with him reflect on what it was like and the lessons they learned.

Dorette Schlidt was WvB's secretary in Germany: *"He wanted to devote his life to the work and development of powerful rockets, guidance systems to one day go into space and to the Moon. It was his dream already from the beginning on, but nobody was actually believing that at that time."*

WvB had commented years before: *"When I was 17 years old, I had no doubt that man would land on the Moon in my lifetime. But, of course, at the age of 17 I had no idea what it would really take to do it, nor did I know what a billion dollars was."*

Frank Williams served as an assistant to WvB in Huntsville and also worked closely with him on plans for future space travel at NASA Headquarters in Washington. Williams: *"When that happened, it was like having a ton lifted off of WvB's shoulders. I'm out of the military business. I'm in civilian spaceflight, peaceful uses of outer space. That was a big, big load off of him and the rest of the team because there was a stigma there. You could tell that was a big point in his life. He was my boss; he was my teacher; he was my mentor. I make no bones about that. During the first two and a half years we became very close personal friends. He knew how to motivate people. And when I say people, he motivated the designer, the technician, the engineer, anybody he came in contact with. He bubbled with newness and excitement."*

WvB invited NASA Headquarters and congressional leaders to Huntsville to witness a test firing that demonstrated the immense power generated by his mammoth rockets. These rockets were designed, assembled, and often tested daily in Huntsville. Subsequently, they were shipped to and launched from Cape Canaveral, Fla. He hosted a wide range of high profile guests, including: multiple presidents, vice-presidents and first ladies. Additionally, he hosted First Lady Ladybird Johnson's family reunion at the Marshall Center in the mid-'60s. He never turned down a speaking engagement invitation from a member of the U.S. Congress. He understood the importance of maintaining a close relationship with those who controlled the funds.

Finally, a reporter asked WvB the sixty-four dollar question:
Interviewer: *Would you like to go?*
WvB: *"I'd love to."*

Interviewer: *Do you think you will?*

WvB: *"Well, in Apollo, they only accommodated astronauts. Being an astronaut was a full-time job. In the Shuttle we are going to do away with that and the Shuttle will accommodate scientists as passengers and the role of astronauts will be confined to the cockpit, pretty much like an airline. Anyone who can ride in the back of an airline can ride in the Shuttle.*

Question: *Have you bought your ticket.*

WvB: *"I hope they give me one, one fine day or thumb a ride at reduced fare."*

Ruth von Saurma: *"He was such an enthusiastic pilot. It really would have been the dream for him to be in space.* He made one statement. He said, *"I wish when I die, that I could die with my eyes open to see the wonders of whatever comes afterwards."*

WvB believed in communications with his people. His office door was always open. Early in 1961, he initiated a particularly effective management tool, the Weekly Notes. Every Friday program managers and laboratory directors were required to submit a brief one page description of the events and activities that occurred within their area of responsibility. This was email 1960s style. His notes could be categorized as programmatic, strategic, institutional, political and sometimes humorous. For example, before a potential maintenance contractor strike, WvB commented: *Get me a broom! I'll sweep my own office.* He thoroughly read and checked every paragraph and made notations on the margin, returning them promptly to the people involved.

Frank Williams recalls the importance of the Notes: *"The Weekly Notes were an excellent management tool and team builder. Everyone wanted to be mentioned in WvB's Weekly Notes. He would take those home and bleed on them over the weekend. They would come back with red pencil. He would give them to Bonnie, his secretary. She would run copies, and they were distributed by noon on Monday to everybody, with all of his comments. Comments might be: 'I don't understand this; come tell me more about it – set up an appointment and come tell me. Or, I think you're right; we ought to do it this way. Or why don't you and Ernst, or Hans, etc. get together and set up a meeting as soon as you can.' Anybody had a problem, they had better go solve it."*

To Karl Heimburg – *"Can we conduct the static firing during Dr. Paine's (NASA administrator) visit?" WvB (Buckbee 2009)*

To Hans Gruene – *(Response to management concerning desire by the Cape to terminate national TV coverage of early Saturn flights) "This was President Kennedy's personal suggestion. Will be hard to turn down, particularly now, after the tragic event (Kennedy's assassination)." WvB (Buckbee 2009)*

Don't let anyone tell you that we didn't have a failure during Apollo. We lost three astronauts, Gus Grissom, Ed White and Roger Chaffee—on the launch pad in the Apollo 1 fire. This accident brought NASA to its knees. It was nearly a showstopper. Many of us thought the program was over and we would never go to the Moon. But WvB rallied the troops, brought the team together, supported the other center's efforts to correct the problems, and ultimately we went on to fly the redesigned Apollo spacecraft on the Saturn IB launch vehicle.

Many of the astronauts believed this accident was a wakeup call, reminding the manned space flight team of the highly dangerous and unforgiving business we were engaged in of launching people in mammoth rockets into space. WvB's leadership and sound management style were important factors in rebooting the program after the tragedy of Apollo 1.

WvB was a key player in the NASA Management Council meetings. These monthly meetings consisted of the NASA Headquarters director of manned space flight, Dr. George E. Mueller; Kennedy Space Center Director Dr. Kurt Debus; and Dr. Robert R. Gilruth, director of Johnson Space Center. He realized to accomplish a dramatic space exploration venture three criteria had to be met: (1) a focused task and well defined mission (2) an exceptional manager, a doer, man of action who is supported by a capable team, managing and executing the given task and (3) a customer is required, an entity who has a sincere interest, much to gain, a real demand and—most important of all—is willing as well as capable of financing the project.

WvB stated, *"I have learned to use the word 'impossible' with the greatest caution."*

It was a challenge in the 1960's to have a management style that worked with government as well as private industry. WvB developed and executed a management style for a fast track technology program with high visibility. The keys points were:

- High degree of trust between von Braun and staff.
- Ability to make everyone of his team feel proud and privileged to work for him.
- Defined the technical projects he wanted to bring to life, worked tirelessly until they were accomplished.
- Ability to form a team, nurture it, keep it challenged and together, even after his death.
- Gift of feeling equally at ease whether he conversed with his driver, or with the president of the United States.
- Often gave the problem to three or more parties in order to have more then one solution to consider.
- Conducted joint technical discussions so that all participants could understand what the problem was.
- Continued such discussions until a reasonable solution had been reached.
- In technical disputes with a declared winner, he was careful the other party did not feel to be the loser, but the second or third winner.
- Gifted scientist with a knack for inspiring people and managing huge technology projects.

- Established an organization designed to develop and produce Saturn rocketry systems and engaged industry on a large scale basis.
- Implemented a direct personal communication channel with his lab directors and project managers referred to as the "Weekly Notes."
- A one page report of the week's activities was expected on WvB's desk by Friday from each of his key organizations.
- Subjects covered were categorized as programmatic, strategic, institutional, political and sometimes humorous.
- If an employee had a subject they felt needed to be brought to the attention of WvB, it was submitted to the "Weekly Notes" editor in that organization.
- WvB carefully and diligently read each note, writing in long hand his thoughts on the margin and signing with his initial "B" and the date.
- Notes were distributed to all persons named in the notes by WvB and others by courier on the following Monday
- Notes were often directed to others to take action or returned to the sender with specific instructions.
- Provided encouragement and suggestions regarding progress on a given project.

To Board Members Ref: NASA Headquarter Control: *"The old creeping NASA disease again! The 'headquarters octopus' strangling all activity if not checked continuously." WvB*

- Constantly advised his staff on changing strategies and politics of the day
- Delegated decision-making to others. When timing was critical he encouraged his staff to go to a higher level of management than himself in order to expedite the issue or problem.
- Kept himself and his staff abreast of new and innovative technologies in the field of rocketry and space flight.

To Werner Kuers: *"Request a briefing on this subject by the most knowledgeable people we have. Please arrange." WvB*

- Assigned personnel to NASA HQ for experience and to Harvard and Yale business schools.
- Firm believer of man-in-the-loop systems, both on the ground and in space. Conducted simulations of such. Developed a Lunar Rover Vehicle cockpit simulator enabling the design engineers to understand the effects of 1/6th gravity while driving the rover on the Moon.

WvB had a fascination for machines and man-in-the-loop technology. He wasn't much for robots or unmanned remotely controlled vehicles. He believed humans at the controls enhanced the outcome. He loved to fly aircraft, drive weird machines and be challenged by sophisticated simulators. It became well known in the space community if you had a new prototype vehicle that would float, fly, or drive and you wanted to sell it to NASA, bring it to Huntsville. If WvB sat in the driver's seat, it was almost a sure thing—NASA would buy it. We went through a period in the '60s when an incredible number of aerospace companies appeared at Marshall's entrance with new prototype toys for WvB to test. He drove Moon buggies of all sizes and shapes, flew airplanes that looked too big to be in the sky, and rode on river barges that looked like blimp hangars. The most interesting one was the Pregnant Guppy, a modified Boeing Strato-cruiser that Carl DeNeen found in California. It had been enlarged to transport rocket stages across country. WvB flew the strange-looking craft and immediately decided NASA needed one. Most vendors were successful in selling their innovative machines to NASA, because WvB loved innovation and enjoyed serving as the pilot and test driver.

To Bill Lucas: *"If possible I'd like to participate in one of these flight tests to get familiar with the problem." WvB*

To Fred Cline: *"Does this mean that astronaut initiated abort in case of pad fallback is not possible during first 25 seconds? What's the rationale?" WvB*

- Encouraged the development and use of simulators and mockups of new vehicles or procedures.

To Lee Belew: *"Are we preparing a neutral buoyancy mockup to verify feasibility? I think we should." WvB*

- Valued the talents and skills of his staff, often giving the same problem to more than one to solve, thus acquiring more than one solution to choose.

To Werner Kuers: *"No, I'm not. A lifetime in rocketry has convinced me that welding is one of the most critical aspects of our whole job!" WvB*

- Sought opinions and advice from his people and other sources.

To Saturn Project Manager: *"Do you recommend we cut back on some of these reviews (meetings) so your people have more time to work?" WvB*

- Apologized for being harsh, short or too abrupt with colleagues in meetings

To Willi Mrazek: *"I'm sorry for my harshness. We just had to come to grips with that launch operations problem." WvB*

To Helmut Hoelzer: *"Hope you are still alive. How is Slidell & Michoud?" WvB*

To Walter Heaussermann (Regarding no report for several weeks): *"I guess I haven't had any notes from Astrionics for 3 or 4 weeks. Have you stopped working, has your place burned down. Or is that you simply have no problems." WvB*

- If sloppy workmanship was reported, he took care not 'to shoot the messenger' but rather strongly suggest improving the procedures to prevent a reoccurrence.

To Fred – *"Who goofed? Please see to it that procedures are tightened. I'm not interested in name of culprit. I am interested in steps to prevent recurrence." WvB*

To Bill Lucas – *"Why does this alarming info come in so late? 501(first Saturn V unmanned launch) are ready for launch. How critical is all this?" WvB*

To Production Manager – *"Let's find out why the man dropped it and did not report it! If people start hiding goofs like this, we'll find ourselves in endless trouble!" WvB*

- Congratulated those organizations and individuals who succeed or surpassed milestone events.

To Karl Heimburg – *"Test deserves a big pat on the back for setting up and running a very efficient transportation system all these years." WvB*
To Walter Heaussermann – *"Congratulations! A splendid record! Congratulation to the Saturn V Program Office and the entire Saturn V Team." WvB*

To All Civil Service and Contractors: *"My heartfelt congratulations to everyone who contributed to the stunning performance of the Apollo/Saturn V on Nov. 1967. This was a critical test, and the Saturn V looked like a thoroughbred." WvB*

- Wrote letters of condolences to families who had lost loved ones.

To Frank Williams – *"Please prepare a condolence letter to widow and other relatives – inform Boeing." WvB*

- Encouraged and requested briefings from staff on new developments.

To Helmut Hoelzer – *"Please call me when you'd like to give me a little demonstration." WvB*
Frank Williams makes this observance about WvB: *"He loved to walk the floors – of the shops, the technical offices, the computer building and all of that. If he was talking to a janitor sweeping the floor or a lathe operator, and they were talking and he asked them a question, you know. What kind of tolerance do you have, and they were talking to him, that person had his total attention. The rest of the world, forget about it. He was focused on what that person was telling him. That's when he would say, 'You know, you can learn something from everybody.'*

To Frank Williams – *"I'd like to have a personal appraisal from you on merits and objectives of the Chrysler study. No formal briefing, just a chat between the two of us." WvB*

- Regarding the urgency of a matter, "How can I help you with this problem?" was one of VB's favorite phrases.

To Bill Huber – *"I understand that Keith Glennan is now Chairman of the Board of Aerospace. My relation to him is very cordial. Shall I grease the skids?" WvB*

To Ernst Stuhlinger – *"Misunderstanding! I don't mind having these problems brought to my attention. I only feel that the deputy for research should always be called to help resolve them before I am brought in the act. No. Let's leave the Weekly Note system as it is. I like it." WvB*

- When prime contractors fell behind delivering key components of the Saturn rocket, WvB often wrote letters to the company presidents reminding them they were on the 'critical path' and their failure to perform would greatly impact the established timeline of landing a man on the Moon within the decade.

To Saturn V Project Manager – *"What's going on here (regarding a million dollar change order)?? If Boeing keeps operating like this, we'll be broke in no time!" WvB*

To Harry Gorman – *"What do you think of a letter from me to the president of AMF telling him of these difficulties and explaining to him that he is squarely on the critical path of our lunar landing program?" WvB*

To Karl Heimburg – *"Just to remind you, you are on the critical path."* WvB

To Dieter Grau – *"We should put our foot down and declare this as unacceptable. If you run into any difficulty, just let me know and I'll raise cane." WvB*

To Dieter Grau – *"I share your concern. But if this Center is to have a lively future, we just got to get some or our most experienced people into the development of new programs such as the shuttle and space station. I'm sure you understand". WvB*

To Hermann Koelle – *"We need both: long range planning and Apollo add ons. The trouble at the moment is, that for the momentum built up with the mainstream Apollo program, there has been too little of the latter with resulting immediate dangers to the stature of the program." WvB*

- Reminded his key people to pursue cooperative efforts with other governmental agencies who had unique capabilities.

To Hans Maus – *"Please see to it that this doesn't bog down in bureaucratic difficulties. I think we should use these tasks as 'icebreakers' to get this cooperation with the Air Force established. It may be helpful and useful in 1000 ways in the future." WvB*

- Maintained good relations with other manned flight centers i.e., NASA-Houston was strongly encouraged even though fierce competition reigned.

To Hans Maus – *"Do we get reimbursed for this work? Houston misses no opportunity to put its hand in our pocket. I think we should reciprocate." WvB*

To Jim Shephard – *"Let's have a nice buffet luncheon on 10th floor with some of our key people present. I think it's very important that we improve our relations (red carpet) with Chris Kraft, director, NASA Houston." WvB*

Question by Interviewer: *Dr. von Braun, in all of this what was your one or two greatest moments, personal satisfaction?*

WvB: *"Well of course, when Neil Armstrong stepped down on the Moon, it was undoubtedly one of the highlights if not the greatest highlights but I must say I was happy when he went back home because in these early Apollo flights you walk around with a lump in your throat for a couple days when you are close to the program you are aware of the fact that a lot of things can go wrong."*

WvB's secretary, Bonnie Holmes said, *"When he came back (following the Apollo 11 launch) and walked in the office with a big grin on his face, he said, mission accomplished."*

What Was It like To Be WvB's Boss?

Arthur L. Slotkin, the author of "Doing The Impossible", an in depth look at George E. Mueller and the management of NASA's Human Spaceflight Program, discussed with Dr. Mueller his relationship with WvB and others in the 1960's. (Slotkin 2012)

Dr. Mueller was WvB's boss from 1963 to 1969.

Question: *How would you characterize Robert (Bob) Gilruth, Director of Manned Spacecraft Center (MSC) when you first met him?*

Mueller: *"A very good engineer, not a very good manager at all. But he had some good people. I found him more difficult to work with than WvB for example. Wernher and I thought much more alike than Bob did. Our thought process was similar. And to this day, I don't have the same kind of working relationship with the folks at MSC that I had with folks at Marshall."*

Question: *You sold the idea to WvB that he ought to have a systems engineering contractor?*

Mueller: *"And he thought it was a great idea. When it became clear that NASA was going to need a better system than they had if they're were going to get this thing built and go to the Moon."*

Question: *It seems that knowing WvB's philosophy that in-house is better and it seems inconsistent with his philosophy that he would agree to having an outside systems engineering contractor.*

Mueller: *"Well WvB had a fair amount of respect for the ballistic missile programs after all he was new to NASA at the time; it wasn't like he was an entrenched member of NASA. And he got over into NASA and discovered that there wasn't any systems engineering in that whole organization. So it wasn't him thinking about NASA needs system engineering. No. It was him thinking that Marshall needed systems engineering."*

Question: *Do you recall WvB's first reaction to the suggestion of all-up testing?*

Mueller: *"Well, no I don't as a matter of fact because it was at a meeting with a lot of people and there were enough astonished looks around that I don't think anyone really jumped on board at the first meeting. It's something that engineers usually do. They want to test it and test it and test it. And I want to test it and test it at a subsystem level. But at a system level you're much better off testing the system because in the end that system has to work. And the system has to work as a system. It was pretty clear that there was no way of getting from where we were to where we wanted to be unless we did some drastically different things, one of which was all-up testing."*

Question: *So your philosophy in advocating all-up testing was not we don't want to test sufficiently; we want to test the system sufficiently. Did you feel that after two successful flight tests that was sufficient to prove that the rocket was man-rated?*

Mueller: *"Well the second one (SA502) didn't work as well as it should. And that caused us a fair amount of consternation. But when we found out why it didn't work we were confident we could fix it and then I was confident that we could go ahead."*

Question: *It has been reported that WvB initially opposed all-up testing, but Gilruth actually supported you on that.*

Mueller: *"Well if he (Gilruth) did I didn't recognize that. He had a strange way of supporting. I think that he had time to absorb it before he committed himself. But the folks at Manned Spacecraft Center were not that enthusiastic."*

Question: *Why was there a Saturn I in the first place?*

Mueller: *"Because that's the way Marshall did things, Saturn I and Saturn IB and left to their own devices in the Saturn II. I expect as well as the Saturn III and IV and V."*

Question: *Wasn't WvB's original plan was to have 20 flight tests? He planned to have the first manned flight, flight 17, so you eliminated about half the testing or more.*

Mueller: *"Well they shouldn't of had the Saturn I. When WvB first discussed this with his 'board of directors', the lab chiefs strongly opposed the idea and the debate continued for several days. This approach in the development of a new vehicle especially a space vehicle that had been designed from the beginning as a man-rated system was against the team's conservative philosophy. At the time, no one thought it was a good idea or that it would work at all."*

Eventually WvB convinced the team it was to be done and he informed Mueller, there is no fundamental reason why we cannot fly 'all-up' on the first (Saturn V) flight.

Who Was this Man WvB?

By Jay Foster

Editor's Note: J.N. (Jay) Foster was an engineer who joined the WvB team in 1955. He worked in the laboratories, on the executive staff and served as WvB's senior assistant at the Marshall Space Flight Center (MSFC). He transferred with WvB to NASA Headquarters, where they worked on America's future space programs. Foster shares his experiences while having had a front row seat with The Rocket Man.

WvB galvanized the nation, the world, the Congress and led America to the Moon. His futurist *Colliers* magazine articles of the 1950's led the way. Even today [2015], the USA is the only nation that has sent humans beyond earth's gravity. The WvB team built an organization that had the complete capability to build large rockets from scratch, from concept through preliminary design, final design, fabrication, assembly, test, evaluation, transportation, and flight. The WvB team could do it all.

For the NASA years, 1960-1970, the Marshall center was created and placed under the direction of WvB. Most of the infrastructure was transferred from the Army. For most of the Marshall years the vehicle R & D, design, development, manufacturing, test and operations were carried out by the Science & Engineering Directorate. The other major elements included: industrial operations (program and project management and contractor management and oversight); program development (long range planning, developing future program proposals with preliminary designs, schedules, and preliminary cost estimates); administration & program support management operations {logistics, transportation, security, etc.}, procurement, facilities, communications, computer services, and technology utilization; and staff functions [comptroller, human resources, public affairs, chief counsel, safety, and equal opportunity).

WvB was a people person. He could and did converse with individuals across the whole spectrum of people involved in the lunar program. They ranged from the project and program managers at MSFC, JSC, and KSC to their counterparts at the various contractors plants. His discussions also went from the janitor to the president, the congress, and his bosses in NASA Headquarters. He conversed with all these people on their level with subjects that were important to the person he was speaking with. To WvB, everyone in America was essential to the Saturn Apollo Program. After all, it was the citizenry who convinced the Congress and the Administration to support the program. He had special friends in the Congress. For example: WvB and Congressman Tiger Teague, from Texas, went fishing in Canada periodically. WvB was convinced that everyone working for the space program was essential to success and it shone through in everything he said and did. He was an exceptional manager, engineer, and scientist, and it showed in many ways.

It is difficult to explain, but perhaps some incidents will help. When the first Saturn I was at the Cape being prepared for launch, WvB wanted to review the status. Several of MSFC personnel were at KSC working on various activities. Several of us planned to meet WvB for the trip to Patrick Air Force Base and the charter flight back to Huntsville. There were about six or eight of us in two cars. WvB spoke of how much he enjoyed the day and how the program was progressing in a good way, and everyone in our group picked up on his mood. The group headed out to Patrick. WvB piped up and said he would like a steak before the flight. So we went to a nice restaurant in Coca Beach and had a great steak and a beer. Everyone was in a jovial mood as we headed to Patrick. However, there were others on the plane who weren't very happy about waiting for us to have our steaks while they ate sandwiches.

We had a King Air airplane that originally was outfitted for nine passengers; three of them were required to sit on an uncomfortable bench seat. WvB recognized that the bench seat was a problem, and he had the plane converted to seven passengers with comfortable seats. He said that the individuals who got on the airplane would appreciate the new arrangement, and those who had to find another ride would not know the difference.

WvB was driving in Washington with several MSFC employees and he was going out to dinner with friends, but we were apparently lost. WvB spotted a cab, so he jumped out and hailed the cab and left us abandoned in the middle of the street. He figured the cab could find the way to his appointment and we could find our hotel.

On the way to review progress at the West coast contractors, WvB was flying NASA 3, a 12 passenger Gulfstream. We passed close to the Mount Rushmore stone faces on our way to a SAC base to refuel and have lunch. WvB wanted the 12 passengers to see the faces, so he tilted the plane onto its side to give us a better look. I do not remember the faces, because it was August and all the tourists below were looking up at the strange sight of the airplane flying low and on its side. NASA Headquarters had recently hired a new senior pilot to manage the agency's aircraft, and he was along as a check pilot to see if WvB was really qualified to fly the NASA aircraft. After this trip, he stated that WvB could fly at any time.

After WvB went to Fairchild, he came to Huntsville on business. He was returning to Washington and was sitting in first class when I entered the airplane. WvB asked me to sit with him and talk. Naturally I sat. Soon the stewardess came by and said, 'Sir, you must move to coach'. Since the plane was not full, WvB got up and accompanied me to coach. We chatted all the way to Washington. That was the last time I saw him. He died in 1977 at the age of 65. These examples illustrate how WvB was a regular guy. He could talk effectively about any subject and usually won the discussion. However, he almost never gave an order. He discussed a subject and if you convinced him that your proposal was rational, he would pursue your proposal with enthusiasm. WvB believed in and practiced knowledge-based decisions, not position-based decisions.

Organization

The WvB organization evolved through the V-2 organization in Germany to the NASA-MSFC organization to include the physics, mathematics, and material sciences required by rocketry and space flight. As the size and complexity of rocketry developed, it was apparent that subtended from the three disciplines mentioned involved almost every conceivable discipline. This in spite of the fact that the simplest rocket every kid plays with is a balloon that you blow up and let go. If you add a few essentials to the balloon you have the beginnings of rocketry. Just add guidance and control, a mechanism to provide a continuous flow of fluid out the back, and make it big enough to provide weight-carrying capability for men and scientific or other payloads. The major disciplines evolved into the laboratories in the Science and Engineering Directorate. They included for most of the 1960's the following labs and directors: Aeroballistics Laboratory / Ernst Geissler, Computation Laboratory / Dr. Helmut Hoelzer, Fabrication and Assembly Engineering Laboratory / Hans Maus, Guidance and Control Laboratory / Dr. Walter Haeussermann, Launch Operations Directorate / Dr. Kurt Debus, Research Projects Laboratory / Dr. Ernst Stuhlinger, Structures and Mechanics Laboratory / Dr. Willi Mrazek, Systems Analysis and Reliability Laboratory / Dieter Grau, and Test Laboratory / Karl Heimburg.

In the early days the complete airframe/structure was fabricated in-house, and many of the internal components were also fabricated and the whole vehicle was assembled in-house. Building 4705 exhibited a real production line. The Redstone vehicle, early Jupiter tanks and a whole row of Redstone tanks, of which eight were required for the Saturn 1 and 1B were fabricated and assembled in-house. With President Kennedy's announcement of the lunar program there evolved many changes in the manufacturing plan. Many more components and whole stages were contracted with industry. MSFC built the initial Redstone and Jupiter vehicles and the Saturn I first stage test vehicle. Later the fabrication and assembly of the Saturn I, and IB vehicles was contracted with Chrysler at the Michoud Assembly Facility near New Orleans. The Saturn V first stage was contracted with Boeing also at the Michoud plant. The 2nd Stage of the Saturn V was contracted with North American Aviation in California and the 2nd Stage of Saturn IB and the 3rd Stage of Saturn V were contracted with McDonnell Douglas also in California. The H-1, F-1 & J-2 engines were all contracted with the Rocketdyne Division of North American Aviation in California. MSFC retained a hands-dirty capability, to effectively work with the large contractor work force. MSFC retained this capability by keeping a major system or subsystem in house. MSFC retained the capability in-house to do anything on the critical path, but not everything.

The Science & Engineering Directorate was initially managed by Hermann Weidner, but later was managed by a number of the American team members: Jim Kingsbury, Jim McMillion, and Georg McDonough. Additionally, WvB encouraged the laboratory directors to bring along by mentoring the young American team members. Some key names that come to mind: Dr Bill Lucas, became the Propulsion and Vehicle Engineering Laboratory director, the Initial Director of Program Development. Later he became deputy center director and center director. Bob Lindstrom was Saturn I program manager then became Fabrication and Assembly Engineering laboratory director, and Space Shuttle program manager. Brooks Moore and Dr. Joe Randall, managed the Electronics Laboratory, formally the Astrionics Laboratory. The list goes on and on.

The Marshall staff grew to 7,000 during the Lunar Program with many thousands in the contractor plants around the nation. The original WvB Team was dwarfed by the American workforce. Recognizing only the small original German WvB team does a disservice to the many American members who were essential to America's triumphs in space. The American members also mentored the next generation who are currently serving throughout the space industry. This mentoring and identifying became more formalized under Dr. William R. Lucas. Once a year each individual who reported directly to the center director was requested to submit the names of ten comers in the organization. It was a plus if an individual appeared on several managers lists.

Management

WvB was required to travel much of the time and this required the creation of an organization systems and techniques to keep him informed and to be available for major decisions. He accomplished this by a series of actions. The major techniques are described below.

The Marshall front office consisted of three principal's: Director – Dr. WvB, Deputy Director Technical – Dr. Eberhard Rees, Deputy Director Management – Initially Delmar Morris and later Harry Gorman. The organization charts were drawn to show these three positions grouped in three overlapping rings after the Ballantine beer logo. The explanation stated that each of the three had individual responsibilities, but they also had overlapping responsibilities. Consequentially, one or more could travel and the MSFC would not miss a beat. Everyone who directly reported to WvB or their deputies attended staff luncheons. This served as an excellent forum to discuss the problem "du jour". The luncheon also served to keep the senior managers fully aware of activity underway across the center. WvB always attended these luncheons when he was in town. You had to be on your toes during lunch, and be prepared for anything.

Staff and board meetings were scheduled monthly and as required. The Staff included all the Staff Office Managers, while the Board consisted of the Science and Engineering Management, the laboratory directors and selected division chiefs, as well as the program managers. After center wide topics were concluded the staff was dismissed and the board continued to discuss difficult technical problems. The way the board operated under WvB is interesting because:

1) Initially an individual would present the problem in sufficient detail to clearly establish the parameters and the presenter would outline potential solutions with pros and cons.
2) WvB would then open the floor for comments by the members.

Depending on the problem and its seriousness, the discussion could last for minutes or for hours. When the discussion wound down, WvB would sum up and state that the consensus appeared to be so and so. But, before he banged his gavel on the decision, he would ask one more time for last minute comments.

WvB rarely, if ever gave a direct order. He always discussed a problem with an individual or with a room full of people, such as the Board. He also would give way if you could swing him to your thought process. Inversely, he expected you to give way if he convinced you of his approach. His arguments typically won out, but he wanted input, particularly from the experts in a particular discipline.

Quarterly reviews were held with the major prime contractors. These included: Boeing, North American Rockwell, Rocketdyne, McDonnell Douglas, IBM, and Chrysler. WvB led these meetings in an analogous way to the Board meetings. The contractor would present his technical accomplishments and his status on meeting schedule, as well as cost projections. After each agenda item WvB would encourage and expect a lively discussion from the MSFC people present and the contractor personnel. The MSFC individuals present would normally include the Project Managers and the Board members involved with the particular contractor's activities.

The Weekly Notes was WvB's idea. He required each organization that reported to him to provide weekly notes by close of business Friday. The notes would present the week's highlights from that organization. WvB's comments were usually available on Monday morning. WvB carefully read and initialed each note and frequently commented in the margins. This was a way to keep WvB informed and it was a way to instill discipline into the organization.

In the early days, the laboratories had automatic responsibility for the quality of their discipline across the total contract structure. This worked well in the early days, but was judged to be too expensive later. Subsequently, the laboratories, once they recognized a problem or concern were required to go through the program managers to redress the concern with the contractor. Most of the MSFC contracts were award fee contracts, particularly all of the major prime contracts. These contracts provided for a minimum and a maximum fee. The minimum could be established at a straight percent, say 24 percent, or a formula could be inserted to incentivize parts of the contract to determine the minimum. Additionally, an award fee board would meet semiannually to review the contractor's performance. Normally, the program manager would present his analysis of the contractor's performance, and then the contractor would present his self-analysis of his performance. The Award Fee Board would then consider both presentations and arrive at a proposed fee percent for the semiannual period of the contract. The proposed fee could be a straight line from the minimum to the maximum, but it also could be a sophisticated curve that required a high grade by the board before proposing a high percentage of the fee. Some contracts held back a significant portion of the fee for on orbit performance. The proposed fee would then be sent to the fee determining official – usually the deputy director, Management. A clause was usually inserted in the contract that made the fee determining official's decision final. A provision was inserted to allow for the contractor to protest to WvB. WvB's decision was final and could not be challenged.

I had the opportunity to work closely with WvB from 1961 shortly after President Kennedy's announcement of the lunar program through the initial lunar landing in 1969. Initially, I was a key member of his executive staff and later his Senior Assistant. I considered WvB my boss, mentor, and friend. WvB did not give orders. He wanted me to monitor everything that was going on at MSFC and brief him on areas that I believed he should be aware. He asked that I attend meetings for him and report the highlights back to him. He asked me to attend meetings with him and prepare summaries for presentations to the staff and board. These presentations usually served to start a conversation and discussion, which would lead to the resolution of a problem or the development of a position to take with NASA headquarters or our sister centers at JSC and/or KSC. The opportunity to work with WvB and his senior managers was the highlight of my career.

The Washington Years

After the first lunar landing Tom Paine, the NASA administrator, was looking to the future, and he could see that the Saturn/Apollo Program would come to an end after several more lunar landings.

Tom Paine was cognizant that the Program Development Directorate at MSFC essentially was the long range planning element for manned flight across the agency. Tom Paine then asked WvB if he would come to NASA Headquarters as the Deputy Associate Administrator (the No. 5 Position in NASA that happened to be vacant at the time) and head up an agency planning office with the goal of defining what the agency should look like in the year 2000. WvB, after taking a long vacation, discussed the option extensively with his family and agreed to move to Washington. WvB agreed to move to Washington for several reasons. 1) He was under a lot of stress with 7,000 MSFC employees and ~100,000 contractor employees working on MSFC contracts, 2) WvB also could see the end of the Saturn/Apollo approaching and he thought that he would enjoy being able to reduce his workload and dream about the future. 3) WvB had some health concerns.

WvB selected two individuals from MSFC to accompany him to NASA Headquarters. One was Frank Williams who was the MSFC liaison engineer located at JSC. Frank had previously been the deputy in the MSFC Future Projects Office, the predecessor to Program Development. He selected me and at the time I was his senior assistant and previously had been the deputy of the MSFC executive staff. Tom Paine gave WvB 20 positions and we began to set up a central planning organization and network across the agency. WvB was the head of our group. WvB's Deputy was Dr. D.D. Wyatt, Assistant Administrator for Planning. He was the existing headquarters long range planner, and I became the Deputy Assistant Administrator for Planning. The organization had three sub-elements: 1) Office of Long Range Plans with Director J. Ian Dodds, 2) Office of Analysis and Evaluation with Director F.L. Williams, and 3) Office of Plans Integration with Director Jim Skaggs.

The Central Planning Office worked on two fronts. Ian Dodds would synthesize projects, Frank Williams would integrate the projects into an overall NASA program, and Jim Skaggs would develop schedules and cost estimates for the projects, in parallel, WvB would schedule trips to each NASA center and ask the Center Director to outline his vision for the year 2000 for his center. Additionally, we brought in all the aerospace companies and government entities, such as the Weather Service and the Military Sealift Command and ask them the question of how they would see their use of space in the year 2000. Central Planning also developed the agency cost projections for the Space Shuttle and helped obtain funding to move the Shuttle out of paper studies into Hardware Design and Development. This process produced a wealth of information that we could use in synthesizing the total plan.

The NASA Administrator, Tom Paine and WvB wanted to gain a wider understanding of the long range thinkers in the agency, so they set up a long weekend meeting at Wallops Station, Va. NASA Conference of the Year 2000 – took place June 11-14, 1970. There were twenty-three NASA participants plus Arthur Clarke, the fiction space writer as keynoter. Arthur Clarke made a good opening presentation that served as a good kick-off. Arthur Clarke subsequently sat in the audience as others presented their thoughts on elements of future plans. WvB presented a very large booster he called "Orion". Orion's propulsion was nuclear and consisted of a large pusher plate on the back, which emitted aspirin size atomic pellets out of the center of the pusher plate. The pellets crossed a laser field and exploded, giving the vehicle a continuing set of pushes to accelerate to extreme speeds.

Subsequent to the conference, Tom Paine took the developed long range plan to President Richard Nixon in the White House. On his return we received a downer message. President Nixon was up to his butt in Watergate and he essentially told Tom Paine that NASA currently had a lot on its plate; Tom should concentrate on current projects and not bother him about future projects.

Shortly after the Nixon meeting, Tom Paine announced that he was leaving NASA to become a Vice President of GE, to take charge of their nuclear program. George Low became Acting Administrator. A few months later, Dr. James Fletcher became the NASA administrator and subsequently WvB decided to leave and become a vice president of Fairchild. I called Dr. Rees at MSFC and received an offer to come back to Huntsville.

There has always been speculation in Huntsville that WvB was pushed to leave MSFC and that he was unhappy in Washington. I would like to give the readers my take on this speculation from my perspective. For the reasons mentioned above, WvB wanted to leave MSFC and Tom Paine really wanted WvB's thinking and dreaming about the future to be placed on paper. The negativism arose after Tom Paine's meeting with Nixon. With George Low as acting Administrator, Central Planning's work dried up. While George Low and WvB had a history from the Saturn/Apollo days of competition between MSFC and JSC, George was perfectly within his rights not to provide assignments to Central Planning since the president did not want future planning at this time. WvB and the Central Planning Office were upbeat and challenged during Tom Paine's administration and the Planning Office fell on hard times after the Nixon meeting. WvB said that we had provided a plan that could be implemented as the nation and the congress saw fit. Consequently, we didn't need to stay at NASA headquarters. So Tom Paine, WvB and I jumped ship. A few weeks later, the Planning Office was disbanded. Frank Williams went to Martin Marietta at Michoud, Jim Skaggs joined George Mueller in industry, and I lost track of Ian Dodds.

Fast forward 20 years, during the first Bush administration. President George H.W. Bush appointed the same Tom Paine to chair a blue ribbon committee to define the future of space flight. The Bush committee published a Space Plan that is in the public domain, that is very similar to the WvB plan of 1970, but it flies under the Paine committee name. One of WvB's tasks at Fairchild was to go on the road in India. Fairchild had a contract with NASA to put up an educational satellite in stationary orbit over India and broadcast educational programs into the Indian countryside. WvB toured major Indian cities to prepare the populace for the coming programs. Mrs. Gandhi was in power, and she provided a number of bicycle driven generators to be placed in village squares. Individuals could take turns pumping the bicycle and villagers could watch educational TV programs.

WvB, was a very special individual, and the United States was perceptive and fortunate to bring him and his team to America, first to Fort Bliss and later to Redstone Arsenal and NASA-Marshall Space Flight Center, Huntsville, Ala.

WvB's Management Style

By Stan Reinartz

Editor's Note: Stan Reinartz, who served as a Saturn project manager and laboratory director during the Saturn-Apollo era, shares his thoughts on WvB's management style.

While many of the items listed below relate to WvB's character, others are part of his total life experience. I believe, that in the end, the key feature that led him to rise to the political and public position as the best known, most listened to spokesman for space exploration, was his unique and truly outstanding ability to communicate extremely effectively. He concentrated on using a level and type of speech or writings that most directly fit his audience needs. Whether it was his "gee-whizz" *Collier's* magazine articles, his illustrative work with Disney or using simple explanations of his plans to political figures, including benefits to their constituents, he made sure they could easily understand what he was advocating and why it was a good idea for them to support. I remember WvB using the analogy that we have been looking at the stars "through a dirty basement window" and now we had the opportunity to study the heavens for scientific data without that handicap—a clear concept.

One other characteristic of WvB was being able to accept and proceed to implement plans that were less or somewhat different than his personal long-term goals. He understood that eventually the U.S. would be expanding rocket technology and that alone would be another big step toward his goals in space. The same manner of thinking applied when development tests for his rockets were designed such that, if needed, they could drive a small satellite into earth orbit. This bidding his time for the right moment came after the successful Russian Sputnik orbit and U.S. Vanguard failure. Again in the late 1950s, the military, to some degree reacting to political pressure, directed the Army and WvB to build a "big booster." While an advanced technology booster, as part of an optimized multi-stage vehicle, would have been the desired next step, he accepted the country's priority need for a big booster quickly. It was well built, all flights successful, using available propellant tanks, engines, and current technology in record time. This booster opportunity provided a huge head start toward WvB's goal of getting men into space.

When deciding how best to get to the Moon, WvB advanced the case that it would be better to use Earth Orbital Rendezvous (EOR). If this method were selected, it would again provide a big step toward launching, assembling, and manning an earth orbital space station following the Moon landings. Once again, he ended supporting Lunar Orbital Rendezvous (LOR) the chosen method, but having this tremendous booster laid the foundation for the eventual Skylab and man's weeks in space. While not at the Marshall center for the final phases of Skylab development, at the start, WvB again hoped for a large, all new optimized space station. That was not to be, so he worked effectively with NASA Headquarters and encouraged his team to design a first step space station using mostly existing Apollo and Gemini hardware. Skylab, although a NASA and congressional stepchild program, for WvB it was one more large step on his dream path into space exploration—even if it had to be less than his dream of an even bigger step.

I was fortunate to personally be involved in the work related to initiating and building the Saturn boosters and eventually the Moon program. I worked with WvB in several capacities from 1957 until his transfer to NASA Headquarters in 1970. For eight years, I was heavily involved in the project management from the very beginning of the manned Skylab Program, when its best configuration description was jokingly called "The Kluge", until the end of Skylab operations.

Industry Contractors

To the extent of my knowledge this story represents WvB letting the system work without pressure on participants to bend their views or emphasize selected data points in order to achieve an outcome possibly more to the overall benefit of Marshall and outside parties. I was selected as chairman of the Source Evaluation Board for the Skylab integration contractor. Our team's conclusion, made-up of people from of Marshall, Johnson, Kennedy and headquarters, was that Martin (Denver Division) was best suited to perform this role for Marshall and NASA. While presenting to WvB and a few other key personnel, it was clear that we had not arrived at the conclusion some wanted. The only minor change that I agreed to make was how the contractors were listed on one chart. WvB cut off further discussion, to my great relief, and agreed for the evaluation to proceed to NASA headquarters. After presentations to Dr. George Mueller, director of manned space flight, and Jim Webb, NASA administrator, Martin was announced as the winner. In my mind, WvB played by the book where he could have pushed us to reconsider the importance of factors that were more favorable to a different contractor.

Board Meetings

A good example of WvB's management style is contained in the following anecdote. Marshall's headquarters, Building 4200, was nicknamed the "Von Braun Hilton." On the 10th floor of that building was WvB's conference room or what we referred to as the "board room." A meeting was under way on the major status review of the Saturn booster with about fifty people in attendance. Most of the presenters were in their late 20's or early 30's, like most of the other engineers who made up WvB's now greatly expanded technical and management team. Seated around the large conference table were the laboratory directors and senior division chiefs who constituted an informal review team. For these reviews with WvB, when not presenting, I usually located myself in a position so as to observe any outward reactions by WvB to the briefing. During a presentation on the progress of structural design by one of the younger speakers, I noticed a frown and questioning look developing on WvB's face and knew there was a problem. WvB politely held up his hand and asked the presenter to briefly stop. He then turned to face the chief of Structures Division down the table, and said, "Emil (Hellebrand), is he trying to tell me that somebody goofed?" Emil's face flushed red as he acknowledged WvB's discernment, of an earlier error being corrected, with a positive nod of his head. This episode was accompanied by the good-natured laughter of the observers to this exchange. At this point WvB said, "Just tell me where we are now and let's move on!" No wasted time on looking for whom to blame—just keep moving on with the right data. The greatly relieved presenter completed his material, knowing now that he need not fear just telling it like it was without trying to sugar coat the data.

A similar incident in a similar conference table seating shows his seriousness and dedication to the matter under review. He clearly demonstrated this when one of the seated laboratory directors started talking to another director during the briefing. Without warning, WvB picked up a three foot pointer, fortunately rubber tipped, and accurately slid it 12 feet down and across the table, ending, not too hard, on the talker's chest. After the nervous laughter had ceased, WvB announced that personal conversations interfered with everyone being able to clearly hear the presenter. He apologized to the presenter for his interruption and asked him to proceed. Similar audience-talking ceased to exist through many follow-on meetings.

There was another similar meeting in early '63, when Rein Ise made a presentation, approved by WvB, to Dr. George Mueller on how we were going to launch our Saturn V rockets in typical, conservative, WvB team approach. This meant three flights each of a one live, then two-and finally three-live-stage Moon rocket with the 10th flight to be the first one manned. Dr. Mueller had only a few questions, and then took the presenter's place and launched into his very opposite view of the conservative approach. His daring plan was to launch all three stages live on the first flight and decide on subsequent flight missions depending on the outcome of each prior flight. His logic was that the so called, "All Up" plan had the highest chance of meeting our 1969 Moon landing commitment, and, with any reasonable rate of success, would be substantially less costly.

I can't remember WvB's immediate reaction, but I do remember his team members were generally stunned by this approach to Saturn development testing. I do remember that WvB did not vigorously defend the plan that his presenter had just completed and, in effect, left him "hanging in the wind". For me, this was a real departure for WvB. While the "All Up" was adopted, and very successfully implemented by WvB and his MSFC team, I never was able to learn the rationale or strategy of his very minimal defense of the plan he had presented. I believe that he may have known, or at least suspected, that Dr. Mueller would favor taking an aggressive rather than conservative approach. To me, the most obvious rationale was WvB wanted to accomplish two objectives. First, he wanted to have on the record his preference for the stage-by-stage development approach. Secondly, this higher risk approach, championed by an "outsider", provided an extra challenge to his technical team. I think this was again an example of WvB making the best of what was obtainable at any given time or under the circumstances, living for another day to further advance his goal.

Traveling with WvB

This mid-'60s story illustrates the concentration and focus WvB had toward anything space-related despite efforts by others to disrupt his train of thought. A group of several top Marshall technical and program management personnel including me had accompanied WvB for discussions with Grumman Aerospace Corporation personnel, at their Long Island, NY facility, regarding possible usage of a Lunar Excursion Module (LEM) for other than Apollo missions. WvB attended a private dinner with the Grumman officials while the rest of our party endured a typical motel dinner along with other patrons. Just as we were finishing our dinner, he joined our group and immediately started a rapid fire of space mission ideas. The servers had cleared the few remaining tables and were given permission to clear ours, by WvB's head of the Marshall laboratories, Hermann Weidner. WvB continued explaining his ideas and only looked up briefly when all the dining lights blinked off momentarily. After another 5-10 minutes of WvB's intense, spontaneous burst of ideas, the lights again flickered with WvB's minor annoyance surfacing again as he quickly glanced around and once again returned to pursuing his thoughts. Since the room was now empty, except for a lonely server, Hermann finally interrupted WvB and said the flickering lights were meant to tell us it was time to get out of the dining room now! Looking around the empty room, WvB said OK and we departed. When he was pursuing an idea, it took a bulldozer to get him off a space topic.

This WvB story is about his private life, and not doing enough advance planning as he did in his rocket life. Several of us had traveled with WvB to a meeting at NASA headquarters. Afterwards, we piled into the rental car, with him as the driver, and he headed supposedly to the northwestern section of Washington, DC, for dinner with his daughter.

She was attending the National Cathedral School. Of course, it was rush hour with very heavy traffic. After several unsuccessful attempts to find the school, he pulled up at a stoplight. Seeing a taxi pull beside him, he jumped out of the car, shouted to us, "I'll see you tomorrow," and into the taxi he was gone. Being less than familiar with Washington, we were left to navigate back downtown by heading for where the lights were the brightest.

Flying with Von Braun

When we were with the Army Ballistic Missile Agency we were flying when WvB was at the controls of a C-47. He was landing the aircraft at the Washington National Airport. After touchdown, the plane bounced about 50 feet into the air. On re-landing, it bounced again about 25 feet. Before losing the last of the runway, he successfully set the plane on the ground on a third attempt.

On the plane was a former Air Force pilot who was a close associate of WvB working as a future planner. As WvB came down the aisle to exit, Frank Williams told him that he was short on flying time that month and would he mind if he "logged one of those landings". WvB, by a noticeable "harrumph", with his face flushed, embarrassingly, acknowledged his poor landing. For myself, I was just glad to walk off the plane safely.

This story relates to WvB's sense of humor but still tied to his technical background. In the late '60s, some of us technical and program management types were traveling to Washington on the Marshall Center's Gulf Stream, often referred to as WvB's corporate jet. All of us were having an alcoholic beverage, which WvB enjoyed as well. When the flight steward inquired if WvB would like to have another drink, he told the bewildered steward, "that it would take only about 1 erg of energy to persuade him." Sitting across the aisle, Hermann Weidner eased the steward's understandable confusion by saying, "that means he will have another drink." For the layman, one erg is about the amount of energy expended when a housefly would do one pushup!

Another flying adventure with WvB occurred at the end of a flight from Huntsville that took much longer than usual. The reason—WvB always looking to increase his knowledge and skills—had not yet attained his instrument rating license and he needed more time flying non-visually. Since he had no spare time to just be flying around in cloudy weather, he combined the needed trip to Washington with lots of instrument rating time flying in and out of clouds.

WvB was always ready for adventure whether large or small. Flying home with WvB, on the Gulf Stream from the West Coast, the pilot asked WvB if he would like to get a look at the Grand Canyon as they would be flying nearby. WvB apparently stated that a close view would be great. As was recalled afterwards, everyone really got concerned about how literally the pilot was giving WvB a "really close view", when they found himself looking UP at the canyon wall ridge tops!

First Lady of Marshall

By Nancy Guire

Bonnie Lou Green Holmes graduated from Eva High School in 1947. She received a scholarship to Andrew Jackson Business College in Nashville and graduated in 1948 with a degree in secretarial science. Her first job was at the Morgan County Courthouse. In 1951, Bonnie was selected for a position at Redstone Arsenal with the Army's Post Engineering Office. Coincidentally, WvB and most of his team members from Peenemunde, Germany, who had chosen America for their homes following the end of WWII, were working on Redstone. WvB was head of the Development Operations Division, and his team was eager to fulfill his childhood dream of safely transporting humans to the Moon and back. In 1952, WvB needed a new secretary. Bonnie was among the candidates selected for interviews. She turned out to be the perfect match for the job because she, too, was inspired to the Moon mission by having read a series of WvB's articles on space exploration published in *Collier's* magazine. Some of Bonnie's colleagues were skeptical of the match, because the memories of WWII were still fresh in the minds of Americans who had fought in the war or had family members who were impacted by the war.

During the next eight years, WvB and Bonnie's working relationship developed into one of full trust and mutual respect. WvB was confidently dependent on Bonnie's administrative management. His direct reports regarded Bonnie as a skilled communicator who could facilitate solutions to many issues for them when WvB was not available.

In 1958, President Eisenhower established the National Aeronautics and Space Administration (NASA), a civilian agency to pursue peaceful space exploration. The NASA/George C. Marshall Space Flight Center (MSFC) was formed in 1960. WvB and his ABMA Development Operations Division became the nucleus of the new center. Bonnie was immediately catapulted in stature, visibility, and responsibility in her new position as secretary to the MSFC director.

It would be difficult to describe a typical day for the Center Director's secretary-assistant at MSFC. After all, humans had never traveled to the Moon before, no agency before had a presidential mandate to do so within a decade, and there were no templates to follow.

When WvB was in town, Bonnie would arrive at the office about 7:30 to get ready for his arrival around 9:00 a.m. with a briefcase full of signature and decision packages. Bonne would retrieve the documents and carefully ensure that appropriate actions were followed. They would review the calendar and make adjustments if necessary. Then, WvB would be off to his first meeting, perhaps a program review in the 10th floor conference room.

Oftentimes, Bonnie would then get a call from NASA Headquarters that WvB was required at a meeting in Washington D.C. early the next day. Bonne had a note passed to WvB in the conference room, called appropriate managers for briefing papers on issues to be discussed, scheduled the NASA Gulfstream for late afternoon departure, notified Transportation Office so other MSFC travelers could use any available seats, notified Mrs. von Braun to pack his bag for overnight, and then send a courier for the bag. Then, Bonnie rescheduled all the next day appointments. Bonnie was excellent at logistics, and it was a valuable asset with WvB's hectic schedule.

Bonnie had an assistant secretary—Evelyn Mueting in the early '60s and Molly Payne after Evelyn's retirement. One important task assigned to the assistant was to prepare a daily journal of WvB's activities. A few notations from May 27, 1960:

9:16 Dr. von Braun arrived
11:30 Mr. Shepherd went in
12:06 Talked on phone with Dr. Gilruth (see Encl 3)
12:27 Got off the phone

These journals were great references through the years to recall meetings, action items and phone calls; today, they are a treasure trove for historians. The assistant secretary would monitor WvB's telephone calls and prepare summary of discussion points for attachment to the journal. Bonnie worked closely with the Assistant to the Director (Frank Williams from 1960-65 and Jim Shephard beginning in 1965) in getting summary notes for the journal from other appointments and meetings.

Bonnie knew that she was living history and recognized that the daily journals, speeches, mission documentation, and fan mail were significant in preserving WvB's legacy. She made sure that files were maintained.

WvB's calendar was almost always full, with no allowance for continuing meetings beyond the scheduled time. To keep him on schedule, Bonnie was the polished diplomat in the office by reminding WvB that his next appointment was waiting in the conference room across the hall. This usually worked, but there were exceptions when he was outside the headquarters building. Once, when he was at the neutral buoyancy tank observing an astronaut training for the weightless of space, his curiosity of the moves overcame him. When the astronaut left for the airstrip to fly his T-38 back to Houston, WvB asked him to drop off on his way a note to Bonnie to cancel his next appointment—he wanted to stay and experience the weightless maneuvers himself. This was well before the days of cell phones! How was Bonnie to judge; if WvB was having fun, it would most likely be awhile before she could reschedule the missed appointment.

The staff luncheon from noon to 1 p.m. was a standard calendar event for WvB and his direct reports; it took place even when he was out of town. It was held in the conference room across the hall from WvB's office. There was a small kitchen adjacent to his office that was staffed by the cafeteria contractor. Menus were marked early in the day, and food was brought up from the cafeteria or the attendant would warm up soups or other food brought from home. Secretaries were served at their desks after the bosses were served. The staff luncheon was very popular because it was in effect a break—it was informal chat time, no agenda, but a lot of business could be done because all the important players were there. It was one of WvB's effective communication tools.

Bonnie's last task before leaving for the day was to pack WvB's briefcase—his homework. She was very organized, and sorted materials with urgent on top of the stack, and prioritized importance below that. Reading material was at bottom, just in case he didn't get through the pile, though that rarely happened. Bonnie was careful to ensure that signature packages had been concurred in by all appropriate managers, had been proofed for format and grammar, and covered salient points. The decision papers would have briefing notes on pros and cons and, usually, a recommended action that had been put together by the Assistant to the Director.

Among the more pleasurable assignments for Bonnie were trips to the Kennedy Space Center for Apollo launches. While there, she would be on duty round-the-clock as required to assist WvB including when there were unexpected holds and delays of liftoff.

Dealing with dignitaries was a common event for Bonnie. She had the privilege to escort President Dwight D. Eisenhower to the podium when he came to Huntsville for the unveiling of the bust of General George C. Marshall when Marshall Space Flight Center was formed. President John F. Kennedy visited twice, and Vice Presidents Lyndon Johnson and Hubert Humphery were guests. Senators and Congressmen were frequent visitors to the Center

Director's office. And many other VIPs came through the years including CBS newsman Walter Cronkite, author Cornelius Ryan, pioneer heart surgeon Dr. Christiaan Barnard, aviator Charles Lindbergh, singer John Denver, cartoonist Johnny Hart, and many others.

Bonnie was mentor and role model for all administrative staff. She managed the director's office with remarkable efficiency, and was able to instill many of her organizational skills in secretaries across the center. In the early 1970s, Bonnie facilitated the establishment at MSFC of a chapter of the National Secretaries Association. It focused on the professional development of the administrative personnel. As part of this endeavor, Bonnie was instrumental in getting Evelyn Lincoln, former secretary to President John F. Kennedy, to come to Huntsville for a seminar following the publication of her book about her White House experiences.

In 1970, WvB was reassigned to NASA Headquarters to lead planning efforts for a mission to Mars. He asked Bonnie to serve a temporary assignment at headquarters to assist with the transition and to help select his new secretary.

Following the assignment at Hdqs., Bonnie returned to MSFC to serve as secretary to the next two center directors, Dr. Eberhard Rees and Dr. Rocco Petrone.

There was never a doubt about Bonnie's commitment to the mission, and she spent long hours in the office at critical times. However, she was dedicated to maintaining a healthy family life. She was married to Emern Holmes and had two sons, Ray and Kenny. Mondays in the office included stories about family activities that included sports, church, and meals with extended family and friends.

After a 27-year stellar career, Bonnie opted for early retirement from Federal service effective March 30, 1978. She expressed mixed emotions about the change stating, "I am sad at the thought of leaving the place that has provided me with so many friends and such a rewarding career. My many years here have been interesting, educational and challenging. But I look forward to fulfilling a personal commitment for some special church and church-related activities, as well as some community services. I am not quitting work. I am only changing vocations."

And that is exactly what she did. Bonnie worked at the U.S. Space & Rocket Center, raising funds for the I-Max Theater and organizing WvB's personal papers for easy access by researchers. She served as Eva's town clerk, and was instrumental in the establishment of the Eva library where she frequently volunteered. She continued to serve in many capacities at Hamby's Chapel United Methodist Church.

Bonnie passed away on December 5, 2014, three weeks after suffering a stroke. Her husband and son, Kenny, predeceased her.

Nancy was secretary to J.T. Shephard, Assistant to the Center Director, from 1965 until his assignment to head Industrial Operations. Nancy remained in the director's office as administrative assistant through the remainder of Bonnie's tenure.

AIDES-DE-CAMP

By Annette Tingle

History tends to shine a spotlight on the winners, the powerful, the extraordinary, the leaders, but as one NASA historian has put it, "...history also gathers up in its sweep many ordinary people, not only those who give orders and do combat at the front lines, but those who slug it out and otherwise endure in the trenches."[1] While those toiling in the trenches and behind the scenes are seldom highlighted, they sometimes have the best seats in the house to history as it unfolds. This is especially true of those who serve as confidential secretary or, in military parlance, *aide-de-camp* to the great figures of history. The seven women who served as secretaries to the first center directors of Marshall Space Flight Center (MSFC) had such seats.

The secretarial profession initially was primarily male dominated, but by mid-20th century it had become predominately female. In some ways the center directors' secretaries, as a group, were representative of the general demographic of the several thousand employees of MSFC in its first decade. As females coming of age in the 1950s, they were a minority[2] in a male-dominated workforce, and all except two were drawn from the north Alabama region. In another way they were a unique generational bridge. Most were female traditionalists of the era, but with an added ingredient. On one end they were the inheritors of the transformation their mothers set in motion with raw zeal born of patriotism and dire necessity in the "Rosie-the-Riveter" era of World War II. On the other they were the vanguard for a new transformative generation; many of their daughters would aspire to become engineers or space program managers themselves. It was, however, to be a slowly evolving transformation; not until 1978 were the first females admitted to an astronaut class. Gertrude Conard, a beginning MSFC employee and subsequently secretary to a center director, described an experience that hints of aspirations and changes to come:

> *I'd like to tell you a little story that I remembered...It had to do with Fred Uptagrafft, if anybody remembers him. He was an engineer. When I first came and worked for Dr. Lucas, we sneaked out one day. I don't know how we ever did it, but anyway Dr. Lucas didn't know that we went over and checked the F-1 engine test.[3] I got to go with him. We were in Engineering Materials at that time. That test was so loud. They said Dr. von Braun used to come over and listen to it. It would deafen you. In fact, I had to hold my ears in order to even talk or try to talk to him. After the test, he was so thrilled because he thought it was all good. Not only did we just listen to the test – I mean in an outside area – there were no buffers; there were no places to stand. There wasn't even a hole to get in. You just stood there and listened. Then we went up into the vehicle itself and examined it, while it was still steaming, to be sure everything went well. It was so exciting! It was the most exciting experience I think I ever had as I worked for Dr. Lucas, although he made things interesting. But I understood what was going on at Marshall. That was real! That was*

real! Every time I went to see a launch, I thought about that because we knew that we tested every single part of the vehicle. It was so exciting to me to be able to do that on a one-to-one basis. I'll never forget it.[4]

The space history archives project at the University of Alabama-Huntsville (UAH) Library, fortunately, has provided a unique glimpse into the role of these first MSFC aides-de-camp in the form of a video-taped interview. In May 2006, the seven women who had served as secretary to the first nine MSFC center directors met in an informal session of relating and reminiscing about their experiences and their bosses.[5] All had begun their careers with the U.S. Army or NASA in the 1950s or early 1960s, most directly out of high school. As one native of Huntsville expressed, "It was just sort of a natural thing that when you graduate from high school or college, you go to work in the space program or the defense program."[6] Whether they came from near or far, they were like most of those who came to work at Marshall in the early 1960s; they caught the fire and have never forgotten the experience and its impact on their lives. As more than one of the secretaries expressed, "The space program ... gets in your blood."[7]

It was easy to catch the fire because at mid-century it was a focused flame. A unique alignment of circumstances—a challenging and singularly defined mission, a specified schedule, and an adequate and relatively stable budget augmented by post-war optimism and confidence—needed only the element of leadership that could motivate the workforce to be successful. Fortunately, it was available. Few in that first generation may have been overtly conscious in their daily routines of what was clear in retrospect. As one of the secretaries put it, "We were all part of making history."[8] Each of the seven interview participants echoed in some way the sentiment of most employees who came early to Marshall and NASA; e.g., "...that was the greatest decision I ever made. We were all so lucky."[9]

As the seven reflected on the demanding jobs they had held, at least three major common threads emerged in their reminiscences: dedication, resourcefulness, and technology's impact. As with the role of any competent aide-de-camp, the job demanded full (sometimes almost 24-hour) engagement, alertness, and flexibility. They viewed their own dedication as being naturally inspired by the extraordinary dedication of their bosses. As one of them expressed: "They were all very driven and very dedicated."[10] All spoke of the early reporting and late leaving working hours to adapt to the director's schedule. Decades later, Bonnie Holmes could smile when she related a memorable experience:

Dr. von Braun had a Congressional hearing that he was supposed to attend early the next morning. We were working late. He had been in meetings all day long. He decided that he was going to change several things in there, in his testimony. Everybody had long since been gone home. He said, "Bonnie, could you retype this for me before you go home and send it to me. He said, "I'll go ahead and have dinner with the family." Well, it was long past dinner time. It was snowing outside. When I called the driver to come to pick up the testimony to take to him, he said, "Do you realize that there's a big snow on the ground? Don't you have a long way to drive?" I said, "I have to drive up the mountain to Eva." He said, "Oh my goodness!" That was one time I drove home scared. It was snowing; all you could see was the snow. I worked late a lot of nights, but that one was one that was really frightening. I didn't mind working late whenever he needed me. I was there.[11]

The dedication was not confined to the secretaries; commitment and urgency characterized the Marshall environment in that busy first decade. Bonnie observed quietly as the others who worked for later center directors engaged in animated conversation about the center Christmas party tradition. It began in the dark post-Challenger period of the late 1980s and was exactly the right tonic for the times. At the end, she added: "We weren't fortunate enough to have those in the days that I was working. Everybody was just on the go, go, go, go."[12] Being driven and dedicated, however, did not equate to one-dimensional. Many of the early leaders were Renaissance men who contributed as devotedly to community activities as to their profession.[13] An Antarctica working trip, as made by WvB, or vacations with family were pursued just as passionately as work.[14]

The commitment of time was not the only factor. Resourcefulness was just as important a quality for a personal assistant and came into play in myriad ways.

...I think every Director and every secretary brings something different to the table. I think we all experienced much the same thing, but yet the differences because of our bosses and their personalities. And then what we ourselves brought to the table. I think we probably all handled the job a little bit differently than the other one; doesn't mean that one was better than the other, but we all made it work.[15]

The former secretaries obviously enjoyed retelling some of their stories of resourcefulness in action. These ranged from long-distance arrangements on short notice to have a tuxedo available for the boss for a Washington, D.C., formal occasion to managing a petty cash fund for the director, an experience familiar to Bonnie Holmes.

Oh, yes, I had to keep up with his money. He would go downstairs to the basement [barbershop] *to get his hair cut. I would get a call from them saying you owe me so much for Dr. von Braun's haircut. So I would go down and pay it, or send somebody down to pay for his haircut. But I kept a running – I've still got my books – with all these expenses. When it would get near a hundred dollars, I would say, "Dr. von Braun, you have to bring me a check." He just didn't carry money.*[16]

Some situations demanded quick thinking and response. On Sept. 11, 1962, President John F. Kennedy made a brief visit to Marshall. He received a briefing by WvB on the progress and testing of Saturn booster development sufficiently impressive to invite WvB to accompany him on to Cape Canaveral.

> *...he wanted Dr. von Braun to go with him down to the Cape. He didn't have a change of clothes. He said, "Bonnie, make some arrangements to get me some clean clothes down to the Cape." He said, "I'm going with President Kennedy on his plane." These kind*[s] *of things happened.*[17]

The immediate requirement for the boss was not the only impetus for resourcefulness. Sometimes it was the innovation or intuition of the secretary. Working behind the scenes was routine when important visitors were expected. Gertrude Conard related a charming story about Lady Bird Johnson's visit to Marshall in March 1964.

> *One* [visitor] *in particular that I enjoyed was Lady Bird Johnson. She came and visited Materials Lab when I was down there. I got the joy, I guess you would call it, of getting them to clean up the restrooms and clean up the director's office so that it was presentable so that she and her assistant, Liz Carpenter – remember her. They came and they enjoyed the Materials Lab very much. They toured all over, and I got to serve them cookies and coffee on my good china that I brought from home. They seemed to enjoy themselves very much. That was one incident that I really enjoyed. It was a plus.*[18]

All resourcefulness was definitely not behind the scenes. In the early days, a highly effective hierarchical secretarial structure matched the management hierarchy and chain of command. Each laboratory and program office director had a senior secretary who was well known to Bonnie and worked closely with her. In turn, the senior secretary was responsible for the guidance, training, and protocol of her junior colleagues in the organization. Even though the video-taped interview covers almost another three decades beyond her tenure, Bonnie Holmes appears to have the respect and deference of the other participants. A testament to her influence is that even until her death in December 2014, her professional colleagues and many other early employees still regarded Bonnie as the "first lady" of Marshall.

The secretarial profession and work environment changed profoundly over the period represented in the interview. Similar positions formerly titled "secretary" currently are more likely to be executive assistant, administrative assistant, etc. Social, professional, and cultural changes were partially responsible. Equally influential was the advance of technology. Viewing the video of the interviewees is akin to watching four decades of Marshall history unfold. A common thread in the accompanying discussion is the impact of technology's opportunities and challenges on the secretaries, their bosses, and the work environment.

A prime theme of the interview roundtable discussion was the director's calendar or daily schedule. Even the eventual efficiency of electronic calendars could not alleviate the challenge of frequent rescheduling of the director's

Beginning

Konstantin Eduardovich Tsiolkovskiy

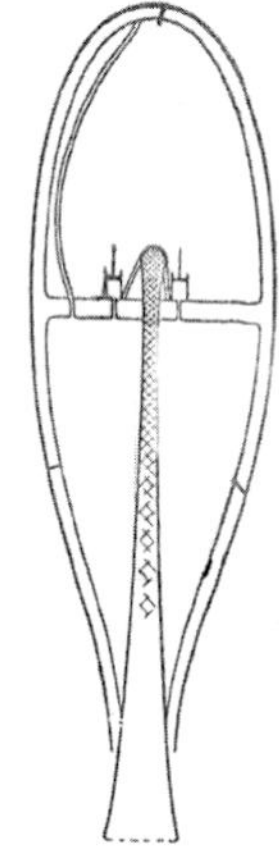

Concept of Crewed Rocket

Robert Hutchings Goddard

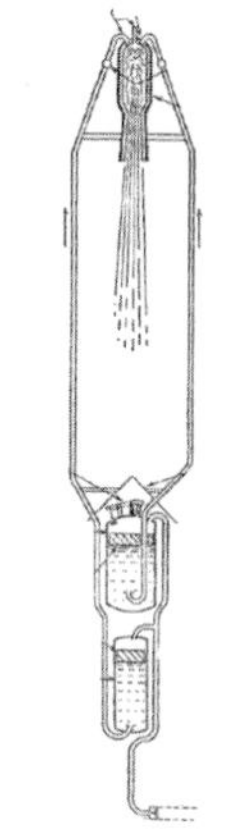

First Liquid Propellant Rocket

Hermann Julius Oberth

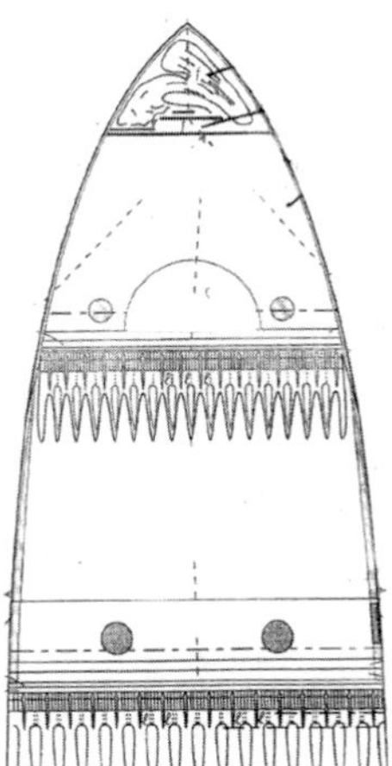

Concept of Piloted Rocket

Dr. William Pickering, Dr. James Van Allen and WvB celebrating the successful flight of the Free World's first satellite, Explorer I. *U.S. Army*

Joachim (Jack) Kuettner, project manager for Mercury-Redstone, standing next to a model of the Mercury-Redstone rocket, describes to the Mercury 7 astronauts how the team will man-rate a missile. Seated at the table from left to right: astronauts John H. Glenn. Jr., Alan B. Shepard, Jr., Walter M. "Wally" Schirra, Jr., M. Scott Carpenter, Donald K. "Deke" Slayton, Virgil. I. "Gus" Grissom and L. Gordon Cooper. *U.S. Army*

Mercury 7 astronauts, from left to right: Grissom, Schirra, Shepard, Glenn, Carpenter, Cooper and Slayton visit WvB shortly after being named America's first astronauts. *NASA*

WvB welcomes Mercury and Gemini astronauts visiting Huntsville in early 1962. From left to right: James McDivitt, Neil Armstrong, Jim Lovell, Thomas Stafford, Neil Elliott, Pete Conrad, WvB, Deke Slayton, Frank Borman, John Glenn, Wally Schirra, Edward White and John Young. *NASA*

The WvB team launched into space and successfully recovered the first monkeys, Able & Baker. Monkeynaut Baker lived-out her final years as a famous space flyer at the U.S. Space & Rocket Center. *U.S. Army*

In 1958, the team launched Explorer I atop a modified Redstone missile. The Space Race was on. *NASA*

On May 5, 1961, Mercury astronaut Alan B. Shepard, Jr. was launched aboard a Mercury-Redstone, the first manned-rated rocket ever to carry an American astronaut, designed and modified by the WvB rocket team. *U.S. Army*

Robert Lindstrom, center, project engineer at the Army Ballistic Missile Agency(ABMA), discusses plans for the project's service tower and blockhouse with his assistants, Stanley Reinartz, left, and Rein Ise. All three men entered the program in the early 1950s as Army Scientific and Professional engineers. They transferred to NASA's Marshall Center and assumed key management positions with the Apollo Saturn program. *U.S. Army*

WvB leads a celebration on the square in downtown Huntsville after Shepard's successful flight aboard the Mercury-Redstone, May 5, 1961. *U.S. Army*

Dave Newby, left, Marshall's first employee, accompanies WvB and Vice-President Hubert Humphrey while touring a laboratory. Newby, associate deputy for administration, represented WvB in development of civic projects such as the Alabama Space & Rocket Center and Von Braun Civic Center. *NASA*

Professor Hermann Oberth and WvB are briefed on satellite orbits by Charles A. Lundquist, June 1958 at ABMA. *U.S. Army*

ORGANIZATION

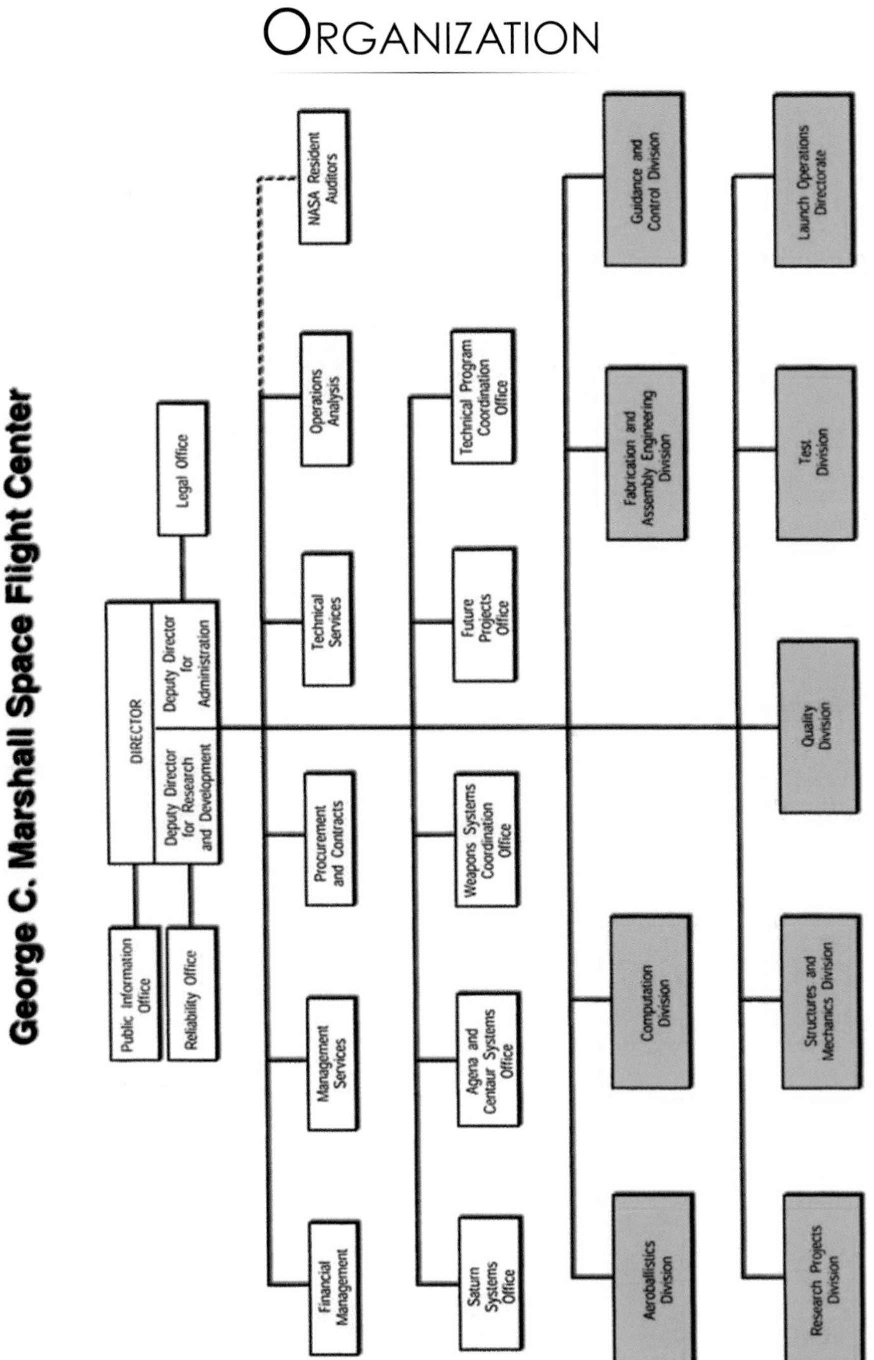

The Marshall Space Flight Center's organizational chart at the time of transfer from ABMA to the newly created NASA agency. *NASA*

WvB gathers the team's board of directors at the time of transfer to NASA. Left to right: Ernst Stuhlinger, Helmut Hoelzer, Karl Heimburg, Ernst Geissler, Erich Neubert, Walter Haeussermann, WvB, Willi Mrazek, Hans Hueter, Eberhard Rees, Kurt Debus and Hans Maus. *U.S. Army*

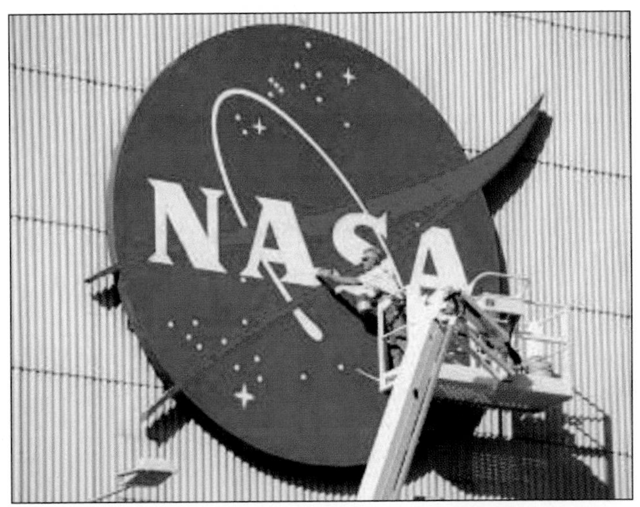

The WvB space crescent represents the organizations in the Southeast that managed manned space flight during the 1960's. Each center was responsible for a mission related to accomplishing manned lunar landings. *NASA*

The roll-out of the first Saturn V booster, S-IC, to be launched was a major press event for the Marshall team. The launch of the all-up AS-501 occurred in November 1967. *NASA*

Saturn I (Block II) booster on the left and the Saturn I (Block I) on the right are shown in final assembly at Marshall's Fabrication Laboratory. Clustering of tanks and engines enabled the team to utilize proven and tested hardware to develop a 1.5 million pound thrust booster stage. *NASA*

The Lunar Roving Vehicle, developed by the WvB team, performed flawlessly during three landings on the Moon. *NASA*

Apollo 17 Lunar Module Pilot, Harrison Schmitt and member of the last crew on the Moon, stands near the U.S. flag with the Lunar Module and Lunar Roving Vehicle in the background. *NASA*

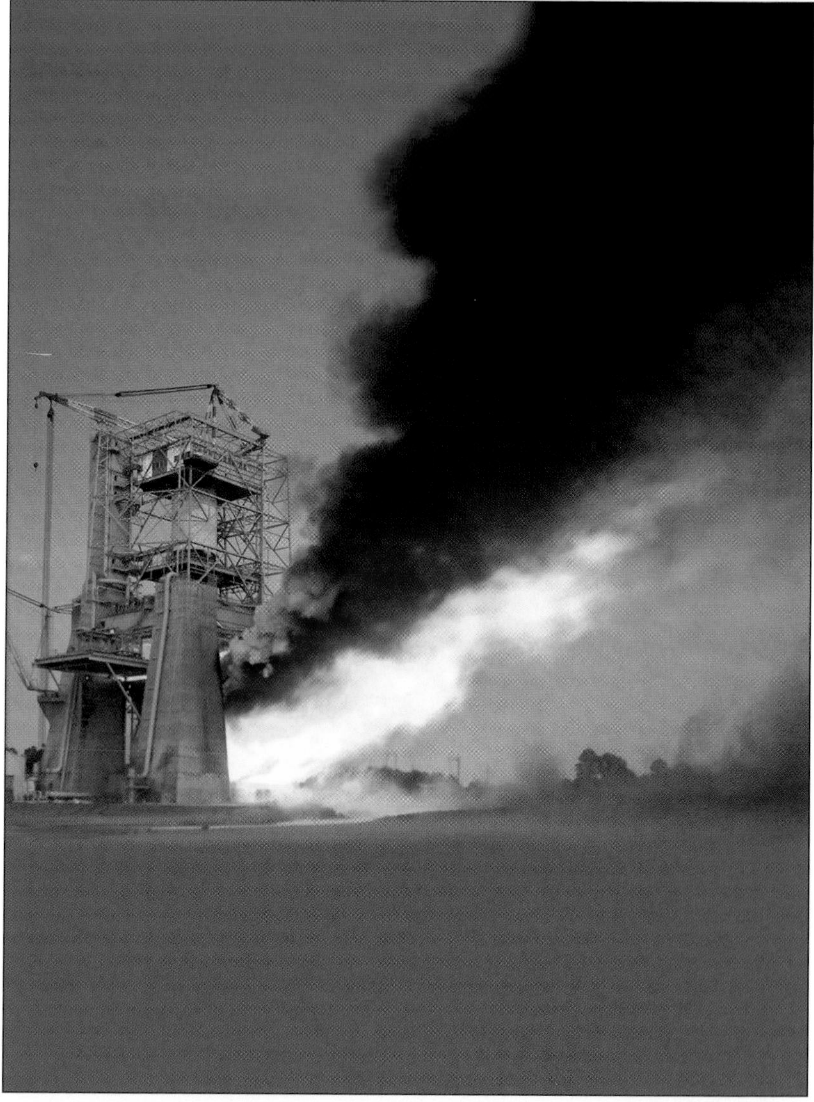

In the late 1960's, ground test firings of the Saturn V booster were conducted in the Test Lab. Twenty tests of the S-IC-T, referred to as the "shop queen," were conducted to verify it's performance. All Saturn stages were ground tested before shipment to the Kennedy Space Center for launch. *NASA*

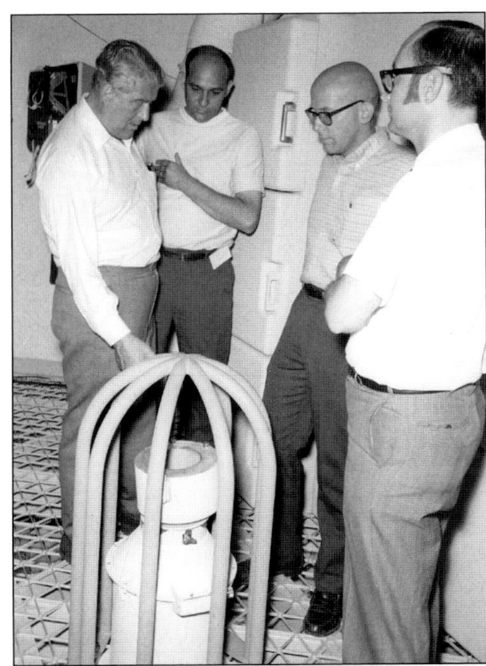

WvB on the left, discusses the Skylab mock-up with young Marshall engineers, J.R. Thompson, Dick Heckman, Jack Stokes and below with George Hardy.

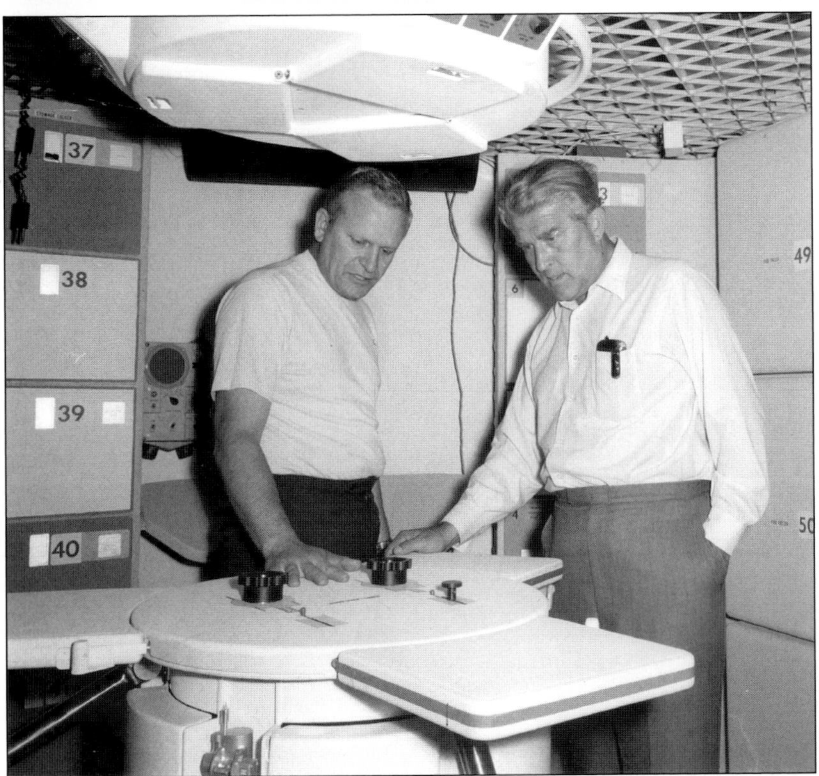

The Neutral Buoyancy Simulator served as a valuable experience for astronauts training to repair the Hubble Telescope. *NASA*

The staff of the NBS headed by Jim Splawn are joined by Skylab Astronauts Pete Conrad and Joe Kerwin, while training in 1973 for the Skylab repair mission. *NASA*

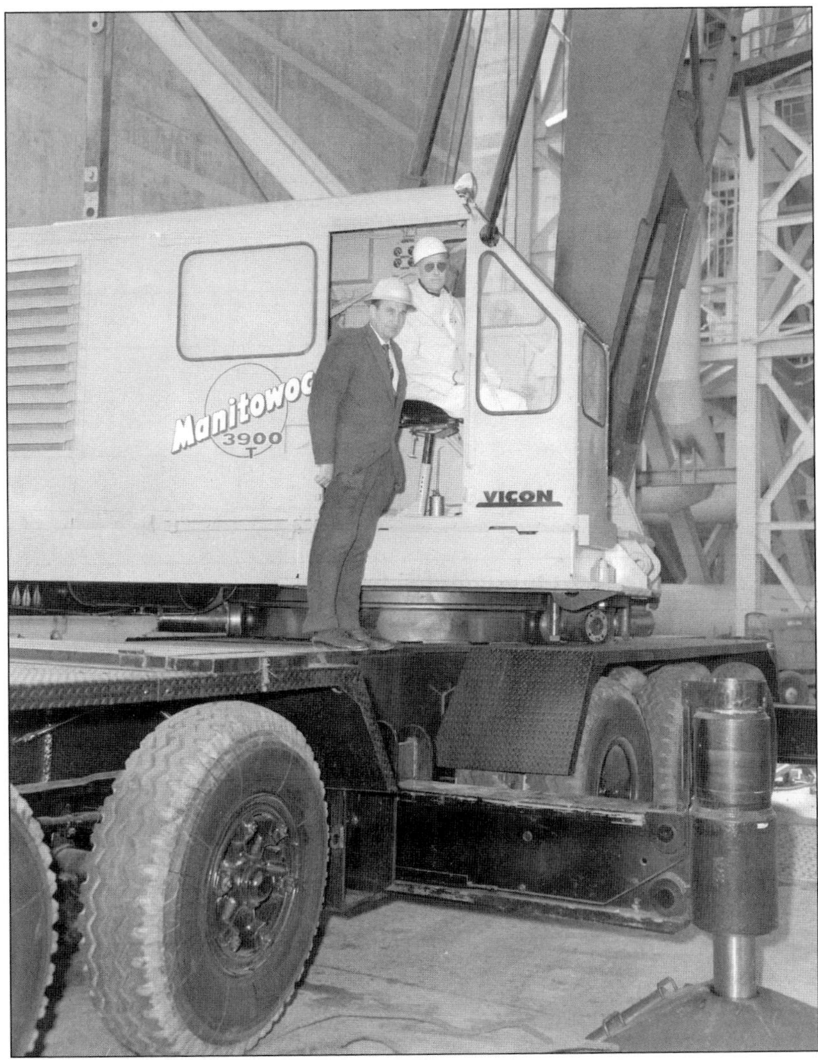

Fritz Vandersee checks out one of the new crane operators in Test Lab. Vandersee managed a shop of skilled technicians and engineers who erected all of the stages for testing at the Marshall Center. He is credited with building and assembling of the Neutral Buoyancy Simulator, with building and grounds funds. In 1969, Vandersee was responsible for transporting and assembly of the rockets at the Alabama Space & Rocket Center, the largest display of its kind in the world. *NASA*

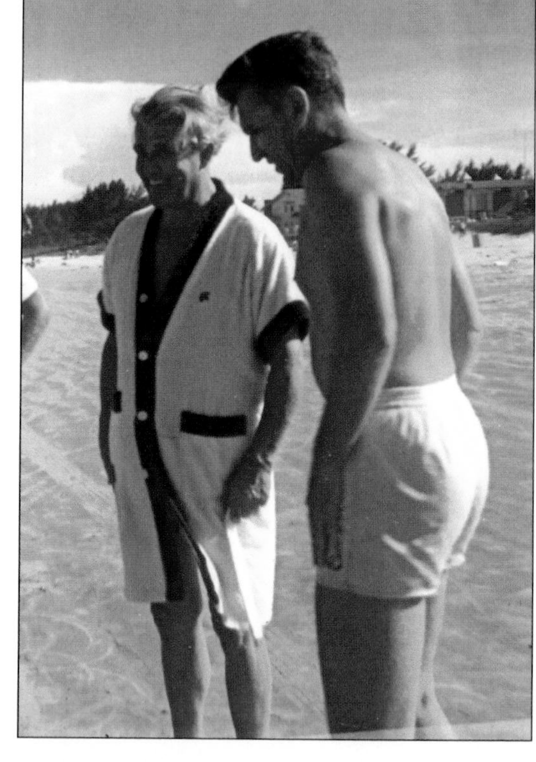

WvB, left, chats with Brooks Moore while relaxing on Cocoa Beach, FL. *Moore*

Below: Astrionics Lab members, seated at left clockwise, Sherman Seltzer, Charley Cornelius, Bill Howard, Walter Haeussermann, Brooks Moore, Fred Digesu, Henning Krome, George von Pragenau, Paul Holland, Stan Carroll, Mike Borrelli and Zack Thompson. *Moore*

In the 1960s, Marshall's Test Laboratory was the most advanced rocket test facility in the world. Rocket engines producing 500 pounds thrust to engines generating 7.5 million pounds thrust were fired in Test Lab. All Saturn vehicles were first assembled vertically and tested before shipped to the Cape for launch. The Test Lab became the team's show place for the press and VIP visitors. *NASA*

Members of the Interim Test Stand front row, left to right: Bill Grafton, Jim Pearson, Frank Rudledge, Jack Conner, Vernon Fesler, Guy Perry and Ed Berger. Standing, left to right: Paul Devine, Roy Whisenant, Red Hamby, Earl Tussel, Bently Erwin, Claud Moore, Jack Troupe, Charlie Gillespie, Gene Miller, Frank Kopera, Hop Hopkins, Eddie Bush, Clark Murphy and Bob Lindquist. *NASA*

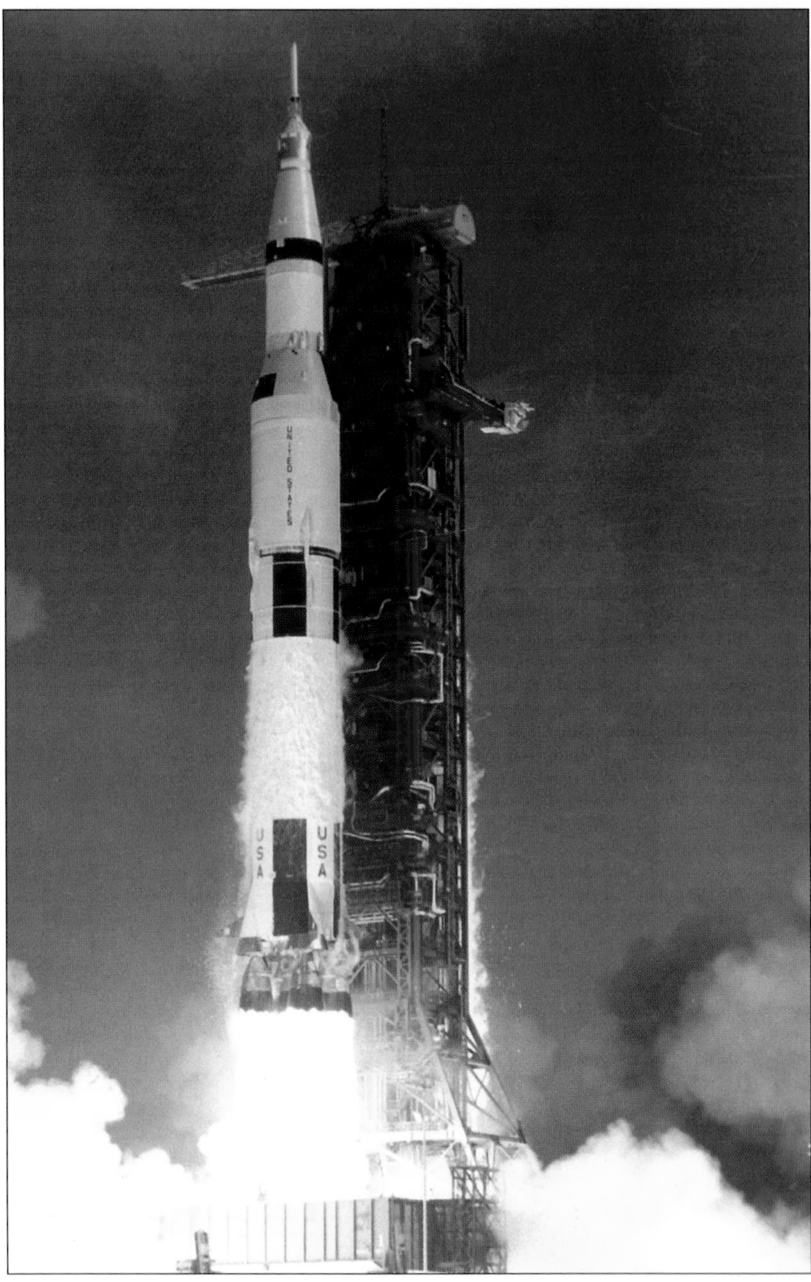

The first unmanned launch of AS-501(Apollo Saturn 501) in November 1967 was a major milestone in verifying the design of the vehicle and proving the all-up concept. *NASA*

Houston's Mission Control sends this message to the Apollo 8 crew in December 1968, the first humans to hear, "You are Go for TLI (Trans Lunar Insertion)." *NASA*

The Saturn Machine
Thirty-three Saturn flights:
- All successful
- All without loss of life
- All without weapons
Saturn sent twenty-seven Americans to the moon:
- Saturn enabled twelve to walk on the moon.
Saturn sent nine astronauts to Skylab.
Saturn sent an American crew to join a Russian spacecraft in earth orbit.

APOLLO 14
Shepard
Roosa
Mitchell

APOLLO 16

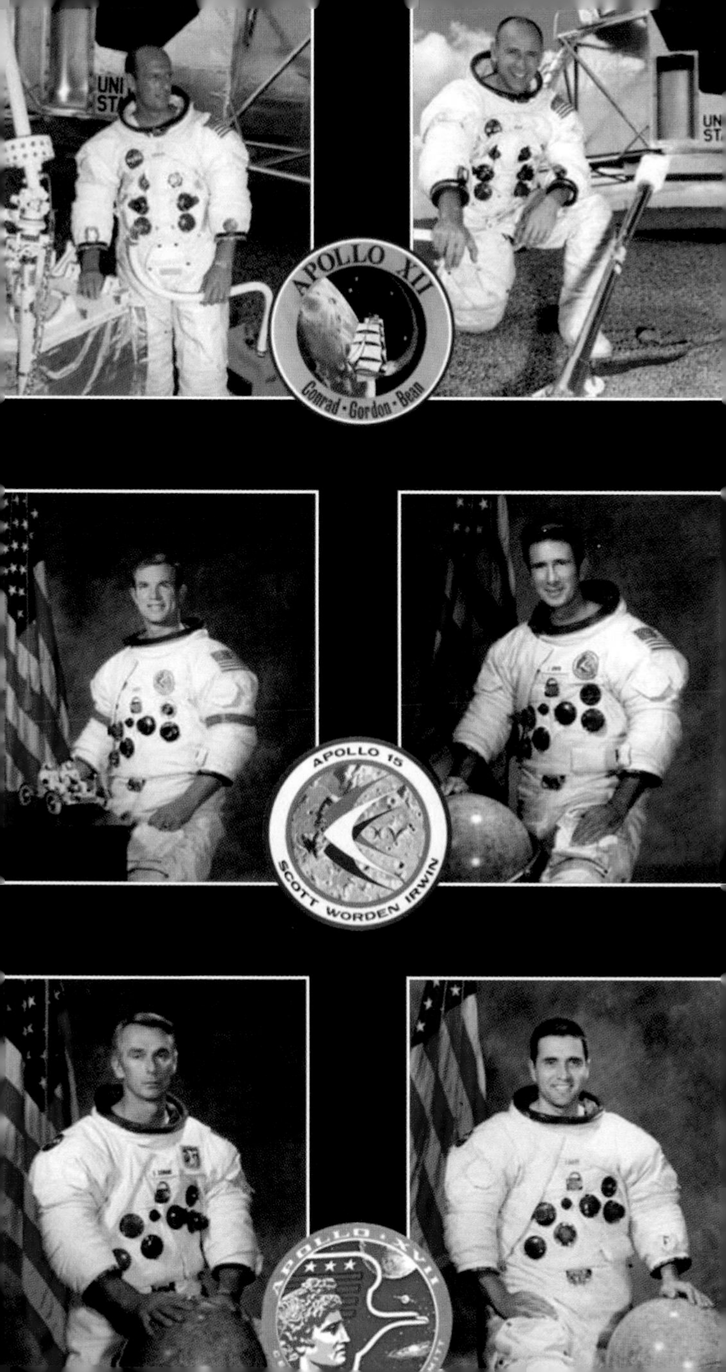

APOLLO XII
Conrad • Gordon • Bean

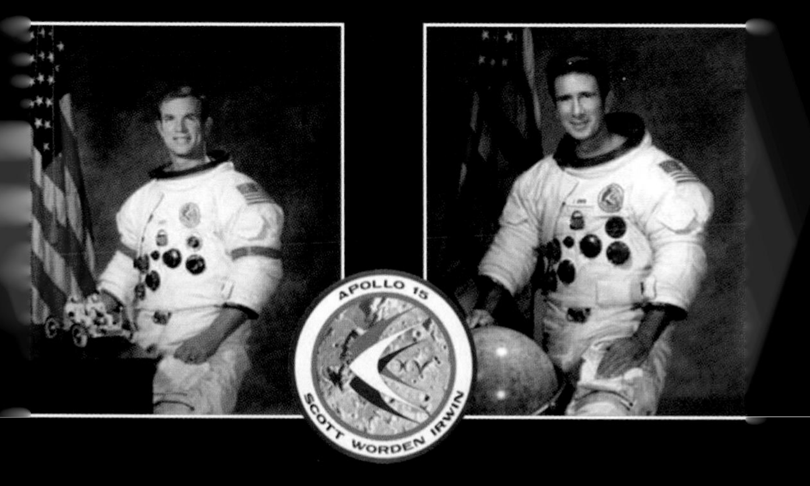
APOLLO 15
SCOTT WORDEN IRWIN

APOLLO • XVII

WvB loved to show this photograph to demonstrate why he supported having "man-in-the-loop." The crater on the right, is where Apollo 11 Commander Neil Armstrong was to land his spacecraft the Eagle. The crater is estimated to have been the width of a football field and approximately 60-feet-deep. At the last moment, Armstrong selected a safer landing site. If he had landed in the crater, it may have been impossible to launch from there or climb the walls of the crater, even in the one sixth gravity field. He landed with only 17 seconds of fuel in the spacecraft. *NASA*

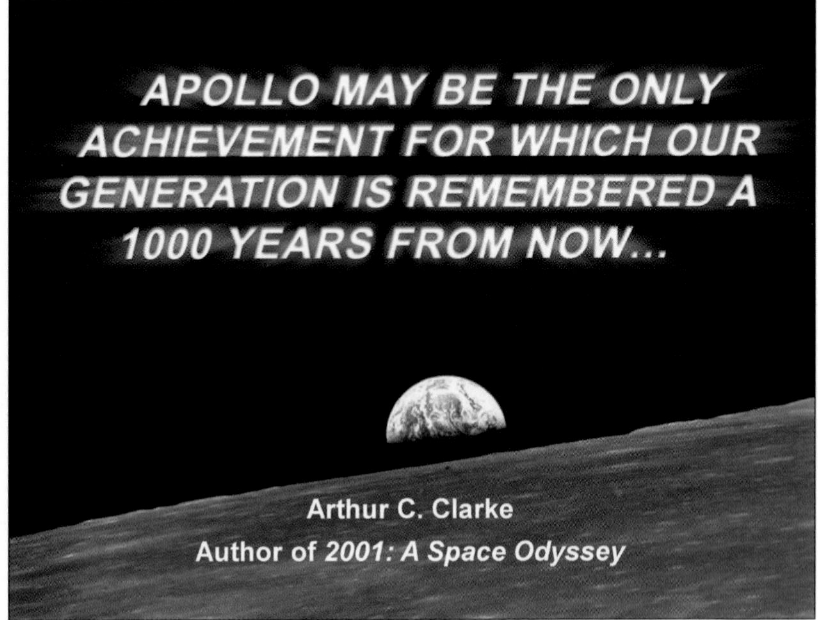

APOLLO MAY BE THE ONLY ACHIEVEMENT FOR WHICH OUR GENERATION IS REMEMBERED A 1000 YEARS FROM NOW...
Arthur C. Clarke
Author of *2001: A Space Odyssey*

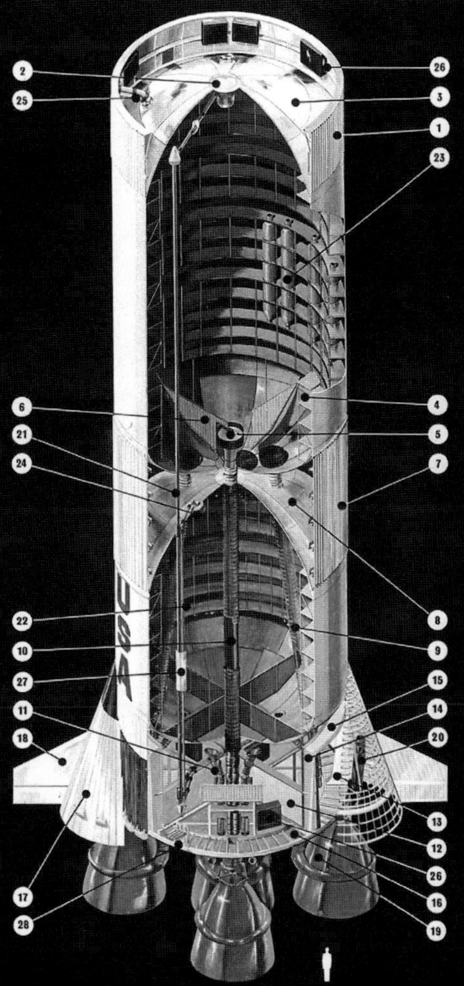

1. FOREWORD SMART STRUCTURE
2. COX DISTRIBUTOR
3. OXIDIZER TANK
4. ANTI-SLOSH BAFFLES
5. ANTI-VORTEX DEVICE
6. CRUCIFORM BAFFLE
7. INTERTANK STRUCTURE
8. FUEL TANK
9. SUCTION LINE TUNNELS
10. OXIDIZER SUCTION LINES
11. FUEL SUCTION LINES
12. CENTER ENGINE SUPPORT
13. THRUST COLUMN
14. HOLD DOWN POST
15. UPPER THRUST RING
16. LOWER THRUST RING
17. ENGINE FAIRING
18. FIN
19. F-1 ENGINE
20. RETRO ROCKETS
21. COX LINE
22. HELIUM LINE
23. HELIUM BOTTLES
24. HELIUM DISTRIBUTOR
25. OXIDIZER VENT LINE
26. INSTRUMENTATION PANELS
27. CABLE TUNNEL
28. UMBILICAL PANEL

S·IC STAGE SATURN V LAUNCH VEHICLE

THE BOEING COMPANY AERO·SPACE DIVISION SATURN BOOSTER BRANCH

MANAGEMENT

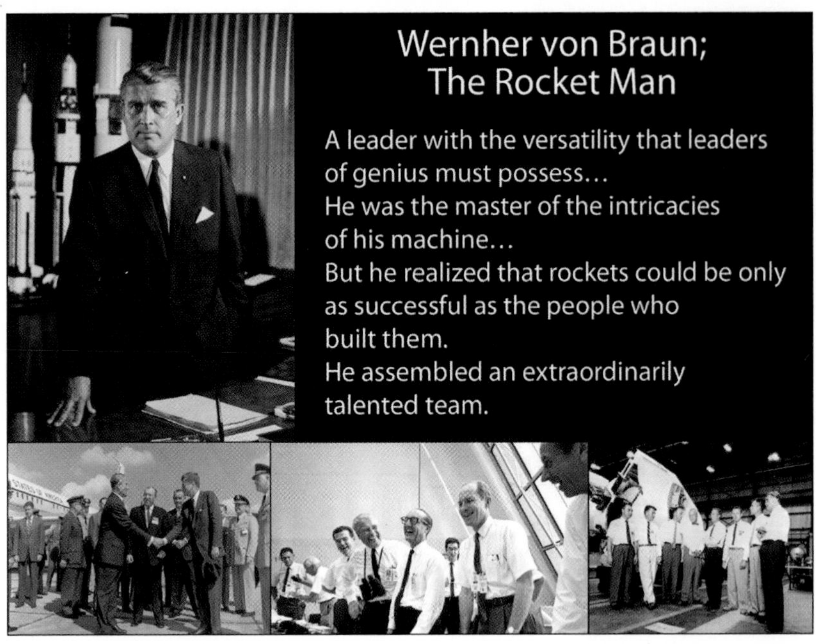
Wernher von Braun;
The Rocket Man
A leader with the versatility that leaders of genius must possess...
He was the master of the intricacies of his machine...
But he realized that rockets could be only as successful as the people who built them.
He assembled an extraordinarily talented team.

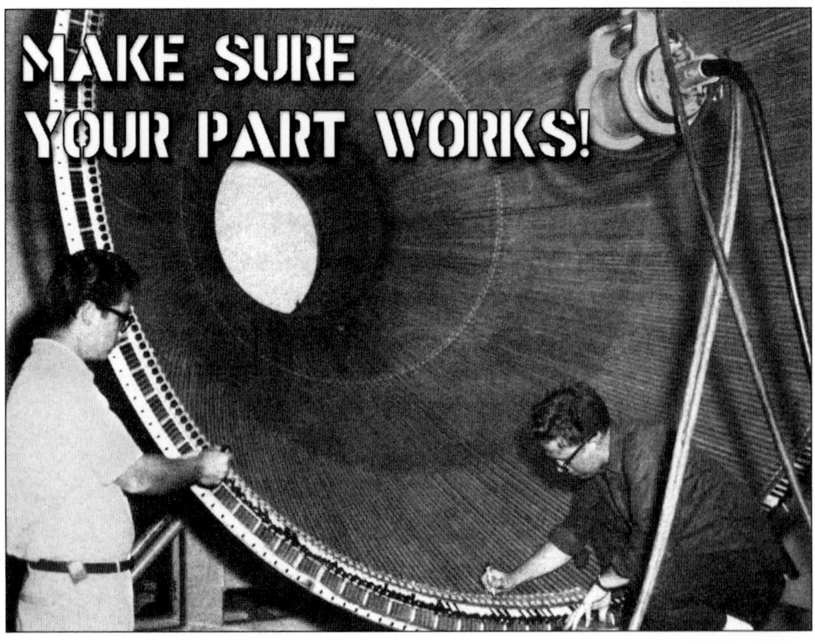
MAKE SURE
YOUR PART WORKS!

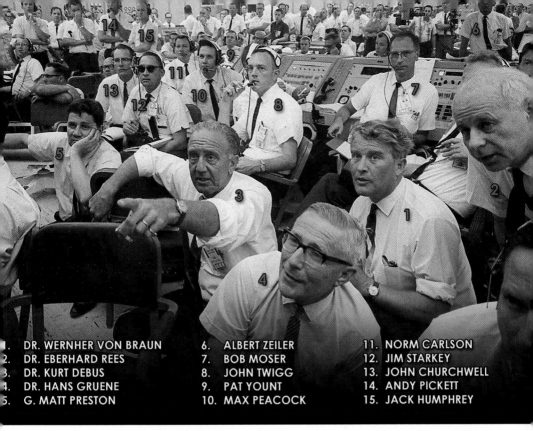

1. DR. WERNHER VON BRAUN
2. DR. EBERHARD REES
3. DR. KURT DEBUS
4. DR. HANS GRUENE
5. G. MATT PRESTON
6. ALBERT ZEILER
7. BOB MOSER
8. JOHN TWIGG
9. PAT YOUNT
10. MAX PEACOCK
11. NORM CARLSON
12. JIM STARKEY
13. JOHN CHURCHWELL
14. ANDY PICKETT
15. JACK HUMPHREY

Everyone's attention is on the rocket in the blockhouse at Cape Canaveral during an early launch of Saturn I. WvB, Eberhard Rees, WvB's deputy; Kurt Debus, director, launch operations center and Hans Gruene, Debus' deputy are grouped together. Others in the background are young American engineers like Bob Moser and John Twigg, learning the skills of launching a rocket from their German mentors. Many of these men were Rocket City residences who began their careers in Huntsville and eventually moved their families to Cape Canaveral and became the next generation of rocketeers assuming firing room responsibilities and retiring there. *NASA*

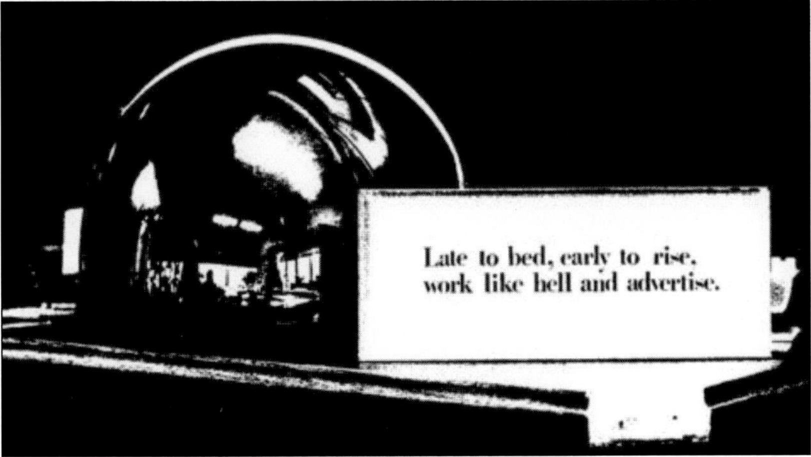

Typical 1960s sign seen atop desks and on walls of the Marshall Center. *Buckbee*

Bill Sneed, Manager of Program Control, seated on the right, briefs Saturn V Program Manager Arthur Rudolph and Harold Price, Control Center Coordinator. *NASA*

View of Saturn V Program Control Center. *NASA*

Ruth von Saurma served as WvB's international relations specialist, handling correspondence, inquires and visits of VIP's and members of the international press. *NASA*

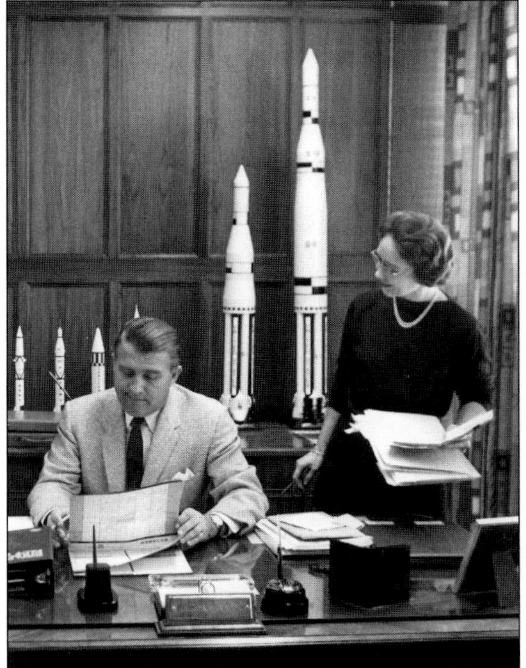

WvB and secretary Bonnie Holmes, who worked for him for 20 years, try to find an opening in his schedule for another meeting. *NASA*

Bonnie Holmes, who was familiar with many of WvB's documents, later worked as a curator of the collection at the Space & Rocket Center. *Buckbee*

The first gift made to the Space & Rocket Center was a collection of a 1,000 letters addressed to WvB from people around the world congratulating him on the landing of Apollo 11. In March 1970, WvB presented the collection to Ed Buckbee, director, Alabama Space & Rocket Center. *NASA*

Bonnie Holmes welcomes Evelyn Lincoln, President John F. Kennedy's secretary, to Marshall for a secretarial seminar held, March 1972. Bonnie often communicated with Ms. Lincoln in preparation of the president's visit in 1962. *NASA*

Nancy Guire, president of the MSFC-Redstone Chapter of the National Secretaries Association, presiding at the annual secretaries banquet, April 25, 1972 when Bonnie Holmes was chosen Secretary of the Year.

Bonnie is pictured with Eberhard Rees, center director. *NASA*

Center Director Jack Lee, presents a special accommodation to Annette Tingle, technical information officer, for her work in managing a Scientific and Technical Information conference held at Marshall in 1991. *NASA*

One of the exciting stops for members of the press visiting the Marshall Center in 1965, was the office of Chief Scientist, Ernst Stuhlinger. Displayed on his conference table was this awesome model of a Mars spaceship, powered by electric propulsion that would propel astronauts to Mars. The famed-scientist points to a Saturn V launch vehicle that would place parts of a Mars spacecraft into orbit for assembly and launch on to Mars. *NASA*

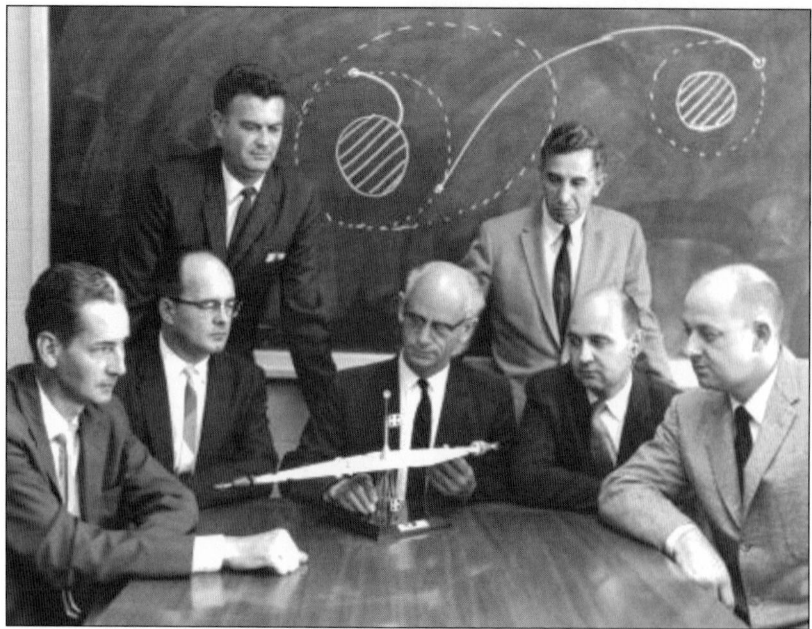

Members of the Space Sciences Lab standing are: left to right, Art Thompson, Gerhard Heller. Seated: left to right, Unknown, Gentry Miles, Ernst Stuhlinger, George Bucher and Russell Shelton. *NASA*

WvB escorts President Kennedy and Vice-President Lyndon B. Johnson on a tour of the Fabrication Laboratory viewing a Saturn I Block 1, completely assembled for the first time. Later, the President's party viewed a full duration test firing of the Saturn booster. *NASA*

President John F. Kennedy answers a reporter's question upon arriving at Redstone Arsenal's airfield, joined by WvB in the President's limousine. *NASA*

Apollo 11 astronauts, left to right, Buzz Aldrin, Michael Collins and Neil Armstrong join Marshall Center Director J. R. Thompson at the full-size Space Shuttle during the Apollo 11 20th anniversary celebration held at the U.S. Space Rocket Center. *NASA*

First five decades, America's Human Space Flights. *NASA*

Jack Lee, left, and Marshall Space Flight Center colleagues, William R. Lucas, Harry Johnstone, Erich Neubert, Bill Schneider, Dick Smith, Jim Shephard, Frank Williams, Eberhard Rees, Bill Schick, Herman Weidner and Lee Belew stand at the base of the SA-512, the Saturn V booster that launched the last crew, Apollo 17 to the Moon in December 1972. *NASA*

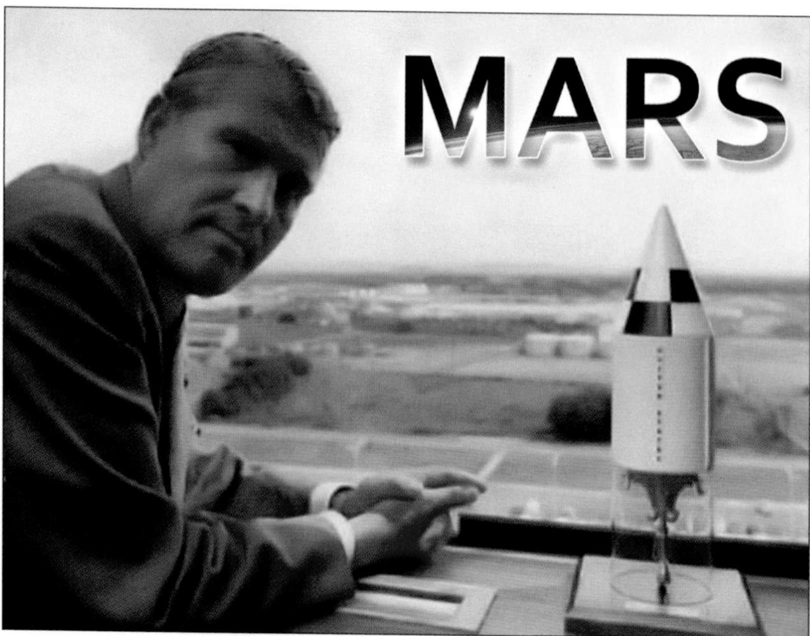

WvB is shown with a model of NERVA (Nuclear Engine for Rocket Vehicle Application) proposed by WvB, designed to power a manned rocket to Mars in 1983. *NASA*

View of the NERVA engine mounted vertically on a transporter prior to being static fired at Jackass Flats, Nevada by the WvB contractor team and the Atomic Energy Commission (AEC). The engine developed 250,000 pounds thrust. *AEC*

President John F. Kennedy is shown entering his limousine while visiting Los Alamos National Laboratory where he was briefed by WvB on the nuclear engine to be used on a manned mission to Mars. *AEC*

Marketing

Marketing space, selling the NASA brand and managing WvB was the task of Marshall's Public Affairs Office. *NASA*

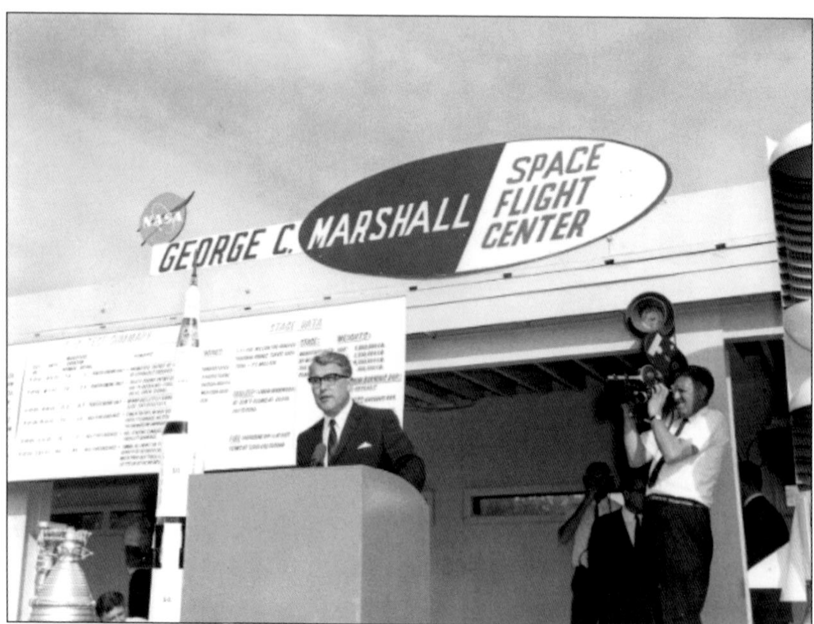

WvB was the star at press conferences whether at the Marshall Center or at the Cape with other manned space flight officials. When his name was announced as a participant, the press sent in the first team. He was known for delivering inspiring quotes. *NASA*

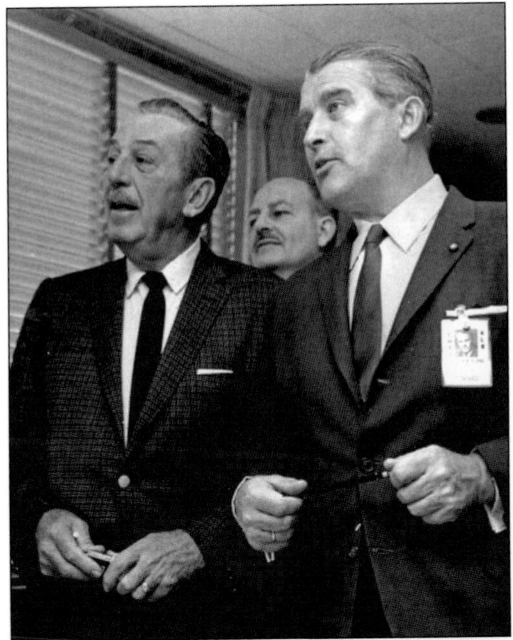

WvB developed a close relationship with Walt Disney producing films about manned space flight. In the 1960's Disney brought his design staff to meet WvB and consider what future space flight plans might be incorporated into Disney's EPCOT project. *NASA*

WvB, wearing the Texas hat given to him by President Lyndon B. Johnson, presents a rocket scientist hard hat to First Lady, Lady Bird Johnson during a visit to the Marshall Center in 1965. *NASA*

WvB was named, "Chief Fire Arrows to the Moon" during a speaking engagement honoring Speaker of the House, Congressman Carl Albert in McAlester, OK. *Buckbee*

Eberhard Rees and WvB, wheeling a gavel toward Bart Slattery, Jr., public affairs officer, during a meeting in the 10th floor conference room. In February 1970, WvB returned from a Bahamas vacation sporting a beard. *NASA*

WvB trained with the astronauts aboard the KC-135, named the "Vomit Comet". *NASA*

WvB drives the Mobility Test Article (MTA) built by the Bendix Corporation for extended exploration of the lunar surface. *NASA*

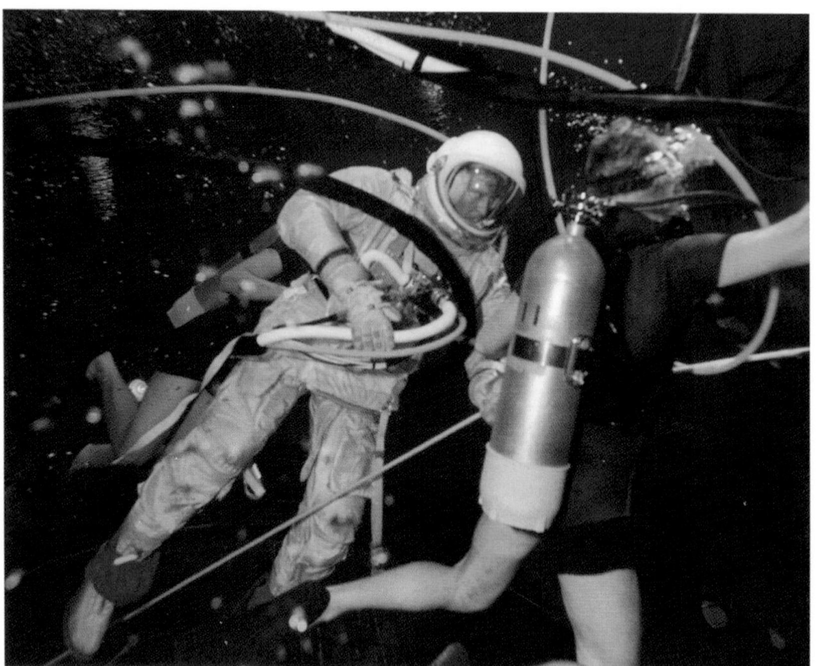

WvB's astronaut training included under water in Marshall's Neutral Buoyancy Simulator. *NASA*

WvB loved flying. He was multi-engine and instrument qualified. He spent many hours flying the NASA Gulfstream to attend meetings at NASA centers and aerospace contractors. He was an accomplished glider pilot seen here in the cockpit of his glass-fiber "Libelle" in 1971. *WvB Family*

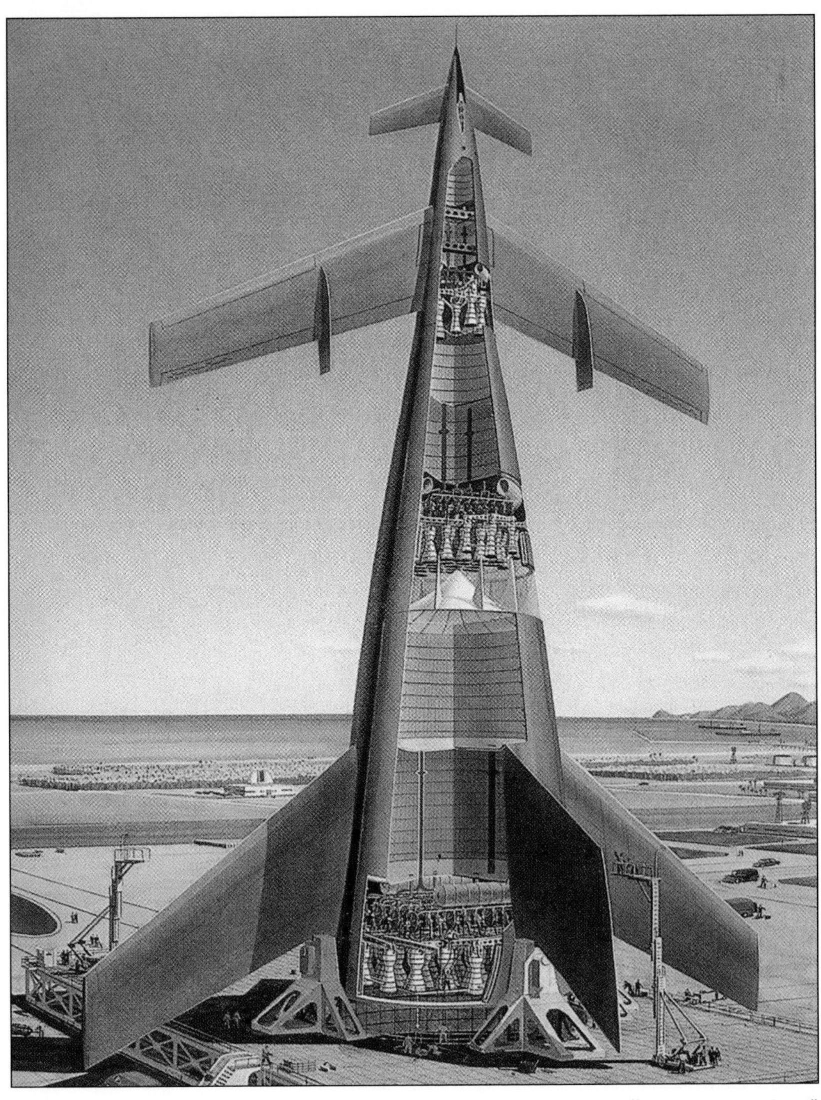

In 1952, WvB contributed to an article in Colliers magazine entitled, "Man Will Conquer Space Soon."
Colliers

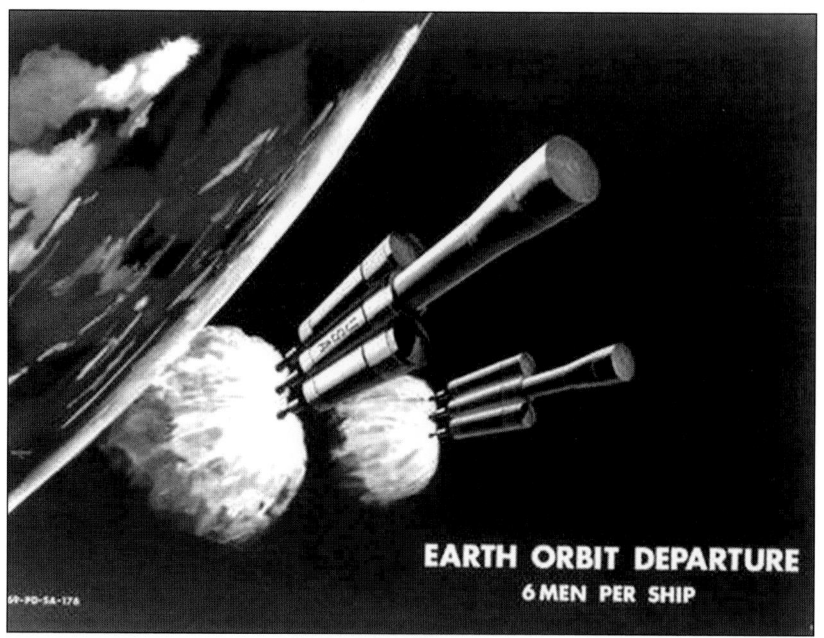

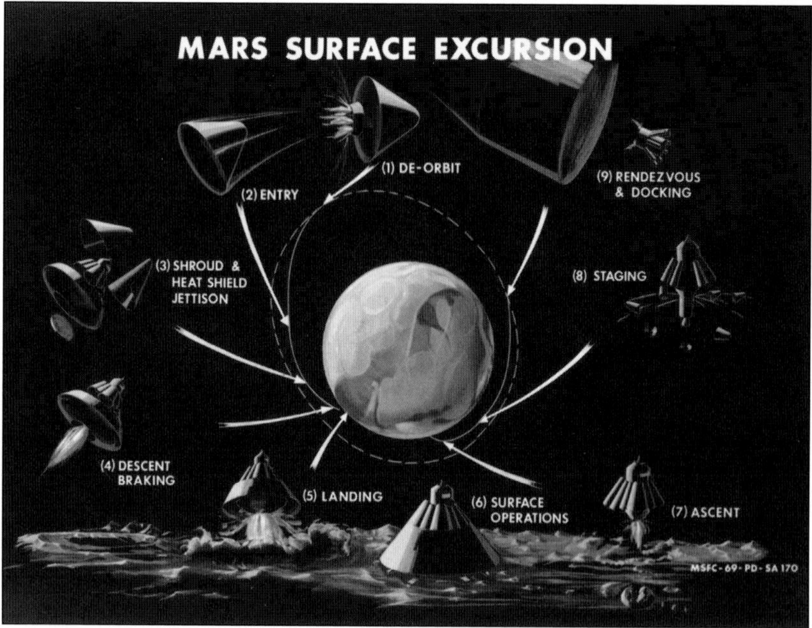

Two Mars-bound spacecraft with six astronauts aboard depart Earth orbit. Landing astronauts on the surface of Mars in 1983 was planned and proposed by WvB and the Future Projects Office. *Buckbee*

"When the craft is ready and the oceans of space are calm… the space-age Columbus and Magellan are presently sitting somewhere today in a public schoolhouse preparing for an adventure. Here are the people to whom we shall pass the baton. But the first lap of the race is ours. And we shall not falter."

-Dr. Wernher von Braun

"Cruel fate denied Wernher von Braun the chance to buy his ticket as a passenger bound for an excursion in space—his boyhood dream and lifetime goal.

Because of Wernher von Braun, however, almost everyone has been brought to the realization that we have been passengers on a spaceship all along—Spaceship Earth. Posterity will not forget Wernher von Braun."

- Eugene Emme, Former NASA Historian

Reflections

Ground breaking of the Alabama Space & Rocket Center took place in March 1969. From left to right are: Ed Buckbee, director Alabama Space & Rocket Center; Jack Giles, vice- chairman, Alabama Space Science Exhibit Commission; Wernher von Braun, director, Marshall Space Flight Center; Martin Darity, director of Alabama Bureau of Publicity and Information and Governor George C. Wallace's representative; U.S. Senator Jim Allen, chairman, Alabama Space Science Exhibit Commission; Major General U.S. Army Charles Eifler, commanding general, Redstone Arsenal and Mayor Glenn Hearn, City of Huntsville. *U.S. Space & Rocket Center*

The Alabama Space & Rocket Center, later named the U.S. Space & Rocket Center, opened in March 1970. It was founded by WvB. *U.S. Space & Rocket Center*

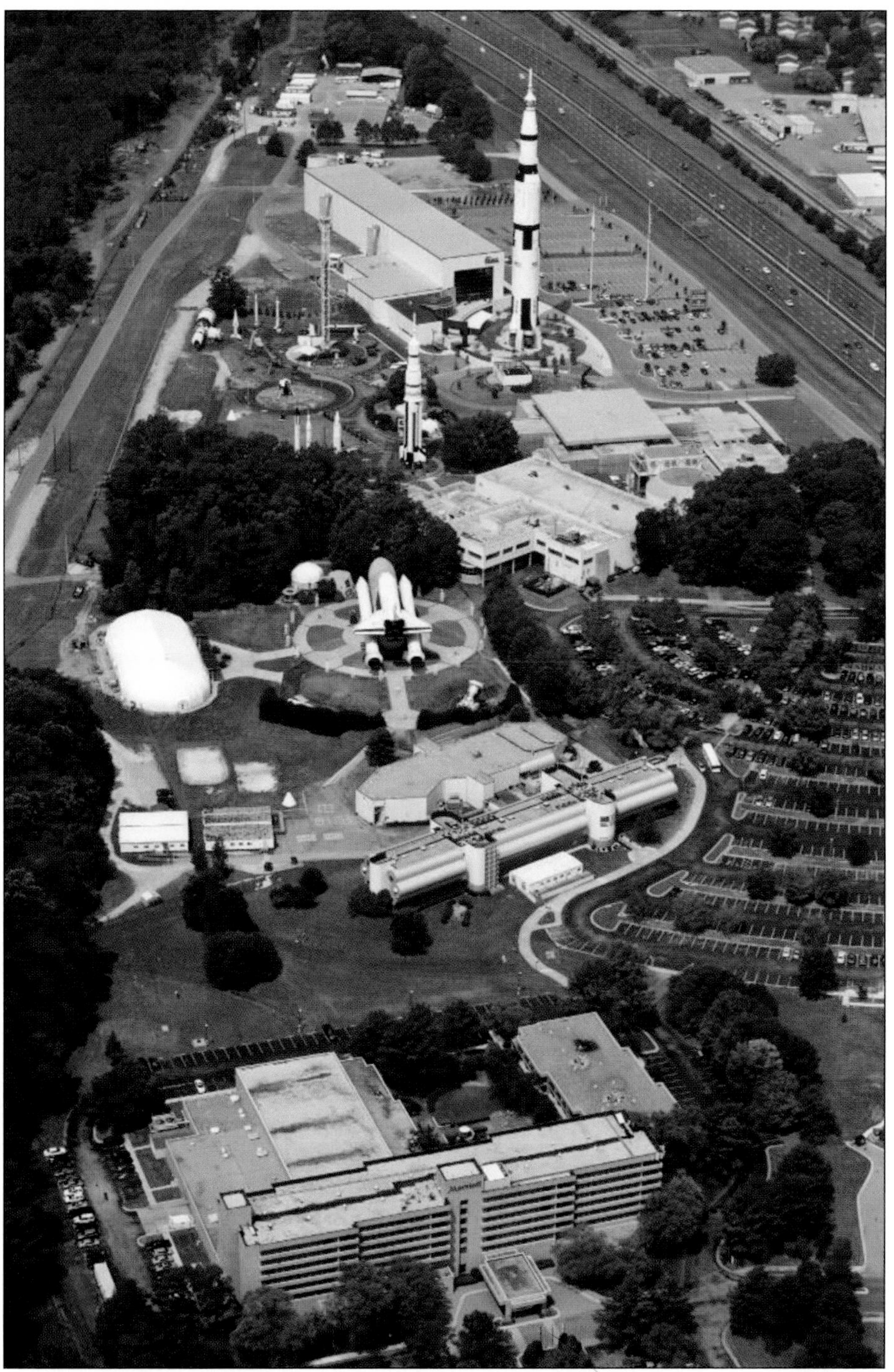

Aerial view of the U.S. Space Rocket Center featuring the world's largest public exhibition of U. S. human space flight artifacts and home of America's Space Camp. *U.S. Space & Rocket Center*

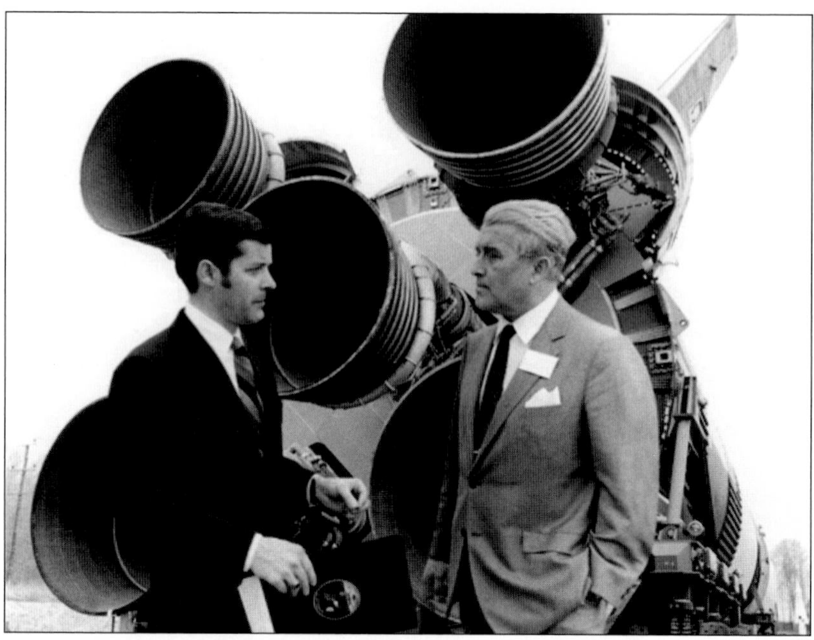

Ed Buckbee, left, stands with WvB beneath the Saturn V booster at the Space & Rocket Center in 1975. While watching a group of students touring, WvB's comments gave birth to the concept of Space Camp. *U.S. Space & Rocket Center*

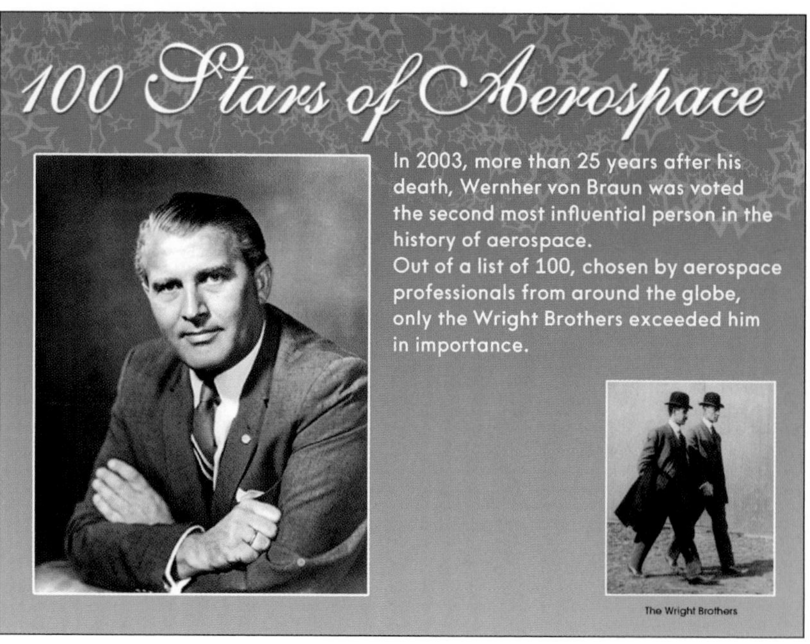

100 Stars of Aerospace
In 2003, more than 25 years after his death, Wernher von Braun was voted the second most influential person in the history of aerospace.
Out of a list of 100, chosen by aerospace professionals from around the globe, only the Wright Brothers exceeded him in importance.
The Wright Brothers

WvB was named second most influential person in aerospace history.

University of Alabama in Huntsville (UAH) students and Space Camp graduates activate experiments aboard NASA's zero-gravity aircraft. *UAH*

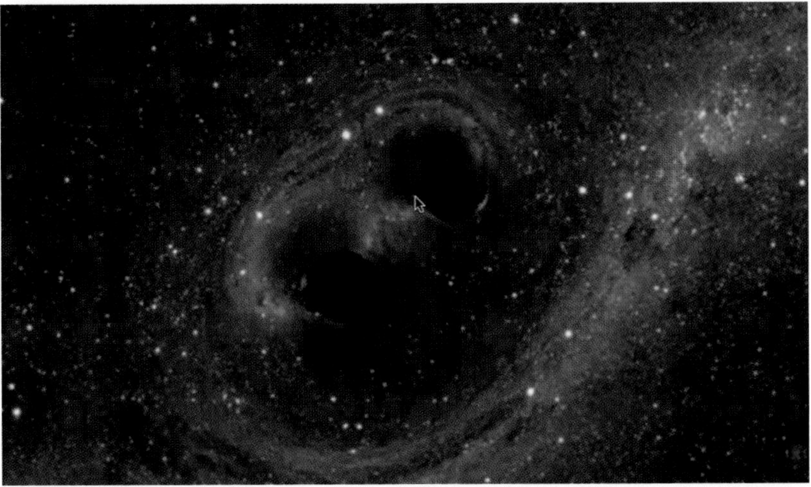

Gravitational Waves and the affect they have on Black Holes is an exciting subject being researched by UAH and studied by Space Campers. *LIGO Team-CALTECH*

The Space Camp Habitat was a highly visible addition to the U.S. Space & Rocket Center campus. A capital campaign, Partners in Our Future, rounded up a million dollar team of aerospace and consumer corporations. As a benefit and recruiting tool, donors' logos were emblazoned on shirts and used to identify Space Camp teams. *U.S. Space & Rocket Center*

Space Camp training center is up-graded annually.

The Space Camp Training Center opened in 1987, enabling growth and clearly identifying Space Camp as a workforce development program for high technology industries. Among programs added that year were the professional educators program and advanced Space Academy. As attendance increased, so did interest in offering Space Camp in other locations. Through licensing, Space Camp was made available in Florida and California, as well as in Japan, Belgium, Canada and Turkey. *U.S. Space & Rocket Center*

Trainees strap into the 5-Degrees-of-Freedom trainer at Space Camp. The 5DF is patterned after equipment designed to prepare NASA's early astronauts for space flight. The simulators were modified and built by Oscar Holderer to teach trainees physics and human response. *U.S. Space & Rocket Center*

Trainees from differing backgrounds learn to work as a team to prove they have the right stuff to succeed. Space Camp scenarios have moved away from the shuttle program and currently focus on International Space Station, NASA's deep space/Mars missions and commercial space initiatives.

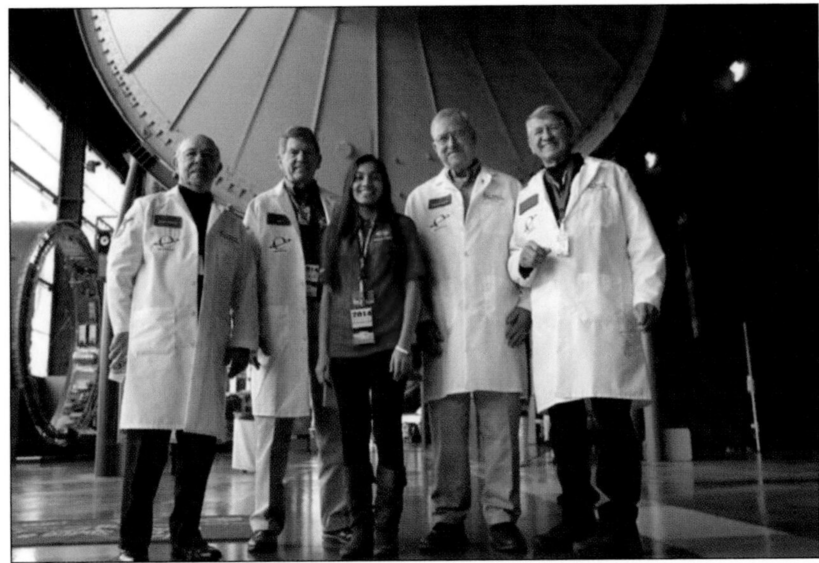

Retirees from nearby Marshall Space Flight Center comprise a robust docent program that supports Space Camp and museum programs. Emeritus members share their passion and knowledge with trainees to ensure the ongoing vitality of America's human space flight program. Space Camp also benefits from current-day scientists and engineers who uniquely share their passion. They are left to right: John Thomas, Jim Splawn, NASA Intern, Gerald Smith, president of the Marshall Retirees' Association and Ron Paulus. *U.S. Space & Rocket Center*

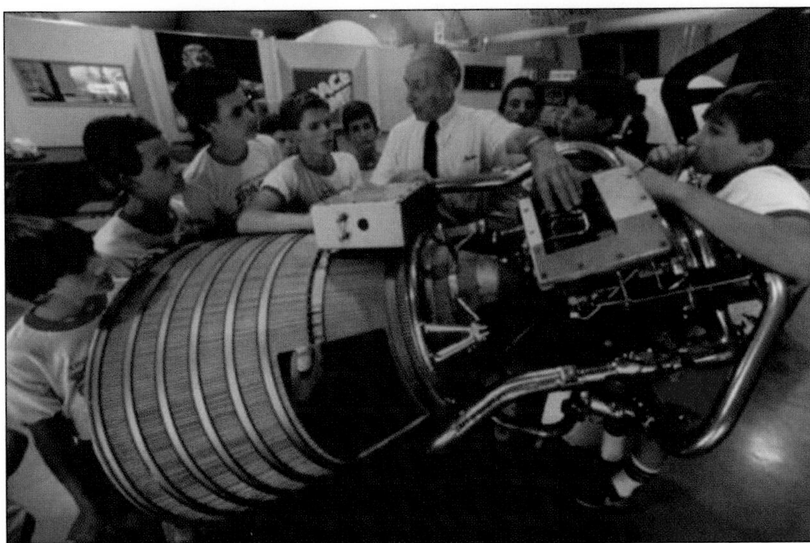

A key member of Wernher von Braun's rocket team, Konrad Dannenberg was a frequent lecturer and the "resident rocket scientist" at the U.S. Space & Rocket Center. The RL-10 engine offered a ready prop for teaching the basics of aerospace propulsion to the next generation of rocket scientist. Space Camp warmly welcomed astronauts from the Mercury, Gemini, Apollo and Space Shuttle eras as special guests and presenters. The tradition continues today with International Space Station astronauts and a variety of aerospace professionals. *U.S. Space & Rocket Center*

International Space Camp was added in 1990. The program hosts students and teachers from around the world and invites U.S. teachers of the year. Today, international trainees comprise 20 percent of overall yearly attendance. *U.S. Space & Rocket Center*

Dr. Deborah Barnhart, Space Camp director, and Ed Buckbee, executive director for the USSRC and Space Camp (standing) often engaged dignitaries, celebrities, politicians and captains of industry to gain widespread support for Space Camp and its sister programs. Friend and fund raising strengthened program content, as well as provided for capital expansion. *U.S. Space & Rocket Center*

Vice President George H.W. Bush experienced what it's like to train to be an astronaut in the Manned Maneuvering Unit at Space Camp. When he became president, he returned to visit the Marshall Space Flight Center. *U.S. Space & Rocket Center*

International Space Station exercises help Space Camp trainees better understand the next great leap in exploration. Research and technology developments aboard ISS benefit human and robotic exploration of destinations beyond low-Earth orbit. *U.S. Space & Rocket Center*

Robotics Camp was added in 2013, offering more variety to trainees and allowing them to explore their potential as entrepreneurs. Trainees engineer robots of the future, test air, land and sea robots, and consider the components of a business plan. *U.S. Space & Rocket Center*

What happens when an actual or want-to-be Space Camp trainee grows up and has children of his or her own? Family Camp is an increasingly popular program that parents, grandparents or adult mentors attend with children as young as seven years of age.

Space Camp launched with 747 trainees in 1982. By the end of the 2016 summer season, the cumulative enrollment surpassed 750,000. Today's trainees arrive in Huntsville from around the world and include students, professional educators, families and corporate teams. *U.S. Space & Rocket Center*

The future is bright for today's Space Camp trainees. How many more astronauts, engineers and scientists will we be able to trace roots in Huntsville? For current U.S. Space & Rocket Center CEO Deborah Barnhart there is more to shining the light. "It's not all flying jet airplanes and spacecraft. There is a complete industry full of people who do contracting, business operations and communications in addition to the scientific and engineering work. At Space Camp, they begin to see there is a place for them in this industry of vision and moving humanity forward." *U.S. Space & Rocket Center*

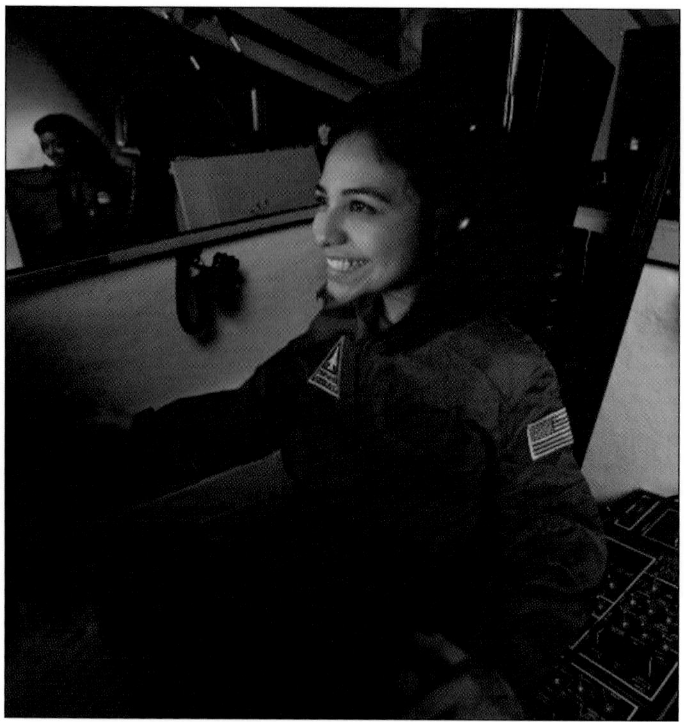

Military aviation was a common thread for America's first astronauts. Aviation Challenge, first offered in 1990, follows this thread and uses the excitement of flight and survival training to provide a different experience for trainees seeking an alternative experience. *U.S. Space & Rocket Center*

A trainee rides a zipline in water survival training at Aviation Challenge. *U.S. Space & Rocket Center.*

The Space Camp brand has long had cachet with the media. Ed Buckbee, USSRC executive director, made numerous appearances on news shows, late night and lifestyle television. Buckbee visits with Late Night host David Letterman. Today. Space Camp continues to trek across the nation, making appearances and engaging personalities with entertaining and educational simulations experiences. *U.S. Space & Rocket Center*

Mercury astronaut and Moonwalker Alan B. Shepard, Jr. addresses Space Camp graduation in the training center. When asked, "Who wants to go to Mars?" all hands went up, including Shepard's. *U.S. Space & Rocket Center*

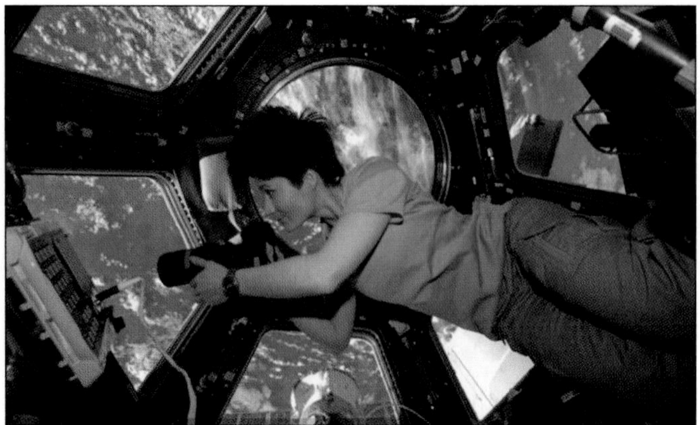

Space Camp alumna Samantha Cristoforetti is floating in zero-gravity aboard the International Space Station. Space Camp has 750,000 alumni across the globe, including five astronauts, two of whom have flown in space. *NASA*

Astronaut Alan B. Shepard, Jr. rode two of WvB's rockets, the Mercury-Redstone and the Saturn V to the Moon. He was one of the first astronauts to endorse and support the Space Camp program. *Fricker Studios*

2013 Alumni Survey: Space Camp Inspires

88% planned to take **more STEM classes** after attending Space Camp.

71% are **now in a career field** related to space, technology, energy, or defense.

66% claimed **Space Camp inspired** their decision to enter that field.

Space Camp Brings the World to Huntsville
In 2015, Space Camp hosted
65
international locations
accounting for
23%
of all Trainees
Compass UAE – Dubai, United Arab Emirates

Science In Orbit, a new exhibit open to the public, teaches Space Campers the Marshall Center's role in managing science aboard the International Space Station. *U.S. Space & Rocket Center*

NASA brings their astronauts home. *20th Century Fox*

It's time for another walk, and this time on MARS! *NASA/Pat Rawlings, SAIC*

meetings to accommodate an unexpected priority. The cascading effect down the line on the calendars of others still had to be personally worked. Bonnie described WvB's calendar as "A nightmare every day."

We always tried to get one more appointment in. Every morning when he'd come in, he'd dump his briefcase on my desk, empty it out all over my desk, everything he'd written to me to take care of that day. Then we'd talk about calendar. It was a challenge to keep all the appointments going. I had a lot of people that got very irritated at me for going in and interrupting a meeting: "Dr. von Braun, you've got ten people waiting out here." And they weren't finished with him. I got fussed at a few times for breaking up a meeting.[19]

Few of those working in 1960 could have imagined the magnitude of technological change they were setting in motion in response to the requirements of the Apollo missions. The advancement of existing technologies or creation of new ones in materials, computers, electronics, and telecommunications was incredible. How quaint it seems today that we were fascinated by meals prepared in microwave ovens at the 1964 New York World's Fair. The Marshall office environment was soon being changed as were many households in the United States and beyond. Near the end of his time at Marshall, WvB attempted to prepare Marshall employees for the advancing Information Age. In a presentation in Morris Auditorium he used impressive graphs to depict the exponential growth of information already in progress. Today we are immersed in an ocean of electronic communication that most take for granted as the norm, at least for this day or week. The future promises more.

Once again, Bonnie generally listened as her successors commented on the advantages of computers, e.g., Evelyn Staples, who was in a program office at the time:

I had the first computerized typewriter at Marshall, in the Skylab Projects Office. It was about the size of this [conference] *table. Actually the console was about the size of the table, and the typewriter was in the center. I had it because we did a lot of experiments and we had all these experiment folders. I would input them. Then I would run to the restroom. I would print it out, and it would be printing out while I went to the restroom. They rented it from Xerox. That was in '76, '77? I think it was like $1,000 a month. But it was huge!*[20]

In a few years, the first desktop computers were appearing on everyone's desk. Eventually, cell phone and computer miniaturization became available and ultimately provided ubiquitous communications capability, allowing the director and secretary to communicate instantly. None of these aids were available in that first decade. Bonnie self-identified as being "pre-computer age." This probably meant, of course, more challenging, more personal interaction, and more hands-on tasks. This extended even to available office aids. In response to a question, Bonnie indicated that WvB's accent was never a problem for her:

> *He sounded different in recordings. The accent seemed more prominent in recordings than in talking face to face. I never had a problem. He would not use a Dictaphone; he hated the Dictaphone, so I had to take everything in shorthand....*
>
> *Dr. von Braun would always bring me notes. When he would be on a business trip somewhere and run into somebody, he would have names and phone numbers scribbled down on paper napkins, on brown paper sacks, tear off the corner of a file folder, etc. I would have to keep up with all these notes because he'd say, "When I go to Los Angeles next time, I want to... make me an appointment with him." I would have to keep a tickler file of all these contacts, all these notes that he'd written to me, whether it be in New York, or Washington, or the West Coast, wherever. When he would schedule a trip, then I'd get out all these little notes and try to work in appointments. Sometimes I would have to say, "If you'll meet him at the airport and drive him to where he's going, you can have a little conversational meeting with him in the car on the way." So I arranged a lot of meetings that way when there was not enough time to schedule something. If they'd pick him up at the airport or have a taxi ready; meet him in the taxi or something like that. It was something to keep up with all of these little notes that he brought to me.*[21]

The interview session was lengthy. Once into sharing stories, the former secretaries were reluctant to close. The facilitator tried three times before successfully bringing the session to an end. There was an obvious sense of bonding and camaraderie among the participants, not unlike the atmosphere of family reunion that surfaces when a group of veteran Apollo retirees gets together. With the Moon landing, the leaders, aides-de-camp, and foot soldiers in the trenches shared with public supporters a sense of pride in mastering the improbable to do the impossible. The Apollo program was a unique aligning of objective and people, a rare achievement that marked a 20th century historical milestone not likely to be soon surpassed. Indeed, we were all part of making history.

Resident Management Office

By Jim Splawn

Editor's Note: As the Marshall center began to contract out launch vehicle components to industry, it became necessary for management to develop a monitoring system known as resident management. During the Apollo Saturn pro-

gram, WvB's deputy for technical, Eberhard Rees, refined and introduced this process of management to all major contractors. He referred to the process as, Marshall management penetration and monitoring. Rees said, "In other words, you can expect our team to be in your 'knickers'." The program offices had a representative at major contractor plants that operated as a "mirror image" of the project manager in Huntsville. The company that had the longest resident management relationship with the Marshall team was Rocketdyne, a division of North American Aviation. The relationship began in the 1950s at Ft Bliss, TX when Rocketdyne was invited by the team to study the German produced V-2 engine. That led to Rocketdyne becoming the Marshall team's propulsion partner and supplier for rocket engines for Redstone, Jupiter, Saturn, Space Shuttle and continues today with the development of the Space Launch System. The following is one resident manager's view of Marshall's extended management into the aerospace contractor world.

This chapter will address the Resident Management Office (RMO) at Rocketdyne, the Marshall center's propulsion partner for over six decades. Rocketdyne was established as a division of North American Aviation in 1955. The Rocketdyne facility was located in Canoga Park, Calif., north of Los Angeles in the San Fernando Valley. At that point in time, Rocketdyne was already supplying engines for the Redstone and Atlas rockets.

In October 1957 the Soviet Union placed Sputnik I in orbit and changed the entire landscape for U.S. space pursuits. Four months later, the U.S. launched Explorer I into orbit, riding atop a Jupiter-C rocket powered by a Rocketdyne engine. Rocketdyne was involved in the first step of the new era of space travel for the U.S.

In the 1960's, following President John F. Kennedy's challenge of…landing man on the Moon and returning him safely to earth…Rocketdyne's mission went on "fast track". Facilities expanded tremendously with accompanying personnel approaching 10,000 by end of the decade.

First came the Mercury manned suborbital flights for Alan Shepard and Virgil Grissom; then orbital flights for John Glenn, Scott Carpenter , Walter Schirra and Gordon Cooper . All these manned flights were placed on-orbit by Rocketdyne engines. The Saturn I vehicle was the first of the "big" rockets. The Saturn I was 162 feet in length and weighed just shy of 1 million pounds. A cluster of eight Jupiter engines, which were upgraded to 200,000 pounds of thrust each, and later designated as H-1, powered the Saturn 1. Ten successful flights occurred during this next logical step to place man on his journey to the Moon.

The Saturn V vehicle utilized an engine with 1.5 million-pound-thrust, and designated the F-1 engine. This single F-1 engine equals the combined thrust of all eight H-1 engines used for Saturn I. Five F-1's, with a combined 7.5 million pounds of thrust, were required to launch the Saturn V. The Saturn V stood 364-feet tall and had three stages, S-IC, S-II, and S-IVB, an Instrument Unit, lunar lander and the Apollo spacecraft. All propulsion engines powering the three stages of the Saturn V were Rocketdyne engines.

In July 1969, Rocketdyne employees received their huge reward as they witnessed 10 years of hard work climaxed by cheers and pride as Apollo 11 lifted from the launch pad for its rendezvous with the Moon, an event hailed throughout the world as the greatest exploratory feat achieved by man since Columbus discovered America.

The follow-on to this historical event was the Space Shuttle Program. The Space Shuttle Main Engine (SSME) was a totally new engine that featured many innovative characteristics, including reusability, liquid hydrogen and liquid oxygen as propellants, higher internal pressure, and a reduced size envelope. SSME was the most advanced rocket engine ever designed in the United States.

The Resident Management Office at Rocketdyne was implemented during the Saturn program. The principle was to have multiple skills representing the variety of laboratory skills at MSFC, but be physically located in the contractor's plant. Thus, it was basically a mini-Marshall Space Flight Center, and only working on one major contract.

Each independent laboratory would determine its need to have an on-site staff person, determine the qualifications necessary to fill the position, select its representative, and ultimately assign that person to the contractor's plant for a negotiable period of time. Obviously the chosen individual must have demonstrated characteristics of self-starter, integrity, stability, communication skills, strong engineering, management skills, good and balanced judgment, steady inter-personal skills, and be a team player.

Following the successful Skylab Program, MSFC initiated a fresh evaluation of the Space Shuttle elements to identify any realignment needs and/or an increase in RMO staff. The SSME program had been recognized and accepted as a huge, demanding and critical element of the Space Shuttle. Further, it was generally accepted as the most challenging engineering system of the shuttle program. Consequently the SSME received considerable attention.

Rocketdyne was a strong contractor who had a solid record of outstanding performance and on-time delivery of engines with 100 percent flight success for many years. MSFC's expertise in the propulsion discipline, however, recognized the complexity of this new engine and the magnitude of the tasks associated with its development. They understood this engine was pushing the engineering envelope to the brink of man's knowledge and experience.

By 1974 the fundamental design of this complex engine was complete; individual critical components were being manufactured for testing. After considerable review, MSFC management made the decision that increased participation by MSFC personnel should be implemented in the Rocketdyne Resident Management Office. Accordingly, this decision was implemented in the Fall of 1974. The RMO resultant staffing follows: Jim Splawn, resident manager, SSME Project Office; Ray Tjulander, deputy manager, SSME Project Office; Ann Bowden, secretary; Lillian Bright, secretary; Bob Swaim, Materials Laboratory; Jerry Kidd, Propulsion & Vehicle Engineering Laboratory, Turbo-machinery; Bob Wagner, Propulsion and Vehicle Engineering Laboratory, Combustion

Devices; Jack Weil, Propulsion & Vehicle Engineering Laboratory, Engine Systems; Denny Holweger, Facilities; Parker Counts, Quality Laboratory manager; Benjamin Pagenkoff, Quality Engineering; Claude Buzbee, Mechanical Hardware; Charlie Pierce, Electrical Hardware; Jack Freedman, Finance; Ken Fisher, contracting officer; and John Chaffee, contracting officer. The disciplines mention above aligned with Rocketdyne's major organizations and their products. At MSFC the center director was William R. Lucas. The SSME program manager was J.R. Thompson.

The Rocketdyne management staff was competent, committed, energetic, and demonstrated strong engineering capability. The comradery of the entire organization was contagious. Their mission was focused. The entire organization knew the mountain to success was steep, but believed deeply that their ingenuity, stamina, and hard work would conquer that mountain.

Dom Sanchini was the SSME program manager and ran a tight ship. His engineering and management skills produced a confidence that permeated the entire SSME staff. J.R. Thompson and Dom Sanchini made a winning combination for the SSME program. Both were strong technically, hard-driving, yet showed compassion and deep appreciation for respective employees.

For the operational posturing of the RMO staff, the environment established by these SSME program managers created a homogeneous atmosphere that permeated all organizations with full-and-open coordination without concern of reprisal. It was a real gift for the future of the SSME Program.

At MSFC the SSME Project Office and its organization structure, and each Laboratory with its organization structure, were heavily involved in planning, scheduling, technical studies, documentation, solutions to operational problems, solutions to hardware test failures, briefings to higher management, and the list goes on and on. At Rocketdyne very similar tasks were being pursued. Consequently, and as you'd expect, there was significant interchange between MSFC personnel and Rocketdyne personnel on an hourly and daily basis, all of which was focused toward production and flight of a monstrous rocket engine. Also, funding for technical and management staff travel was significant.

A couple of differences: the manufacturing shops, and the Santa Susana test facility. The huge manufacturing area produced flight configured components and subsystems for testing and ultimately complete engines for flight. Test hardware would be transported to the Santa Susana mountain, a test facility of 2,635 acres northwest of the Canoga Park facility. Several test stands accommodated hot-fire testing of turbo machinery, combustion devices, and limited full engine tests. Major full-up engine hot fire tests would be conducted at National Space Technology Laboratories (NSTL) in Bay St. Louis, Miss.

The continuous test program was the real proving ground for the final configuration and operational SSME. Along the way and as planned, much component hardware was destroyed in test as threshold life was determined for all piece parts contained within the assemblies. The strategic location of multiple sensors (temperature, pressure, flow rate, etc) provided critical data for engineering analysis. After a test failure, an iterative process would be followed: Redesign was based on instrumentation data that showed how, why, and where the failure initiated, revised new hardware would be fabricated, hot-fire test again—repeat, and repeat again, until success was achieved. Turbo-machinery was the most difficult to fine-tune because of pressure, rotating parts, temperature, and clearances that changed with temperature variations. The components produced by this testing process resulted in minimum loss of a complete engine during test firing at maximum operating conditions. Testing was key.

The mission of the RMO was to actively monitor, on a daily basis, contractor performance against the many contract requirements. The RMO responsibility included the overseeing of Rocketdyne performance for design, development, manufacturing, test, and delivery to meet flight requirements while keeping the MSFC SSME Program Office and laboratories informed of the progress toward program objectives. This responsibility was a huge undertaking and required acceptance of, and respect for, the government employees, individually and collectively, by the total Rocketdyne management staff and working level employees. This was a day-to-day continuous discipline situation. Executive level managers, Thompson and Sanchini, obviously establish the initial environment and expectations, but it was the daily personal discipline, demonstrated professionalism, and inherent integrity that yielded the critical relationships. This relationship could be lost in one day or it could be nurtured every day, on both sides of the equation. The key was stating the facts, accurately, fairly, honestly, and straight-forward. The balance of praise and short-comings had to be handled carefully and in good taste.

Fortunately the MSFC staff and the Rocketdyne staff, top to bottom, understood the difficult and critical tasks and conducted themselves accordingly. Forgiveness without individual blame, and praise for the team accomplishments, were great to witness. Don't misunderstand; there were many, many times when significant, very significant problems occurred, and many disheartening hardware failures during test operations, but the joint teams tightened their belts, worked the situation, and pressed forward. This relationship made the MSFC / Rocketdyne working environment fun but pressure packed, rewarding but with a few scars.

The individual staff members in the RMO owed their home organization at MSFC daily feedback on the progress or problems for their particular field of specialty. Many times the home organization would send to California a traveler with specific skills matching the problem area to work with the Rocketdyne specialist. So, the RMO staff had to really understand the problem, articulate the situation clearly, and request MSFC assistance. Once on-site, the RMO staff would work continuously with the traveler and the assigned special team for understanding the process of correcting the problem while simultaneously gaining a good background for later follow-up. This scenario is a typical example all RMO staff would follow for similar situations.

Each RMO person was responsible for making continuous contact with his assigned Rocketdyne counterparts; attending joint meetings, monitoring schedules, following manufacturing progress on hardware, writing reports, and getting answers to questions from his home organization at MSFC.

Each work day at 7 a.m. Rocketdyne held a stand-up meeting in the "war room", a large conference room with no tables or chairs, just status charts covering every wall where each organizational manager (Engineering, Turbo -machinery, Combustion Devices, Manufacturing, Quality, Engine Assembly, Test, etc) would brief their progress of yesterday, the needs of today, and forecast for tomorrow. It was a significant communication tool as each could see the interplay between disciplines and how each was dependent on progress of others. When a problem was evident, those involved worked that situation immediately following the meeting. Our RMO staff attended these stand-up meetings for monitoring progress and understanding problem areas. With this data and if appropriate, communications with their Huntsville organizations was conducted to initiate support work on the problem as well.

The following is a partial list of key MSFC personnel that served as my / our mentors during my assignment at the RMO at Rocketdyne, each with critical skills and unique personalities, and all great friends. I'm indebted to all: J. R Thompson, Larry Wear, Brenda Sutherland, Joe Lombardo,Gerald Smith, Zack Thompson, John McCarty, Otto Goetz, Len Worlund, George Hopson, Bob Schwinghamer, Jerry Thomson, Norm Schlemmer, Bob Ryan, Walt Mitchell, Ed Mintz, Carlyle Smith and Bob Morris.

The following is a partial list of key Rocketdyne personnel that served as my / our mentors during my assignment at the RMO at Rocketdyne, each with critical skills and unique personalities and all great friends . I'm indebted to all: Dom Sanchini, Jerry Johnson, Ted Benham, Ed Larson, Bob Crain, Willy Wilheim, Paul Fuller, Bob Biggs, Erv Eberle, Byron Wood, Jim Hale, Frank Lary, Joe Stangeland, Matt Ek and Don Mikuni.

The opportunity and experience of the Resident Management Office for this writer was significant and rewarding. The huge technical task was humbling, but the people at MSFC and Rocketdyne demonstrated the will and the confidence to conquer the challenge. Together, they created a magnificent machine.

From Fire and Smoke to the Top Man

By Ed Buckbee

When you talk to J.R. Thompson, you learn that this Georgia Tech graduate started his career relatively soon in "fire and smoke." He worked on the RL-10 at Pratt and Whitney, the aircraft and rocket engine company.

He came to Huntsville in 1963 and went to work for Hans Paul, a Peenemuende veteran in Marshall's Propulsion and Vehicle Engineering lab. He worked on the RL-10, H-1, F-l and mostly the J-2, all Saturn rocket engines.

"Hans Paul was the first guy to really influence my thinking," recalls J.R.. "I learned a strong work ethic from him. It was not unusual to hear him say, 'today's work must be done today, ' and we worked on Saturdays. He had a lot of experience with engine values, and he taught me the importance of testing and that it was critical to pay attention to details."

J.R. often visited Rocketdyne, Marshall's first engine contractor and North American Aviation. There he met Dieter Huzel, another Peenemuende veteran. He learned more about design, development, fabrication and how to keep a team focused and engaged in rocket engines and manned flight awareness.

He attended Flight Readiness Reviews run by Eberhard Rees, deputy for technical. J.R.'s recalls: "I grew to respect Rees' technical ability. He knew how to run a meeting with a room full of lab directors and project managers. It wasn't very formal. He went around the room polling people to see if they were ready to launch. In those early Saturn reviews, the Germans were pretty much in control. Rees was big on establishing good relations with the contractor team. He wanted to be sure they were engaged and involved. I remember one review when we had the Saturn V second stage engine shut down; WvB ran that meeting. It was a bit more intense. That same readiness review model was used later for Shuttle."

After working on the J-2 engine, J.R. moved into the Human Factors area and concentrated on early Skylab development. He later was named the program manager of the Space Shuttle Main Engine (SSME), an engine that performed faultlessly on 135 shuttle missions.

J.R. recalled, "Mentors at Marshall that I was fortunate to engage with were Hans Paul, Eberhard Rees, Bob Lindstrom, Bill Lucas, Rocco Petrone and Jim Kingsbury. Also, I had the opportunity to work with some very talented people in Marshall's contractor ranks."

J.R. left Marshall in 1983 and began a new career in research with Princeton University, where he was deputy director of the Plasma Physics Laboratory. J.R. explains, "I applied a lot of what I had learned at Marshall in designing and building hardware and how to keep technical people focused and engaged. They needed to be challenged. I encouraged attention to detail, checks and balances and working as a team. I established a similar structure at Princeton that we had at Marshall, science and engineering separate from program management. It worked and the lab grew to nearly 1,000 employees."

In 1986, J.R. returned to his NASA roots as director of the Marshall center. He recalls, " I really didn't have to change much. I told them to do what you know how to do better then anyone, develop and manage launch vehicles for human space flight. I moved some people around and knocked off the 9th floor luncheons and challenged people to get their hands dirty. I revitalized the Test Lab and we went all out in testing. We had ' fire and smoke' again. It was fun to see the center move forward. I had great help from Jack Lee and guys like George Hopson and Bob Schwinghamer."

One of the developments that occurred at Marshall during J.R.'s day was reconnecting with leaders in Congress. This had not been done effectively since WvB days and J.R. loved it. On more then one occasion, U.S. Senate committee chairpersons, lead by U.S. Sen. Howell Heflin (D-AL), visited the Marshall Center.

Senators Gore, Kerry, Kennedy, and Mikulski were among the Washington leaders who came. Vice President Dan Quayle, chairman of the National Space Council, visited as well. Calls came from NASA Headquarters hinting these visits might be a conflict of interest. The visits of congressional leaders continued.

In 1989, J.R. was appointed deputy administrator of NASA working for Administrator Richard Truly, a former astronaut. J.R. commented on his headquarters position, "Dick, let me run the business at the centers. On one occasion I applied the 'Tiger Team' approach that I had used at Marshall to a problem on shuttle. I put Bob Schwinghamer in charge, and he pulled together a team from the centers and industry. It worked and demonstrated the ability of a carefully selected team of people could identify the problem and come up with solutions."

In 1991, J.R. left government and entered the private sector. He became the 50th employee of Orbital Sciences Corporation. Later, he was named president, chief operating officer and vice chairman of the board. Again, J.R. applied what he learned at Marshall and established a similar structure at Orbital. He hired consultants like George Hardy, a former Marshall engineer and manager. Hardy said, "J.R. wanted to apply what he had learned to grow this technology company. It seemed to me wherever J.R. went, he brought with him some of that Marshall culture." When J.R. retired from Orbital in September 2013, the company had grown to over 10,000 employees.

As a young engineer, J.R. was given the opportunity to engage in a new and rapidly growing field of technology. He had mentors who took him under their wing and introduced him to rocket propulsion. He learned to appreciate the "can-do" attitude of a team of people who were challenged to accomplish what had never been done before—flying man to the Moon. His leadership style developed over the years enabling him to get the job done whether it was rocket engine development, managing a NASA center or leading a technology company. J.R. was one of those people who contributed to "The Greatest Space Generation."

Mentoring

By Rein Ise

My first exposure to the business of space travel was in reading some of the early science fiction books, such as Jules Verne's, "From the Earth to the Moon". This, and my general fascination with astronomy, triggered the interest in anything to do with outer space. This received a significant boost when I ran across the early articles published by WvB in *Collier* and *This Week* magazines, describing various concepts of rockets and living in space on a permanent space station. As fate had it, I graduated from the Army ROTC program and requested active duty at an activity involved with rockets. I was familiar only with the Army missile work at White Sands, so I requested that assignment, but instead I was sent to Redstone Arsenal, as the Army missile activities there were building up at a rapid pace. I was assigned to the Army Ballistic Missile Agency, which had the responsibility for fielding the Redstone ballistic missiles, and developing the advanced Jupiter ballistic missile.

As a young (21 year), inexperienced second lieutenant, I found the Redstone and Jupiter work at Redstone extremely intimidating. But probably the earliest mentoring I received was from my commanding officer, Col. Robert Mullane, who encouraged me to make procurement and deployment decisions based on the best information on hand and not expect any help, since no one else had any better information than what I had determined in my research efforts. And this certainly proved true, as all of us, regardless of age, were working in a new field of endeavor. The confidence I thus gathered carried over later to the technical field where I interfaced with Claude Gage, a lead engineer at the Chrysler Corp., the company contracted to build the Redstone and Jupiter missiles.

I was responsible for coordinating the engineering changes to the structures of these missiles. Mr. Gage, who was nearing retirement age, dealt with me in full confidence, notwithstanding my age or lack of experience, again providing me additional insight into working with other professionals and management. About the same time, I had the opportunity to meet Dr. Wernher von Braun, the director of ABMA's Development Operations. He spent nearly 30 minutes with me in a private conversation, showing great interest in my background and education, and encouraging me to stay with his organization as it started developing the Saturn I launch vehicle. He demonstrated for me several key traits of a good leader and manager by taking the time to show interest and provide encouragement to one of his employees. I always considered that interview as a great learning experience.

When I completed my military tour, I was advised by Bob Lindstrom, whom I had previously met at work, to work for ABMA as a civilian in the Saturn I project office with responsibilities to coordinate certain aspects of the Saturn I development. I decided that this would give me a great opportunity to grow with the space program and become familiar with all aspects of rocket development. My assignments soon brought me into face-to-face contact with the senior technical managers of the development organization. These managers had accumulated invaluable information and experience working on the V-2 weapons system in Germany, the V-2 versions used in space research at White Sands, and the Redstone and Jupiter missiles in Huntsville. The depth of their knowledge and ability to make quick and correct decisions pertaining to complex design problems impressed me and set certain standards in my mind.

The Germans also radiated a team spirit that was clearly recognizable, and led to more effective efforts. This was also demonstrated in a social way, as I remember attending team parties at Dr. Willi Mrazek's home, the manager of structural and mechanical design. The team also frequently met Fridays at Huntsville's Elks Club to celebrate TGIF. However, recognizing that my technical knowledge base needed improvement, I decided to take a sabbatical and spend a year at Purdue University, studying rocket propulsion under the guidance of Dr. Maurice Zucrow, one of the leading rocket experts of that era. Again, Bob Lindstrom gave me his full support and encouragement for accomplishing this goal, and Dr. Zucrow and his staff taught me to feel more comfortable in the complicated rocket development business.

After receiving my master's degree from Purdue, I returned to Huntsville, where the ABMA development group had been in the meantime established as the Marshall Space Flight Center (MSFC). MSFC had just been given the responsibility for developing the Saturn V Moon rocket, and I joined its fledgling project office. The project manager, Jim Bramlet, asked and encouraged me to head the systems engineering efforts within the office, which was obviously a very responsible role for a still very young engineer. Being a major NASA program, the Saturn V activities received a lot of high-level attention, and the program was asked to implement several new systems engineering concepts, such as configuration management and interface working groups, a new and challenging experience for me to manage and coordinate.

Again, I had many opportunities to interface with senior managers and learn their methods. One of the memorable lessons was provided by Dr. Arthur Rudolph, the new Saturn V program manager, in preparing for presentations to higher management at NASA Headquarters. He could not emphasize enough to his staff the importance of "chartsmanship", involving choice of words, structure of sentences, and highlighting important words and data for the purpose of getting the desired point more clearly across. Although he sometimes seemed to overdo this, there was no question that he was very much on target.

I learned one of my most important lessons from Dr. George Mueller, the head of the Apollo Program at NASA. I was requested by WvB to give Dr. Mueller a presentation that explained the MSFC recommendations for the Saturn V flight development program. About halfway through the presentation, Dr. Mueller stopped me and declared that my recommendations were not acceptable. He then stood up and outlined a flight program that was totally different. However, he did this in a very respectful manner, demonstrating that differences in opinion are acceptable without embarrassing the other party, and convinced me that expressing controversial opinions to senior management should not create fear. This experience served me well in many future encounters with management.

The guidance, support, and respectful treatment by my supervisors and higher NASA management were all a form of mentoring and were invaluable throughout my career.

Part IV

Marketing

Marketing WvB and Saturn-Apollo to the Moon

By Ed Buckbee

It is almost inconceivable to look back upon the great marketing efforts of the late 1950s and 1960s and witness a brand—or more suitably, a mission—created by a governmental agency that was second only to Coca-Cola in popularity. That brand was NASA, and its mission was "We choose to go to the Moon." Between 1969 and 1972, we sent 27 American astronauts to the Moon and back. Not only was this a governmental agency, but it was one whose employees had little to no previous experience working for the government. How were they able to accomplish both mission and marketing on such a grand scale with no prior history or experience?

At the time it was difficult if not impossible to foresee that this governmental agency would create a relationship with the press that has not been matched. Why, because we were going where no man had ventured. Transparency of information to the public was critical to the success of the mission. We in NASA operated within the statute of an agency that must "provide the widest possible dissemination of information" to the public. The successful marketing of NASA to the public was a milestone for our generation and a testament to its versatile leadership.

An early challenge we had to face was convincing the engineers who ran the agency that we must be transparent. NASA was not a skunkworks nor a black-ops program. NASA was different. Many thought NASA was not a part of government, but rather a collection of nerds taking us on treks into space. There was a common saying around the workplace: "late to bed early to rise, work like hell and advertise!"

You never turned down a media request while you were working for/with WvB. The media wanted three things: kick the tires of a Saturn booster, witness a close-up Saturn ground test (static firing), and, of course, interview WvB. Every effort was made to accommodate the press. There was no media plan; rather, it was more of a reaction to what they demanded at the time. A typical week at the Marshall center in the 1960s might find visiting newsmen Walter Cronkite, CBS-TV; Jules Bergman, ABC-TV; Jay Barbree, NBC-TV; and writers Howard Benedict, AP; Bill Hines, *Washington Post*. Add a television documentary crew from Europe to complete the week. In 1968, Richard Lewis, *Chicago Sun Times* science writer, was invited to witness a Saturn static test firing. When informed he was coming WvB said, "a very nice, constructive man. I would like to meet him." WvB was a PR man's dream. Houston may have had the astronauts, but Huntsville and the Marshall Space Flight Center had WvB.

We were overwhelmed with the sudden interest in our program. Initially, interviews with the press were a challenge for the team. Most engineers were well trained in their respective fields but lacked communications skills, particularly when relating to the media. Many had come from classified military programs that were off limits to the press. With NASA having an open door to the public and the press, we found it helpful to accompany most NASA spokesmen during interviews. We in public affairs learned, along with the engineers, one can say too much about a program or use terminology that escapes the interviewer. In one interview, the engineer began by stating to the reporter, "We aren't going to be using unsymmetrical dim ethyl hydrazine and inhibited red fuming nitric oxide in Saturn as they did in the Titan, we will be using RP-1." After a couple of those experiences, we developed a checklist on how to prepare and what to say to members of the press. It became known as "Interview 101." It wasn't long before we had a number of managers who could conduct a productive interview with excellent results.

Marshall's Public Affairs Office was organized and established by Foster Haley. He hired most of the staff, many of whom had newspaper and writing experience. Captain Bart Slattery, Jr., on loan from the U.S. Navy, was brought in to head up the office. Slattery reported directly to WvB. Irene Shafer was Slattery's secretary and Wanda Reed was Haley's.

The office was divided into four branches with the News Branch being the center of activity. Joe Jones, who later became chief of public affairs, directed the branch. His office published the center's house organ, "The Marshall Star," prepared all news releases and handled all press inquiries. The other branches were Community Services, headed by E.C. Wooten, Education Liaison by Dick Pratt and Special Events by Ed Riddick. Ruth von Saurma was in charge of international affairs and handled all of WvB's international correspondence and much of his writing. Guy Jackson and Amos Crisp were WvB's speech writers. The staff produced films, operated the Spacemobile and exhibits programs, ran the speaker's bureau, handled VIP visitors, conducted tours and managed open house and special events. Most of the public affairs staff engaged with WvB's office.

At the office's annual Christmas Party, it was customary to roast the boss, Bart Slattery, who enjoyed sharing humorous stories and experiences. The highlight of the party was a personal message written by the staff recorded by WvB to be played at the party. To accomplish this required tracking down WvB prior to the party and that often was done in a car, on a plane or wherever he had a free moment. One of his favorite lines in the script was, "Remember, Bart, public affairs is the last to know and the first to go."

What it was Like Working for WvB

What was it like working for WvB in the 60s? You couldn't go anywhere in a NASA building and not hear three words, "Man, Moon, Decade." We were preparing for the mission as if we were flying with the astronauts. We had a passion for completing the mission successfully, which also meant safely. It was a team effort. If you received a note from WvB, it might include, *"P.S.- Remember you are on the critical path…"* meaning the manner in which you completed your tasks could have an impact on the success or failure of the astronauts' mission to the Moon.

He asked us to pronounce his name "Warner von Brown" in the English pronunciation and not as one might say it in German. He said, "I'm an American with an American citizenship and I wish to be recognized as such." In his presence and with visitors, we referred to him as "Dr. von Braun." When not in his presence, we often referred to him as, "vB." We saw him in his role as a corporate CEO and communicator. He often testified before Congress and spoke to public and technical societies and toured dignitaries. When he visited the White House and President Kennedy visited Marshall Space Flight Center, the Secret Service gave him the code name: "Rocket Man."

WvB never turned down an opportunity to reach out to the public. He made films, gave speeches to the general public and technical audiences. He was the star at press conferences. He felt an obligation to the American people to tell them how their space program was doing. He was unique in that as a scientist and manager of a major national technical project he had the ability to communicate with all of the media. He developed a connection with the press that made him the darling of the press corps covering the Moon landings and the space program. He seldom turned down an interview or an opportunity to speak with the media one on one. He could write wonderful articles about his vision of flying humans into space and on to the planets.

He had the most interesting group of friends you can imagine, some of which had no relationship to rocket science or space flight. Walt Disney asked his advice on the production of space films and brought his design staff to visit WvB to seek ideas for the Orlando EPCOT project. I think WvB got the idea for the Alabama Space & Rocket Center from Disney. Disney said to him, "If you want to have public support for your programs, you must let the public experience what you are doing. Almost every kid in America wants to be an astronaut, so take advantage of that."

Ruth von Saurma, WvB's assistant for international affairs and public affairs specialist said:

> "WvB was so terrific in inspiring people and bringing out the best of everyone who worked for him. I was envious of the ease with which he could write. He had written several books. He wrote, I don't know, one piece each month, or even each week, for *Popular Science*. He was the best communicator you could possibly think of, and not just in written communication, but also in oral communications. He was always happy to talk with people, especially if they were the ones who could also, in turn, spread the word about what he said and help to publicize the idea, which was such a novel idea for people, that you could really fly beyond what he called the 'dirty window' of the atmosphere and look further beyond into the great universe."

Over the years, von Saurma experienced a unique insight into WvB's ability to meet the public and communicate. One of her memorable moments was witnessing the launch of the Apollo 11. She climbed onto a helicopter with WvB and writer Cornelius Ryan to meet some French VIP's. WvB welcomed the guests with remarks she had prepared for him in French. She escorted the European group to the Saturn launch viewing site. She recalled:

> "It was an exciting time. At the launch pad, the huge Saturn V was illuminated by searchlights, forming a brilliant star against the sky. To be able to launch something to the Moon, especially our astronauts, you really could not describe the excitement that we had. And the relief when the huge Saturn V took off, it was mind-boggling. That's when I realized I worked for a man, WvB, who would make history."

WvB was outstanding with the media and we PR people were relieved to have such an impressive spokesman. He answered their questions without confusing them with technical jargon. The press liked, trusted, and believed him. With his handsome appearance and German accent, he became the darling of the space program's press corps. At press conferences he was the star, always getting most of the questions. However, this came with a backlash. Occasionally this disturbed others present, especially if some were his bosses. WvB always had at least two superiors immediately above him in the change of command. They were the director of manned space flight and the administrator of NASA. To counteract WvB getting all the good questions, we encouraged friends in the press to direct their questions to others presence at the press conference.

One of the most interesting subjects discussed in press conferences during the early days of Saturn was the subject of solder joints. Would solder joints of critical components withstand the rigors and g-forces of launch? Why solder joints had such a high level of media interest is still a mystery. If that question was asked, it created a lively exchange between the members of the press and the launch team. Once while in California for the roll-out of a new rocket stage, a press conference was held with the top leaders of manned space flight from NASA and private industry. After three questions by the press, complete silence fell over the proceedings and the PR people who had devoted many hours planning the press event were on the verge of panic. Since I was one of them, I noticed WvB passing a hand-written note across the stage to where I was standing. I opened it and in that very clear and bold WvB scroll it read, "I think it's time for the solder joint question."

When you traveled with WvB, you quickly learned he never had money on him. Here's a man who managed billions of dollars of government contracts and when it came time to pay the bill for his soup and sandwich, he began searching his pockets for cash. As the search continued with no results, you knew that was the signal to cover WvB's tab. Upon returning to the office, it was customary to send Bonnie Holmes, WvB's loyal secretary a memo, with the amount noted. A check for $6.27 would be prepared, signed by WvB and promptly returned. Most people never cashed the check, preferring to save WvB's personalized signed check to show to their friends. This hampered Bonnie from balancing WvB's checkbook in a timely manner.

Once he mentioned to Bonnie that he felt he was wasting time driving to work in his car. I had learned earlier from a friend who once shared a car pool with WvB that he was dropped because of tardiness. I never determined who the person was who dropped von Braun from the car pool. Now that he was a carpool of one, he took Bonnie's advice and began learning another language in addition to the four he spoke.

He was constantly sought after to make public appearances and speeches, numbering more than 50 appearances per year. More were turned down than accepted. He became so popular that his superiors informed us to forward all requests for WvB to NASA headquarter for review. It was apparent that WvB became too popular to his superiors. We complied and forwarded requests from congressional members who requested WvB speak in their behalf. Not surprisingly, those requests were returned for WvB to accept after headquarters received calls from the congressional members saying, "We requested WvB, not you!"

He never missed an opportunity to sell the program, especially if it was a congressional request to speak. One such engagement took us to McAlester, Oklahoma for a speech at the request of Speaker of the U.S. House of Representatives, Carl Albert. WvB flew the last leg in his NASA Gulfstream aircraft and we landed at an old WWII airstrip. The official greeting party met us, followed by an immediate press conference. The standard questions in those days were, "Are we going to beat the Russians to the Moon? Couldn't this money be better spent on something here in Oklahoma?" Thirdly, "What do you think about UFO's?" He answered all of the questions with clear and concise answers as if it were the first time he ever heard them. After the press conference, we loaded into a limo and were joined by a caravan, complete with motorcycle escort and blaring sirens, and proceed to speed through downtown. As we sped past the courthouse WvB claimed: "These high-speed police escorted motorcades are the most hazardous part of my job!" We arrived safely at the local high school and entered a non-air-conditioned gymnasium, filled to capacity. WvB was taken to the speaker's platform and treated to the standard fare, fried chicken and peas. WvB seldom had the opportunity to eat a full meal due to constant requests for his autograph. From the beginning of his career, to the time when he became quite prominent, he never understood the demand for his autograph. Autographs were for rock or movie stars, not rocket scientists.

WvB gave the heartland speech that evening in Oklahoma, reminding the audience that Americans had always been explorers and pioneers. If Americans want to continue as world leaders, they must establish a leadership role in space. He commended the local congressman for his great leadership in Congress and Washington and his invaluable contributions to the nation's space program. At the close of his speech, the audience gave him a standing ovation and dubbed him: "Chief-Fire-Arrows-To-The-Moon." He also received a beautiful, full-length Indian headdress. Donning his headdress, WvB posed for pictures, said his good-byes to everyone, climbed in the limo and returned to the airport with the high-speed police escort. Onward to the next mission.

A significant day in WvB's early NASA career was the visit of President John F. Kennedy to Huntsville in 1962. It was huge for WvB and the Huntsville team. He had advised the president to challenge the nation to send men to the Moon and defeat the Russians in an all-out space race. The young President learned on this trip that if America was going to the Moon, then that journey would begin in Huntsville, Ala. He came to see for himself the progress being made and asked WvB, "Can we beat the Russians to the Moon?" He answered confidently and concisely: *"Yes, Mr. President, we will beat the Russians to the Moon and we will do it within the time frame you set."*

He understood the importance of proper political jargon. As Von Saurma noted: *"It didn't matter so much whether you beat the Russians; it was the idea that you were able to get to the Moon. But, of course, there was a political background that was especially important for Congress and for the White House that something would be done to beat the Russians in space who had in previous years often held the first spot in some of the space projects."*

Kennedy's challenge to the nation and to the team was well received. Von Saurma: *"It was absolutely fabulous, and WvB was elated. So was the entire team. We also were delighted that we had a project that was not geared to a military concept but what we hoped was really the exploration of another planet. It was just awesome to think of that. It was a tremendous encouragement."*

WvB was a true crusader for manned space flight and always looking to the future as can be seen by reviewing his daily journals, documents and notes between him and his managers. While developing the Saturn family of space vehicles for the peaceful exploration of outer space, WvB and his team were looking to the future. They studied space stations, missions to Mars, and Nova, a mammoth rocket with enough power to go directly to the Moon and return. His dream and concepts of a manned mission to Mars dates back to the early 1950s, when he collaborated with the famous artist, Chesley Bonestell on a space travel series for *Colliers* magazine. Just a decade latter this was a serious project being studied by WvB and his Marshall center team of future thinkers.

A nuclear engine was under development in the '60s for use as a third stage of the Saturn V. Called RIFT, Reactor In Flight Test, this program was started by WvB in cooperation with the Atomic Energy Commission. Additionally, Aerojet General, Westinghouse and Lockheed were involved in various aspects of the program. The tests were conducted at the Los Alamos test site in Nevada, at a place called Jackass Flats. WvB and others witnessed several test firings. I learned from reading his notes and studying other documents that on Dec. 7, 1962, WvB met President Kennedy at Los Alamos. During that visit, the president learned of WvB's plans of a mission to Mars. Unfortunately, President Kennedy was assassinated in Nov. 1963. WvB was convinced that had President Kennedy lived, he would have been a strong advocate for a follow-up program to the Moon landings that would have kept the NASA team challenged and committed to a manned landing on Mars in the 1980's.

His support of the efforts by Vice-President Lyndon B. Johnson to build NASA facilities in Louisiana and Mississippi surprised and concerned some of his Huntsville team members. Why build facilities elsewhere that duplicated Huntsville's? WvB's future planners had been working on a 25-year plan of manned space flight that involved an Earth-orbiting space station, bases on the Moon, deep planet probes and a manned Mars expedition. His planners estimated it would require 50 Saturn IB and 100 Saturn V rockets to meet the launch requirements. Consequently, additional assembly and test facilities were required to support the increased production. These additional facilities—located in the Southeastern manned space flight crescent—would enable WvB's Huntsville team to concentration on developing new cutting-edge technology.

WvB loved to orchestrate events, especially when VIPs were coming to visit. Ground test firings of the Saturn rocket were incredibly popular. In fact, Marshall was the only show in NASA in the 1960s. In Huntsville we called it the "fire and smoke" show. Initially, we put everyone in the dark and drab blockhouse where they could observe the firing, protected behind portals of several layers of glass and thick concrete walls. Although quite close to the firing, they couldn't hear the sound nor feel the heat or vibration of the firing. WvB didn't like it. He felt the whole point of the demonstration had been lost. Bonnie Holmes agreed: "If they were in the blockhouse, they didn't get the full feel of the test firing. Dr. WvB suggested they put them on the roof of the blockhouse so they could get real close to the test stand, feel the vibrations, the heat, and all of that. So they were very impressed by getting to witness some of these Saturn firings right there in the Test Lab."

This worked for awhile until the safety people determined it unsafe for visiting dignitaries. He suggested they observe from an open bunker on what became known as " Heimburg Hill", named after the director of the Test Laboratory, Karl Heimburg. The bunker had a roof, and a concrete wall with a slit for viewing and open in the back. It was affectionately referred to as the "cow shed." Upon completion of the next firing, the VIP's were not only impressed, they were startled and some scared out of their wits. Suddenly, we had a flood of Washington politicians coming to Huntsville to witness one of "WvB's fire and smoke shows." Standing 2,000 yards away in the "cow shed," first timers watched a giant rocket ignite generating 32 million horse power, shaking the ground, feeling the vibration hitting their chest like a hammer, and heat flashing directly upon their faces. It turned out to be a crowd pleaser that WvB took great pride in orchestrating.

In 1972, while I was driving WvB to the airport, a CBS-TV reporter conducted an interview. WvB was the ultimate multi-tasker. Asked by the reporter, would he like to go in space, he commented, "I'd love too." Being a pilot, he understood space flight was the next step in the challenge. His plan was to fly as a scientist aboard the Space Shuttle, so it would be deemed safe for other non-astronauts. The fact that he, a non-astronaut could safely fly in space, would set the stage for the president of the United States—the leader of the Free World—to address the world population from space. Unfortunately, events did not unfold in time and von Braun was taken from us in 1977 at the relatively young age of 65.

Congressional Hearings

WvB was often invited to appear before congressional committees in Washington. One such invitation was from Congressman Oli E. "Tiger" Teague-(D-TX), chairman of the House Space Committee, a big fan of WvB's. The congressman had many friends in Texas including a radiologist, Dr. James R. Maxfield. Maxfield was well known to us as the NASA supporter who chartered plane loads of people to visit WvB, view ground test firings of the Saturn and attend launches at the Cape. Once we received the call, we began making preparations for the visit. First, the Gulfstream aircraft would be reserved followed by collecting photographs and Saturn models to be given to members of the committee. Biographic information on committee members was gathered and dry runs of WvB's presentation were conducted. Once onboard, WvB would review his comments, make notes on the margin, become acquainted with committee members and memorize names of wives and family members. When finished, he headed for the cockpit and flew the Gulfstream to the Washington airport.

When we arrived on The Hill, we headed for the hearing room. My job was to make sure the projector for the visuals worked and the models and photographers were available for WvB to present at the appropriate time. We waited for several minutes until women and children began to arrive. I thought to myself, either they or we are in the wrong room. I quickly learned from an Intern these were family members of the congressmen who had been invited to meet and be photographed with WvB. This continued until committee members arrived and asked to be photographed with him. My job was to get names of people noting from previous events they wanted WvB's autograph. Soon it was standing room only in the hearing room. The congressmen were happy to have brought WvB to meet their friends and family.

Chairman Teague dispensed with WvB's presentation that we had spend days preparing. The chairman asked WvB: "*Are we on schedule with the Saturn V and are we going to beat the Russians to the Moon? Do you need any more funding?*" Even though they were not invited to attend, NASA Headquarters always sent someone on loan from Houston to observe these hearings and report back what WvB said. WvB presented a four-foot tall model of the Saturn V to Chairman Teague and smaller models were presented to the committee members. One more group photograph was made of WvB with the committee. Before departing, WvB extended an invitation to the chairman and his committee to visit Huntsville to observe a ground test firing of the Saturn V booster. That event occurred a month later.

WvB's NASA Headquarters colleague, Alan Lovelace said it well: "*In the tradition of Newton and Einstein, he was a dreamer pursing visions, and at the same time a creative genius. He was a 20th century Columbus who pushed back the new frontiers of outer space with efforts that enabled his adopted country to achieve a pre-eminence in space exploration.*"

WvB and the Press
By Bob Ward

Bob Ward is a veteran space-program journalist with 40-plus years as managing editor and editor-in-chief of the Huntsville Times. *He interviewed WvB numerous times and covered the team as a science reporter before the title became popular in the newspaper world. He's the author of Dr. Space (*Naval Institute Press*), the story of the private and professional life of WvB. As a journalist who covered the manned space flight program during the 1950s and 60s, he shares his thoughts of the man.*

He was an absolute master of the media. Early on, he unashamedly courted the American press in spreading the gospel about the coming Space Age. He anticipated that accomplishing his and like-minded believers' visions depended on strong public and political support. It was in stark contrast, of course, to the complete press secrecy he had faced in leading the revolutionary advances made in German rocketry during the 1933-45 dictatorial reign of Adolf Hitler.

Dr. Wernher Magnus Maximilian von Braun's message and his manner appealed widely to journalists in his adopted country. Beginning in the early to mid-1950s, while working for the U.S. Army, WvB gained a reputation for honesty and truthfulness with reporters and their editors, despite the skepticism of many in those pre-*Sputnik* days. It was a reputation also recognized later for the most part by Congressional committee members and others in government.

Space supersalesman WvB came across generally to members of the press as a passionate, articulate engineer-scientist-leader who respected them as professionals—and who fully appreciated their ability to inform American citizens and influence public opinion.

He gave every indication that he welcomed—even enjoyed—most interviews and press conferences. He saw each as an opportunity to advance his team's program, along with the space quest in general. He seemed to have a natural desire to please people—reporters included. He was practically always enthusiastic and optimistic—"just naturally upbeat," as one observer put it.

An exception to those traits came into play around 1966-67, when cutbacks in NASA funding and personnel began—well before the triumph of Apollo 11 and its successor missions. Through the news media, he made sure the public and political leadership knew of his great disappointment and opposition to such retrenchment. But even during the national mourning of the fatal Apollo command module fire at Cape Canaveral in January 1967, he displayed a strong, forward-looking, let's-press-on attitude.

The famously voluble WvB had the reputation also of being a perfect gentleman in his dealings with journalists, male and female alike, and not one given to belittling or arguing with his questioners.

To top it all off, he had what Apollo 11 astronaut Mike Collins termed "a marvelous knack for explaining his machines in simple, understandable, human language."

My own observations of him as the consummate "Mediaman" stem largely from covering WvB & Company, to one extent or another, over a span of fifteen years in several journalistic capacities and as author of two books about him.

One role was as a reporter for *The Huntsville Times*, especially as its first full-time missile-space writer, in the exhilarating Project Apollo years of 1962-66. And beginning back in late 1957, about the time the Soviets launched the Space Age with their *Sputnik 1* satellite, I also served for more than a dozen years as the Huntsville area correspondent to technical and trade periodicals in New York. In the early '70s, in my role at *The Times* as an editorial page writer-editor, I continued to write about the rocket-*meister*. I authored a WvB biographical anecdotage published in 1972 in Germany and the 2005 biography "Dr. Space: The Life of WvB," published in the U.S., in two foreign editions, and in an online version.

All through the 1950s and beyond, WvB cultivated relationships, some close and long-lasting, with journalists and other communicators. The more notable included writer/space visionary Arthur C. Clarke, newscaster Walter Cronkite, entertainment wiz Walt Disney, writer-publisher Erik Bergaust, author and magazine writer-editor Cornelius "Connie" Ryan, and television's Hugh Downs, among others.

The charismatic, opportunistic WvB was known on occasion to take technologically challenged members of the press under his wing and mentor them on the intricacies and history of rocketry and space flight.

One especially noteworthy case was that of the Associated Press's Howard Benedict. In October 1957, the New York-based AP reporter just happened to catch the assignment to write a follow-up newsfeature story after the Soviet Union's *Sputnik* went aloft. Among the people he interviewed by phone around the country was a fellow named WvB.

After that, Benedict was AP's "expert" to cover any "space story" that came along, he recalled to me. The wire news service moved him permanently to Cape Canaveral in 1959. After an early news conference there, WvB discreetly offered to tutor the admittedly "befuddled" Benedict privately on the history and technology of missiles, rockets, satellites and such.

Benedict told me in 1998 that, aside from just wanting to be helpful, "WvB knew that one way to sell his ideas, to get people to start thinking about them, was through the popular press, especially those with worldwide circulation, like the Associated Press."

And so, over "a couple of beers" from time to time in WvB's motel room at Cocoa Beach, Fla., the two met for evening tutorials. Through the years that followed, Benedict managed to maintain his journalistic objectivity toward WvB and his pursuits, despite what the writer told me was "a lasting friendship" with "this great man." Benedict retired after 30 years as AP's man at the Cape—the dean of the press corps there.

In rare cases, WvB's dealings with the press didn't go nearly so well.

As the Moon landings of Apollo approached, a feisty Italian female journalist and author with a penchant for going for the jugular scheduled an interview with WvB in his offices at NASA's Marshall Space Flight Center. Ex-naval officer Bart Slattery, who was WvB's public affairs chief at Marshall, personally handled all requests for interviews with him, in coordination with WvB's longtime secretary, Bonnie Holmes. On this day, Slattery escorted the Italian lady into WvB's inner sanctum and took a seat there. The writer objected, asking Slattery to leave. He refused. She insisted that no one would *monitor* her interview. Slattery didn't budge.

The writer appealed to WvB, who as a rule had no objection to going solo with interviewers. But he didn't intercede in this case. Protesting all the way, the woman proceeded with her interview. She later wrote in detail about Slattery's intransigence in her book on the U.S. space program.

And then there was the cantankerous reporter for one of the Washington, D.C., newspapers. He cared little for WvB or anything to do with NASA. In interviews and at news conferences, he accentuated the negative. His caustic comments and writings earned him the reputation as the No. 1 sourpuss of the space press corps. Still, WvB never turned down his request for an interview.

In fact, he never declined to be interviewed by any American journalist, to the best of longtime secretary Bonnie Holmes' knowledge. He also welcomed most opportunities to speak with foreign reporters and television journalists, and honed his linguistic skills in advance of a number of such interviews so he could respond in their languages.

My first one-on-one, sit-down interview with the Marshall Center director, in the mid-'60s, was pretty much a disaster, through no fault of his. Public Affairs' News Branch chief Joe Jones, escorting me to WvB's office, offered the use of a tape recorder, which few reporters had in those days. I declined the offer, for reasons now lost to me.

WvB proceeded to talk nonstop at 90 miles an hour as I tried vainly to get it all down in my notebook. I missed half of what he said, including some highly newsworthy remarks about the need for Congress to make multi-year authorizations for funding the space program, as opposed to the uncertainty of year-to-year budgeting.

Before long I bought a tape recorder and never made that mistake again.

The charismatic WvB tended to be the star of the show among government officials and others at press conferences, whether in Washington, Houston or at Cape Canaveral. For one thing, he was the best known, even in the early days. The landmark series of cover stories about future space exploration in the popular *Collier's* magazine in the early and mid-1950s, and the equally high-profile, space-themed, Walt Disney television programs in that decade, had spread his fame.

At the news conference in the nation's capital after the Army Ballistic Missile Agency's (ABMA's) *Explorer 1* reached Earth orbit the night of Jan. 31, 1958, practically all the questions were directed to him. That was true despite the fact that the main men behind the trailblazing American satellite itself and its scientific payload, as well as key Department of Defense and Army officials, shared the same spotlight as the rocketeer from Redstone Arsenal.

Something of a reprise occurred exactly one year later, also in Washington. The occasion was the presentation by the Army of a mockup of the satellite to what is now the National Air and Space Museum of the Smithsonian Institution. With members of the press present, ceremonies began with brief remarks by Maj. Gen. John Bruce Medaris, ABMA's commander—and WvB's immediate boss—followed by Secretary of the Army William Brucker. WvB spoke next.

"When WvB stopped speaking, there was pandemonium," space historian Fred Ordway recalled years later in an interview with me. "People all rushed down to get his autograph. Secretary Brucker, who had also been governor of Michigan, looked at me and said with all seriousness, 'Nobody gives a [expletive deleted] about the secretary of the Army!'" (Note: Brucker and WvB remained on very good terms with each other.)

As usual, the Huntsville rocketeer garnered the most attention at the July 1969 news conference at the Cape after Saturn-Apollo 11 had lifted off toward the first manned landing on another celestial body. A journalist asked him how footprints on the Moon would rate as an historical event. WvB's answer: "Of about the same importance as when aquatic life first crawled onto land." It was the best quote of the day.

From the very beginning, WvB showed sound instincts where the news media were concerned. For an important, futuristic 1952 "Symposium on Space Travel" in New York, he prepared a paper espousing an aggressive, "great leap forward" approach to space exploration. Another presenter planned to take issue with him, advocating a slower, step-by-step program. Some thought that unwise, that it would create controversy and do damage to their common cause. But WvB argued that a sharp difference of opinion would be more newsworthy than having everyone sing the same song.

He was right. The event made headlines in the New York newspapers, and *Time* magazine also took note.

WvB later showed restraint when drawn into a dispute that some of his Marshall center officials had with *The Huntsville Times* and one of its reporters. Relying on a respected source at Marshall, the reporter had written a story about a coming reorganization there. The article ran before the center's employees were told about the changes and after the paper's publisher had supposedly taken that information in confidence. The reporter knew nothing of the latter.

One or more officials had proposed in a memo to WvB that the reporter and his paper be punished. The suggested plan was to pull the reporter's MSFC press visitor's badge, dry up his news sources, and see to it that all of the center's news releases, and those of local interest or impact emanating from NASA Headquarters, be issued at times of day favoring the morning competitor of the afternoon *Times*.

In the margin of his copy of the memo, WvB wrote his reaction to the "dry them up" proposal: "This doesn't seem to work too well. Now they [*The Times*] are getting their stories from Sen. [John] Sparkman or NASA Hqs." He signed the note with his trademark monogram "B."

The aftermath was that while the timing of a few subsequent news releases may have suspiciously favored the morning *Huntsville News*, no other known punishments were dealt *The Times*. The Marshall director's negative reaction had put a chill on retaliation against the reporter or the paper.

In a totally unrelated matter, WvB once indirectly sent word to *The Times* that the "v" in "von" should *always* be lower-case, never capitalized, even at the beginning of a sentence or paragraph. The paper was sorry but, well, it chose not to comply and violate standard style.

At the space leader's death in 1977 at the age of just 65, *The New York Times'* John Noble Wilford described WvB on Page 1 as "the master rocket builder and pioneer of space travel." In Wilford's full-page article inside, he noted: "Dr. WvB's name, perhaps more than any other, [is] synonymous with space travel."

London newspaper headlines termed him the "Moon Rocket King" and "Father of the American Space Age."

Best-selling author Tom Wolfe of "The Right Stuff" fame declared repeatedly in later years that, "WvB was the only philosopher NASA ever really had." Appearing one morning in autumn of 1998 on NBC Television's "Today" program, Wolfe elaborated:

> ...(He) *was* the philosopher, and he said: "The Importance of the space program is not surpassing the Soviets in space. The importance is to *build a bridge to the stars,* so that when the Sun dies, humanity will not die. The Sun is a star that's burning up, and when it finally burns up, there will be *no Earth, no Mars, no Jupiter.*" And he said, "You *have* to find a way, because humans are the only thinking creatures that we know of in the entire universe, and we have to build a bridge to *save this particular species.*" I think that is a grand thought, and it should be the thought that everybody has in supporting the space program.

WvB is remembered, certainly in the Rocket City and in this country, as the father of America's Moon landing program.

Part V

Reflections

What Was It Like to Grow Up with a Rocket Scientist?

Articles by Heidi Weber Collier, Margrit von Braun, Carol Foster, H. Christoph Stuhlinger, Uwe Hueter, and Mike Sneed

By Heidi Weber Collier

Heidi Weber Collier reminisces about growing up with parents Lissy and Fritz Weber, who was a member of the original German WvB rocket team.

Is it in my DNA? It seems so ironic that my father's and my career paths have mirrored each other. We both worked as U.S. Army civilians for over 13 years and both at Redstone Arsenal, AL. Fritz-Horst Weber's career as an American rocket scientist started with the U.S. Army in Ft Bliss, Texas, in 1948, and he continued to work for the Army as a civilian at Redstone Arsenal from 1950 to 1960 when the majority of the WvB team moved over to NASA's newly formed Marshall Space Flight Center adjacent to Redstone. I, Heidi Elizabeth Weber Collier, continue to work for the U.S. Army at Redstone Arsenal in training and leader development.

After World War II my parents, both German citizens, had entered the United States under Operation Paperclip. My father reached Ft. Bliss, Texas, in January 1948; my mother and sister followed in September of that year. So then it also came to pass that I, Heidi, arrived in July 1949.

The WvB team of approximately 120 rocket specialists had been brought to Texas starting in 1945 to educate Americans about rockets. In 1950, this group of rocket scientists and their families were moved to Huntsville, Alabama, to work on rocketry at Redstone Arsenal.

My childhood, looking back on it, was somewhat unique. Of course, at the time I felt I was just a regular kid growing up in America with my prime responsibility to perform well in school and be a dutiful daughter. Yes, I knew that my parents spoke with a German accent and that we seemed somewhat foreign to many in the Huntsville, Alabama, community. But mostly folks were very kind and respectful.

There was very little housing available in Huntsville in 1950, as hundreds of families began to move into the area. However, there was one small development of new, very modest two and three bedroom homes called Colonial Hills on Meridian Street where about twenty of the young German families purchased homes. With little or no credit, the families each contributed to come up with a lump sum of $5,000.00 to show adequate cash assets for securing a home loan. This single quantity was moved from one family's bank account to another's. The banks, of course, knew exactly what was going on but felt that the Germans could be trusted. My parents bought a little two-bedroom home on Delaware Circle. They were thrilled to be homeowners, yet quite uneasy to make such a commitment. They preferred to pay for things they needed in cash.

The German children played together, rode tricycles and bicycles around our safe little two circles of neighborhood. The Vowe twin girls, Gisela and Ursula, babysat my older sister and me on the rare occasion that our parents went to a movie or someone's house for a party. My best friend was Margot Beichel, but her family moved on to California in the mid 50s, and we lost touch for over fifty years but eventually rekindled a wonderful friendship.

In Huntsville in the early 1950s, there was no busing available to take the school age children to school and each family only had one car, which the dads needed to get to work. Several families came up with the unique solution of hiring a taxi that daily took five children, back and forth, to a few schools in downtown Huntsville.

By about 1955, the majority of the German families had either built homes on Monte Sano Mountain where a group of families bought 40 acres together and divided it up, or many also chose to live in the valley just below the mountain called Blossomwood. This is where my parents built a comfortable three-bedroom home in 1955, because a visit from my grandparents was imminent. They were coming by ship to visit from Germany for several months. Travel was not quite as efficient back as it is today. In the springtime, my sister and I stayed with the Pfaff family while our parents drove up to New York City to pick up Oma and Opa who, after their arrival in Huntsville, soon enjoyed the local hospitality and restaurants. My grandfather loved the dogwoods and redbuds in the spring and took long walks daily back and forth into town during their extended visit.

One particular memory from about this time is that my parents were planning to surprise my sister with a new bicycle for her birthday from Grants Five & Dime store in downtown Huntsville. My father had a coworker drop him off after work, since the plan was to purchase the bicycle and drive it through downtown and on to Blossomwood about two to three miles. What he didn't know is that they would sell him the bike in a box. So after closing time, Fritz was left on the sidewalk in downtown Huntsville with no tools and no way to get home. Of course, he did not have easy access to a phone to call home to notify our mother. Thus, well after dark and late in the evening, he finally came riding up at our house with my sister's birthday surprise.

Since there were so many of the German team members in our Blossomwood neighborhood my father started carpooling to work with four other gentlemen, among them were Dr Walter Haeussermann and Hans Hosenthien. Mr Hosenthien had baptized several of us German babies back at the Ft. Bliss, Texas, Main Chapel on 24 September 1949.

The German families continued to be very supportive of each other and helped celebrate special occasions and milestones. Generally, we children led a rather sheltered, easy going life. Our worlds revolved around a relatively small area environment that consisted of home, school, the local swimming pool, friends and birthday parties. I remember that when I was graduating from high school and again when I was getting married, my mother was told by several of the other German wives and mothers that we had better throw a very good party because they would be bringing a nice gift. My sign-in wedding book reads like a who's who of the German rocket team.

In sixth grade Iris, WvB's oldest daughter, and I were in my same class so our teacher arranged for WvB to come and speak to us. I happened to sit right across from the teacher's desk, which is where our illustrious speaker spent over an hour keeping a bunch of 12-year-olds spellbound. This professor of rocketry could inspire people of all ages!

In 1958, my parents had carefully saved enough money, and my father was finally cleared to travel to Germany to visit since they had not been back in ten years. My father would stay four weeks while my mother, sister and I would stay longer. We drove up to New York City from where we flew over on a long flight, eventually landing in Frankfurt and renting a small gray Opel Record automobile. There were not many cars on the German highways in those days, so this car albeit small was certainly noticed wherever we parked. First, we visited my mother's parents in Schleswig near the Danish border, and then my father's family who had been refuges after the war in the Heilbronn area of southern Germany. My parents left my sister and me with grandparents and aunts and uncles and cousins and took off to take a long delayed honeyMoon to Italy. What I have since found out from my uncle is that my father had to find a phone, usually in a bank or the village community center, and call in to the U.S. Army each day to verify that he had not been nabbed by the Russians. On Saturdays he had to appear in person to an Army civilian who was assigned to track him. Well, my mother had such a good time in Italy that she talked my father into staying over until Sunday instead of returning to Germany on Saturday as originally planned. My uncle tells the story of one very excited U.S. Army civilian dancing around his living room extremely distressed that my father had not returned as scheduled. Fortunately, this all worked out once my parents returned to Heilbronn on Sunday.

The highlight of our European vacation was to be a three-day visit to the World's Fair in Brussels, Belgium, Expo 58. Just the year before in 1957, the Russians had launched two satellites, and the U.S. finally responded with the WvB-team-built Redstone rocket carrying the Explorer I satellite into space in January 1958. Yes, the Atomium was the main attraction in the center of the World's Fair Park, yet my father dragged us into the Russian Pavilion several times. A replica of the Russian satellite was hanging from the ceiling, and what came next left a lasting impression on me; he grumbled, complained and muttered to himself that we could have easily beaten the Russians into space with a satellite.

Back home in Alabama in the 60s, the German families were starting to feel much more settled and started enjoying various outdoor activities. Since the Tennessee River is quite close to Huntsville, many of the families bought boats, water properties and built small, simple cabins. I spent a lot of time with my good friend Gerda Poppel at her family's place on the river at Pine Island. Our family held a very serious discussion, and each of us voted on whether we would like to take regular vacation trips or would prefer to invest in lake property. We all came to quick agreement to purchase a water lot that was in the Skyline Shores development. I have since found in my parent's papers that they actually borrowed the money for the property from Anni Lindner, the widow of Kurt Lindner who had died in 1960 of a heart attack.

Two other German families also purchased property in Skyline Shores. My sister and I traded off riding with Erich Neubert and his son in a sailboat, whereas two of his children enjoyed ski rides behind our motorboat.

Are you starting to notice the recurring theme? The Germans maintained a tight-knit community; we were family for each other since we didn't have grandparents or aunts and uncles nearby.

I moved away from Huntsville in 1967 to go off to college, got married, and yet returned for one year from 1971-72 to take care of my mom who was dying of cancer. Then I spent the next 23 years moving around the world with my U.S. Air Force officer husband. Eventually I did end up back in Huntsville, in 1998, to work for the U.S. Army at Redstone Arsenal. I found community again with my old German friends, started gathering email addresses to help set up a communication network and have thrived. But the best prize I have found in visiting with both the German and American rocket team members is the amazing amount of mutual respect they have for each other. So this helps explain how they managed to get to the Moon in such short order!

By Carol Foster

In my earliest memories of my father (Jay Foster), it is Sunday morning and my sister and I are running into our parents' room to "play Daddy," romping and tickling and laughing a lot, then settling in while he reads us the funnies, one daughter on each side so we can see the pictures.

For my fifth birthday, Dad gave me what became a favorite stuffed toy, a long-haired white kitten with felt flowers on its collar and its very own red woven basket. That was the beginning of my cat collection, when the new kitten joined Anda Sue, a squeaky toy cat that I had been attached to for a long time already. A few years later, searching for the perfect Christmas gift for Dad on a low budget, I wrapped up my beloved Anda, with her pink rubber bow and the tip of her tail chewed off, and gave her to Dad. Luckily, he let me borrow her whenever I wanted.

It was always pretty much assumed that Sherry and I would go to college and be scientists of some sort. There was never any question of girls' abilities; it didn't come up. Sherry excelled in biology, and I was drawn to math. Numbers were much simpler to me than literature or history, where answers and expectations often seemed vague and subjective.

Dad is definitely a science guy—orderly, organized, practical. He had a pegboard in his workshop, and everything had its place. He built desks for Sherry and me out of boards hung from brackets like bookshelves, painted turquoise to match the trim on our dressers and topped with glass for a smooth surface. On the wall above my desk were pictures, including my kindergarten diploma; a photo of Ham, the first chimp in space; and a photo of WvB, signed "To Sherry and Carol." In a house where we lived later, we had a 7-foot pool table in the garage, and Dad made a ping pong table out of plywood to fit on top. He hung some ropes with a pulley above the pool table so we could crank the ping pong table up and down. That garage became a frequent hangout for our teen crowd.

Dad loves the water, and in his younger days he was an enthusiastic water skier. We spent countless weekends camped out by Guntersville Lake, swimming, boating, and "dragging him around the lake," as he would say. He was the athlete of the family and had a set style of teaching others to ski: Line up your skis with the tips sticking out of the water, rope in between the skis, then give the signal to the driver to go rapidly from zero to 25 miles per hour. That way the skier doesn't have to push all that water in front at slow speeds. Soon, ideally, she is skimming along the surface, enjoying the breeze. Wiping out at that early stage and at that speed, however, was not fun, and I had bruises to prove it. We had a pair of long, wide beginner skis, red plaid, with a lot of surface area to aid flotation. My little legs and ankles could barely hold them upright. I did not learn to ski until I was older and living on my own. I went out with friends who advocated starting the boat slowly, with the skier pushing the weight of the water while getting the feel of the skis as long as she liked, and only then giving the signal to speed up. I had never thought of trying another way before that—Dad is so often right in his theories and calculations, and he appears very confident. But there was one time…

One day our neighbors got a fancy new clock—these were some of the neighbors who shared myriad practical jokes with my family. It had started on a Sunday when we came back from the lake to find a large, professionally painted sign on our porch railing, "Hamburgers To Go, 5 for $1." Next to the doorbell was another notice, "Ring bell and walk in." We were confused, then amused. No one rang the bell, but many other pranks followed. On the day of the clock, our friends showed us their new timepiece, which had a small metal arm with a chain at the end connected to a small gold ball. The arm would periodically shift to one side and the chain would wrap and then unwrap itself around a post at one side of the clock. Then the arm would swing over to the other side of the clock where there was another post, and the wrapping and unwrapping would be repeated, the chain forming a candy cane pattern on the post. We were mesmerized. A day or two later the subject came up over dinner and we discovered that Dad—he who is rarely wrong—thought there were two balls and two chains. He was so sure that he was right even though Mom, Sherry, and I all swore there was only one ball—okay, two posts, but only one ball and one chain—that he bet the house, the car, the boat, everything he owned and everything he would ever own. We raced down the hill to our neighbors' house, pounded on the door, then rushed past them. They did not know what to think as the four of us stared intently at the clock, at the one arm with the one chain winding and unwinding, over and over again.

Of course NASA and space were the weft and warp of our lives and the lives of much of the community. Sherry and I watched Gemini and then Apollo launches on television sets on stands rolled into our school classrooms. There was no U.S. Space & Rocket Center back then, so we took out-of-town visitors on a tour, with Dad as tour guide. We went to see the Saturn V test stand, the Redstone and some smaller rockets lined up in order of height, like nesting dolls; the neutral buoyancy tank where the astronauts trained—possibly my favorite stop on the tour; and Building 4200, where Dad and WvB had offices on the ninth floor. There were exhibits on the first floor of the building—I was fascinated by a model of the LRV (Lunar Roving Vehicle) rolling slowly across a simulated lunar surface well before the real LRV navigated the real Moon.

Occasionally Sherry and I would go to the office with Dad if he had to work on a weekend—he put in a lot of overtime in the early days. He had one of Sherry's school art projects, a clay ashtray, on his desk until he retired. It was supposed to be solid orange but the glaze must have been uneven, so it came out of the kiln a splotchy orange and white. It looked good, though. I liked it, and I love that Dad kept it on his desk all those years. He still has it on his desk in his home office, filled with paper clips.

Dad went to Florida for many of the launches, and we went as a family to see Apollo VIII and Apollo XI lift off. I especially remember Apollo XI and the crowd gathered at Cocoa Beach to watch the big event, the result of so much intense work and planning. Dad was high enough in the organization by then that we were invited to the VIP debriefing the evening before the launch. Former President Lyndon Johnson and some astronauts were there, up front in a roped-off celebrity section. The next day, from our vantage point in the viewing stand, I was struck by the fragility of the rocket, tiny from a distance, rising against the vastness of the summer sky, dropping one stage, then another, disappearing into the clouds on its almost unimaginable journey.

Also on the beach that day was a small group of protestors led by Rev. Ralph Abernathy, the "civil rights twin," as he was sometimes known, of Dr. Martin Luther King Jr., who had been assassinated the year before. Some of the group held signs, and Rev. Abernathy spoke briefly about the amount of money spent to go to the Moon while so many were suffering in poverty here on Earth. I don't think I had thought about it that way before. I was 13 years old. When the countdown came, the group turned and watched silently. They were very respectful. Later, Dad, who had been snacking on mixed nuts, made his way through the group to shake hands with Rev. Abernathy. I was impressed and slightly embarrassed. Dad offered the reverend some mixed nuts, which he declined. That night, back in the motel room, we watched news coverage of the launch. The crowd of protestors looked much bigger, maybe due to the angle and location of the camera. I was learning a lot.

When Neil Armstrong and Buzz Aldrin stepped onto the Moon, we were at a Foster family reunion in my Great-Aunt Reba's farmhouse in Bloomfield, Indiana, watching on TV. Dad was the hero of the moment, not surprisingly. He always brought packets of NASA information to give out to the various cousins, many of whom gathered that evening to view the grainy images, the only time we ever watched TV at a family reunion. Dad's parents were there, too. They had always been very proud, doting on him, their only child. When we went to the Museum of Science and Industry in Chicago, for example, my grandfather would chat with the guards, telling them about his son who worked for the space program. As we drove back home to Huntsville, we probably sang folk songs, as we often did when traveling. We each had a solo in *Puff the Magic Dragon*.

Growing up, I didn't usually think of my father as a rocket scientist—he was just Dad, who could be fun-loving or serious, who knew a lot about a lot of things, and who happened to work for NASA, as did a lot of people in our town. But I guess there weren't too many dads who would linger outside on a warm evening, look up into the sky and say, "There's a beautiful target out tonight."

(Carol Foster is the daughter of Jay and Betty Foster. She has a PhD in Computer Science and works for the Computer Science Research Department at the University of Massachusetts, Amherst, Mass.)

By Margrit von Braun

When I left for UCLA as an undergraduate, my dad suggested I buy a street atlas when I landed at LAX. "It's a big city," he said somewhat somberly. I realize now that, like most parents, he was worried; but at the time all I felt was his confidence, his trust, and the high expectations he had for my education. And that I should not, could not disappoint him. Two years later he sent me a telegram that I had been accepted at Georgia Tech, where I graduated in 1974, using a slide rule (the new calculators were too expensive). It's one of two telegrams I have in a scrapbook. The other one he sent to his parents on the day I was born.

I still feel his pride, confidence and inspiration to do well. And I learned from him, that to do well, you must have a supportive team. Our team was my family and my childhood was in Huntsville, Ala.

Now that the statute of limitations has expired, I can confess that he trained my team of silly teenage girls to roll yards on Big Cove Road. As children of Rocket Scientists, we had to understand the aerodynamics and physics of the activity to get it right. Wernher taught us how to roll the toilet paper around a small rock to improve the aerodynamics so we could get the streamers higher into the trees. He was not advocating delinquency; this was a teachable moment.

There was early Astronaut Training, too. We attended Randolph School, originally in an antebellum home. One Sunday night, my sister Iris realized she had forgotten her homework. The solution was obvious. Wernher drove us to the school, launched Iris, then me, through the back window (it *was* open), we grabbed the goods and he drove us home. Anything for our education.

He used a similar technique for our house on Big Cove Road. When we were accidentally locked out of the house—no, there was not a hidden key—it was another opportunity for a Launch of the Kids! He would hoist me through the bathroom window in the front. I'd land on the toilet seat, and go open the front door for my waiting family.

Another part of our astronaut training was food related. One night he brought home colored pellets that looked like dog food. We kids immediately judged them as inadequate for their intended purpose—astronaut food—but quite suitable for our mongrel puppy.

He taught me to balance a checkbook. I clearly remember him comparing his meticulous balance to that of the bank and then rectifying with a line that says,

"Bank says add 3 cents." For some reason, I loved helping him pay bills. I would fill out the checks and he would sign them. In the '60s, we had trouble balancing the checkbook because people would save the checks for his autograph.

Our only big car trips were in the 1950s to Cocoa Beach. After that, he was creative at substituting family car vacations with a way he could fly. One time he arranged to meet our mother and us in Toronto, Canada. Maria was to drive the Oldsmobile Delta 88, and he would fly to Toronto from a nearby business trip. I am not sure how he pulled that off: "You drive 800 miles with the three kids and I will meet you there"? As we were leaving Huntsville, the Oldsmobile broke down and the dealer gave our mother a huge, fancy Cadillac to take as a loaner. During the long car trip we schemed and plotted. When we finally met dad in Canada, we told him the sad tale about the broken down car and pointed to a rusty VW bug "our loaner." The guilt began to overtake him—and probably some fear of his wife's coming retribution. When we could no longer contain our laughter we showed him the Cadillac.

In the 1960s, he drove a Mercedes roadster, a flashy two-seater with red leather seats and a tiny bench in the back. One night he came home proudly announcing that a policeman had pulled him over for speeding. Upon seeing the "Dr." on his driver's license, the policeman apologized for interrupting what was certainly a race to do surgery at Huntsville Hospital. He liked to speed. He was often late to work. He didn't think anything important ever happened in the morning.

Although he travelled on business almost every week, he also took trips with us, individually, just one kid at a time; those were the best. At Christmas 1969, just a few months after the Moon landing, he and Iris sailed through a hurricane in the Bahamas. He took my brother, Pete, on numerous hunting and fishing adventures in Alaska and Canada. I was recruited as his dive buddy. He knew he wouldn't make it to outer space himself, but found scuba diving allowed him to experience another world. In 1967, the two of us travelled to Cozumel, Mexico for a week of sleeping in hammocks and diving.

The race to space was busy but not *that* busy. Sundays were Family Day—a whole day spent swimming, water skiing and fishing in Guntersville, the closest body of water. We had to get special permission to include friends. Our 1950s boat was a small wooden Chris Craft, named Orion, his favorite constellation. It was replaced with a houseboat, Surfside Six, named after the 1960's TV show, and a speedboat, called AOK, borrowing Alan Shepard's phrase.

At the lake, the briefcase stayed home. This was probably the best thing I learned from my dad. When we were with him, we were the center of his universe. He told me once that the good stuff happened only after I was born. But I think he made everyone feel that way.

His other family was the NASA team he led. As immigrants, they were welcomed in Alabama and becoming U.S. citizens was often recounted as their happiest day. The team had a high bar—250,000 miles high. To get to the Moon took determination and an expectation of consistent, excellent performance. There were not A's for effort. I don't remember them ever talking about failure.

I know they had what-if scenarios, tests and backup plans, but the focus was on success.

The famous picture of The Rocket Team at White Sands, New Mexico does not include the team of women behind the scenes—wives, sisters, daughters, granddaughters, staff, scientists who were connected to the men in the picture. They could not have made it to the Moon without them. For example, our mother, Maria, is still the strong, independent person she always was. When we travel together, she insists on being at the airport three hours before the flight. I think she was a big reason Apollo 11 landed early, in July 1969, instead of December, the "end of the decade" deadline established by President Kennedy.

My dad was open-minded, funny, focused, determined, curious, and a consummate team builder. When I look at the Apollo 8 image of Earth Rise, I feel his eternal optimism. His spirit lives on in everyone who is an explorer—whether they are exploring culture, religion, art, science or outer space. His spirit is alive in every person who wonders how to reach beyond perceived limitations.

That enduring spirit of exploration is in all of us. His dream was that space exploration could bring the people of all nations together in the peaceful pursuit of knowledge and wisdom.

On to Mars!

Editor's Note: Dr. Margrit von Braun, the daughter of WvB, is an environmental engineer with degrees from Georgia Institute of Technology (B.S.), Washington State University (PhD), and University of Idaho (MCE), where she taught and served as dean of the College of Graduate Studies.

By Mike Sneed

My childhood was probably not unlike other kids growing up in the 1960's. We were focused on school, summer vacations, and sports. The biggest difference was growing up in Huntsville, Ala., the home of NASA's Marshall Space Flight Center (MSFC). There was not as much an appreciation of what was being developed at MSFC, as the fact that rockets were being designed and tested. It seemed like the whole town was full of engineers working on the space program.

From the Redstone rocket (which carried Alan Shepard and Gus Grissom) to the Saturn family of rockets, there was a constant air of excitement about engine testing at the center. I remember the dining room chandelier in our home rocking back and forth during engine tests of the Saturn V F-1 engines. I did not have an appreciation of what Dad specifically did at work, other than he was working on a rocket that would put men on the Moon.

The fact that we grew up in the heyday of America's space program was a way of life in Huntsville. Our parents went to work, like other parents, but they seemed in go in a little early and stayed a little later. It was not unusual to see my Dad come to my early evening baseball games, removing his tie as he got out of the car. He might miss the early innings of the game, but he was always there. The opportunity to attend open houses at MSFC and see what was going on only added to the interest in the space program. The opportunity to visit WvB's office and see a model of the Saturn V rocket was not something every kid could tell his friends about. The fact that ceiling panels had to be removed in order to accommodate the model really made an impression!

Going to school in Huntsville during this time was also not unlike other schools across the country. From the first launches of America's manned space program, television sets were turned on for the class to see history being made. There was a high level of interest starting from the Redstone launches through the Saturn V / Apollo Moon launches that seemed to grasp everyone's attention. I guess because space travel was so new there was excitement that seemed to wane after the initial Moon landing in July 1969. It seemed that once the country achieved its goal of placing "man on the Moon by the end of the decade", and the later launches seemed to be just another space trip for the astronauts.

My career came full circle in 1989 when I had an opportunity to return to Huntsville and work for Rockwell International on the Shuttle program. Huntsville was the last place I thought I would ever return to for a job. I received my degree from the University of Tennessee in marketing and business, and without an engineering background Huntsville and NASA were never on my radar. That changed when Mr. Ed Nicks offered me a job working business development with Rockwell on the Shuttle program and other advanced space transportation programs coming out of MSFC. Ed stated that he wanted someone that had "big picture" vision versus a more technical nuts and bolts approach for his marketing person. I can't thank him enough for the chance to get back home and complete a career in the space business.

By Uwe Hueter

Moving to Huntsville, Alabama

After living in El Paso, Texas from 1947 until 1950, WvB and his team of rocket scientists were moved to Redstone Arsenal in Huntsville, Ala. My father, Hans Hueter, and mother, Ruth Hueter, packed up their belongings as well as their three children (sons Eike and Uwe and daughter Wendula) and loaded them into an old Nash Ambassador that had the distinction of fully reclining seats, allowing it to become a bed-on-wheels for me and my brother. It was a hot, long drive across the never-ending state of Texas and then the Deep South. When we finally reached Alabama, my parents decided to stop for supper. The first thing my dad ordered was a cold beer after a long day on the road, only to be informed that alcoholic beverages were not sold on Sundays in Alabama. Welcome to Alabama and a new way of life!

When we arrived in Huntsville, there was a severe housing shortage from the large influx of Germans, contractors, and Army personnel. The city at that time had less than 15,000 residents, and the city limits were much constrained compared to the Huntsville of today. After moving several times within Huntsville, we finally landed in a rental house on Locust Avenue. Even as I cycled through several schools in the Huntsville public school system, I never felt like a stranger in a new environment. I knew we were different from the typical Southern kids, but we felt accepted. Every day was a new adventure.

After settling in the rental house on Locust Avenue, my parents decided to build their permanent home on property they had acquired prior to arriving in Huntsville. Before the move, several members of the WvB team came to Huntsville to scope out the surroundings. They decided to buy a parcel of land on the north side of Locust Avenue, between Lily and Owens Drives, in what would become the entire block bordering McClung and Locust Avenues. The land cost approximately $1,500 for a lot and a half. In those early years, the neighborhood evolved into a brain-trust of sorts, composed of the many members of the German rocket team: Dr. Wernher von Braun, Dr. Hans Gruene, Gerhard Drawe, Bernhard Tessman, my father, Karl Heimburg, and Dr. Martin Schilling. The Locust Avenue block later became known as Sauerkrauthügel ("Sauerkraut Hill"), after the November 10, 1958 article referencing it as such in *Bunte Illustrierte*, a popular German magazine.

We moved into our new 3-bedroom, 1-bath home in 1951. The lack of both air conditioning and a garage didn't deter my parents, as they acclimated to a new life in the United States. In fact, my dad loved the heat, commenting that he must have been reincarnated from a previous life in Africa. The house at 1409 Locust Avenue signaled the end of the gypsy lifestyle of our first years in the United States and in Huntsville, as we planted our feet firmly in Huntsville soil.

On Being Raised German in the Deep South

Family Life

The rules of the house, I'm sure, varied from family to family during those early days, although our collective German backgrounds almost promised a more formal approach to parenting than many other Southern families. In the Hueter home, several rules remained constant around mealtime: Dinner was eaten together, German was spoken, no one sat before mother was seated, and you ate what was put before you. On rare occasions, we had to forfeit supper due to 'unacceptable behavior'. My mother usually cooked German food, but sometimes we were treated to a Mexican meal, a habit that followed us east from our time in El Paso. We never ate Southern cooking, why I'm not sure but I imagine that since my mother didn't have that many southern American friends, she was simply not around the deep-fried goodness of chicken or French fries.

Classical music was preferred in our home. My brother Eike's lifelong love of classical music was probably borne from those days, listening to such famous composers as Mozart, Hyden, and Beethoven. Occasionally, my father would introduce new music, such as the Kingston Trio or Sons of the Pioneers. My sister and I rebelled against authority and embraced the likes of Elvis Presley, Fats Domino, Chuck Berry, Bill Haley and the Comets.

Weekend time was chore time for the kids; I helped my dad cut the grass or repair things around the house. The value of taking care of and holding onto your things, now it would be considered 'green' but back then it was just a way of life. Only things that absolutely couldn't be fixed would be thrown away. Economics drove my father to build furniture and stone walls, as well as sprig the entire backyard in centipede grass. This mindset became one of the most important life lessons my father passed on to me.

Saturday morning German lessons also filled my weekend time. Apparently, my parents felt that, despite the 24/7 German being spoken at home, I was not progressing well enough, so I required weekly tutoring. Every Saturday morning, I rode my bike several miles to Frau Hertha Heller's for German lessons. Needless to say, it was not the way I wanted to spend my Saturdays.

The Social Scene

When the families settled into life in Huntsville, social groups formed around common interests such as card, playing ("Skat", the national card game of Germany), fishing, boating, and the arts, or by neighborhoods. Most of these groups consisted primarily of Germans, with an occasional American.

Thanksgiving and Christmas were the major holidays my family celebrated. The Thanksgiving holiday was always spent with our next-door neighbors, Ruth and Karl Heimburg, as well as their three children, Klaus, Stephan and Ruth. Our families' rotated houses each year, and it was always a really nice event, with outstanding food and good company.

Christmas was celebrated with my family on Christmas Eve rather than Christmas day. My parents made us take a nap during the late afternoon while 'Santa Claus' came. Our tree was always a spruce and instead of Christmas lights on it, we lit real beeswax candles. The festivities began around 6:00 or 7:00 p.m., when we came out of our bedrooms to the living room where Stille Nacht (Silent Night) was playing on the record player. The candles were lit and we began the celebration. Christmas morning was spent socializing with the neighbors, mostly hosted by the Heimburgs. There was plenty of food and drinks, and the celebration normally continued well into the afternoon.

As my German buddies and I became teenagers, we started having more opportunities outside the German community and school that broadened our horizons. One such activity was involvement with St. Mark's Lutheran Church and their sponsored Boy Scout troop. I attended church, Luther League (a Sunday evening activity for young adults), catechism classes and then confirmation and involved myself in various scouting activities (camping and hiking).

In general, the German children tended to mingle more with the surrounding community than our parents, mainly due to school and the surrounding neighborhoods being filled with non-Germans. Although a large percentage of time was spent playing with our German friends, we always could find time for sandlot baseball and football with our American friends. These were new sports to the German community, and our parents never embraced or discussed them. In my family, it was education and hard work that were rewarded, not the ability to hit a home run.

Around the age of 14, even though my friends and I had migrated upwards from the Boy Scouts to the Explorers, we soon let that slip because dating and socializing moved quickly to the fore. We started throwing parties at our homes, where you'd find a group of both American and German kids, with the Germans outnumbering Americans only by a small margin. During one party at my house, I told my parents that it was customary in the United States for the parents to leave while the party was going on. Not knowing that this was a complete fabrication, they went to a neighbor's house. Being teenagers and, so, our own worst enemies, we lost control of the festivities, the party got loud and rowdy, and back home our parents rushed to bust us all. Our ruse was discovered, and we never got away with that again.

During high school, cars, dating, drive-in theaters, football games, dances, and, of course, school were the things our world revolved around. We were no different from American kids except when we were home and speaking German, listening to my parents' classical music, and enduring other rituals brought over from our homeland.

My father died in September of 1970, after a 3-year battle with colon cancer. My mother spent the rest of her (long) life in the same house I grew up in. She often traveled back to Germany in summers for extended periods, staying with Wendula in Munich, and at home enjoyed daily laps in her pool, the symphony with her friends, and knitting for her children and grandchildren.

By H. Christoph Stuhlinger

My father, Dr. Ernst Shuhlinger, has always been an excellent driver, and he is also a very good teacher. I recall how he taught me to drive, and it started long before I got my driver's license. For a long time I looked forward to the day when I would finally be able to drive a car myself. I often watched how others drove and learned how they manipulated the different controls.

My father frequently had to visit his office at NASA in Redstone Arsenal during the weekend, and usually I was happy to go along because I knew that I would get to "drive". I must have been in my early teens. Once we entered the arsenal, he would stop the car and I would scoot over and sit on his lap. Then I would get to steer the car (usually his VW Beetle) for the five or six miles to his office. My father worked for several hours, and I worked on my schoolwork or played with his automated slide show. Then I couldn't wait until it was time to go home again because I knew I would get to "drive" again. Eventually, I even learned how to shift gears while he stepped on the clutch. Initially, that is how he taught me the basics of driving.

As my 16th birthday approached, my father took me to the observatory grounds in Monte Sano State Park (about one mile from our house in Huntsville), where there is a long driveway. There he let me sit in the driver's seat by myself, and I learned to operate the clutch and shift gears, and steer all at the same time! His encouraging words helped me to quickly master the art of driving a stick shift, and when the big day arrived for my driver's license test, I passed easily, thanks to his lessons!

But I will always remember the times I got to "drive" on the arsenal. And I remember very clearly the moment I sat in the driver's seat at the observatory, with him sitting next to me, and actually drove the car for the first time.

In 1977, my father arranged a special trip for us. He was going to visit Germany that summer, and so was I, but at a different time. We scheduled our trips so that we could meet in Iceland and spend several days driving around to explore the country. He was returning from Germany, and I was headed over to Germany. I flew Icelandic Airlines to Keflavik, which has a military base as the airport, and took the bus to Reykjavik. He met me there, and much to my pleasure and surprise, he had rented a genuine Land Rover 4WD! We spent the next several days visiting many picturesque villages, seeing the glaciers and waterfalls, and driving on both improved and unimproved roads. I am glad that I was able to share that wonderful experience with my father!

During the early years of the space program, our family of five made several trips to Cape Canaveral (Kennedy) to watch the Apollo launches. My father usually had a NASA pass so that we could drive closer to the launch pad than most of the public. I remember well the very loud thundering rumble as the mighty Saturn V blasted off into space. Later, we also made a trip to see one of the Space Shuttle launches. The difference in how the two huge vehicles lifted off was fascinating. The Saturn V lifted off very slowly as it built up thrust and accelerated towards orbit, while the Shuttle, with its solid rocket boosters, literally jumped off the launch pad and was gone in a matter of seconds.

Before WWII, my father owned a BMW motorcycle with a sidecar. Unfortunately, he had to give it up when he was drafted into the German Army, but his affection for those motorcycles never went away. He dreamed of the day when he might have another chance to ride on a BMW. Whenever he saw a BMW motorcycle on the road, no matter what model year, he always pointed it out to us.

As my father's 90th birthday was approaching, I wanted to do something special for him. Then I came up with the idea of finding someone with a vintage BMW to give my father a short ride, just for old time's sake. How in the world would I find someone relatively close to Huntsville? I did a little research on the internet, and found out that there was an Alabama BMW Club. And believe it or not, the president of the club at that time lived in Huntsville! I contacted him and told him about my father's fondness for BMWs. He knew of someone in Huntsville who owned an older BMW motorcycle with a sidecar. The owner immediately agreed to help out with this project.

Originally, I wanted the BMW owner to come up to our house and surprise my father as he drove into the driveway. Instead, our whole family, who was in Huntsville for the big birthday, went to the owner's home to see the motorcycle. It was a white 1957 model, and seemed to be in excellent condition. A big smile came over my father's face as he inspected the motorcycle, and the owner let him sit on it. Much to our surprise, he actually started the engine, and before we knew it, my father was riding down the road! He drove around the block a few times, and then it was my turn to drive. A highlight for me was when I drove, and my father sat in the sidecar, and we drove around the block together two or three times.

My father commented later that his instinct for driving with the sidecar, which requires extra attention when turning, came back immediately. The 70-year lapse since he had last driven with a sidecar did not matter at all. I was very happy to have been able to arrange this thrill for my father's birthday.

The Matterhorn in Switzerland was his favorite mountain. He rarely missed a chance to visit the Matterhorn during his frequent trips to Germany. He climbed to the top of the Matterhorn twice (with a guide). At NASA, his office was on the top (9th) floor of Building 4200. To help train for his mountain climbing, he always took the stairs to his office, often several times per day.

Sometimes, when our family traveled together in the U.S. and Europe, we would hike in the mountains. It happened once or twice that while we were resting, all of a sudden my father would be gone. Many hours later, he would finally reappear. He had quickly climbed a nearby mountain without telling anyone of his plans. And there were many other mountains he had climbed during his lifetime.

My father told me several months before he died that he wished he had more time to travel. Our family traveled together often. We camped in many national parks, visited relatives in Germany, traveled around Europe, and made about 50 visits to one of his and our favorite spots—Sanibel Island in Florida. In Sanibel, he spent many hours on the beach collecting seashells and watching the birds and alligators in the wildlife refuge. He visited Africa, South America, even Antarctica. But there were so many more destinations he wanted to visit....

My father often made the point of how important it was for the German rocket team to work as one unit with the Americans. In fact, the collaboration is what made America's space program so successful during the Apollo years.

There were quite a few Americans that my father worked with closely during much of his NASA career. There was much mutual respect, and each gained from their work together. My father had numerous good American friends within NASA, the overall space program, and the community. Some that I can remember are George Bucher, Chuck Lundquist, Fred Ordway, Dave Christensen, Brooks Moore, Jim Downey, Ed Buckbee, Bob Ward, Ed Stone, Don Thomas and Leland Belew.

Reflections on a Career with NASA

By Thomas J. "Jack" Lee

Editor's Note: The career path of an engineer at the Marshall Center during the Apollo Saturn era was challenging, exciting and rewarding. Young engineers were provided the opportunity to engage with experienced engineers and project managers, providing them the opportunity to become involved in highly technical projects. This enabled the team to grow their own future managers, and keep the culture alive. The following is the story of one of those engineers who learned the culture, contributed to its growth and applied what he learned to become the center director.

The route that lead me through a most interesting and rewarding career with NASA can in some measure be credited to working with and for some of the most competent government and industry aerospace managers, scientists, and engineers, along with timely decision making and recognizing opportunities.

I was scheduled to graduate from the University of Alabama in January 1958 with a degree in Bachelor of Science in Aeronautical Engineering. Even though Sputnik had just been launched, I had no thoughts about a career in the space business, only in joining some airplane company. I soon found the airplane industry was saturated with engineers, and none of the offers I received was attractive. This led me to making the first major decision related to my career, which led to my involvement in the Manned Lunar Apollo/Saturn Program and a path to becoming the director of the Marshall Space Flight Center.

It began in the fall of 1957. My roommate at the university, Larry Wear, a mechanical engineer major, convinced me to accompany him for an interview to Redstone Arsenal where he had spent the previous summer as a student trainee. His selling point was the WvB team was really doing some interesting, yet classified, work at the Army Ballistics Missile Agency (ABMA). Reluctantly I agreed, recognizing it would be temporary until something better came along. I had always wanted to become a pilot, and one of the options for something better was to join the Navy and become a jet jockey. As it turns out, that trip to Huntsville was the best career decision I ever made. The desire to be a pilot would come some 14 years later when I received my private pilot license.

I began work at Army Ballistics Missile Agency (ABMA) on the third of February 1958, just three days after launch of Explorer 1. I was assigned to the Launch and Handling Laboratory, one of nine Laboratories under the direction of WvB. When MSFC was made a field center and the WvB team became the core of the center, the Light and Medium Office was formed under the direction of Hans Heuter. I was given an opportunity to join the office and was assigned to the Centaur Resident Office in San Diego, California, where I was able to expand my knowledge of space systems and to learn more about program management. I had previously decided this career path seemed to be best suited for me. The job brought me in daily contact with pioneers in space system development, the likes of Kraft Ehricke, the Centaur program manager at the time and Karel Bossart, the father of the Atlas, along with others who were instrumental in developing the country's ballistic missiles. They shared numerous stories about the Atlas development program, including when they had five failures in a row commonly referred to as "the bloody five". One of my primary functions was to work with payload developers on the integration of the Surveyor Lunar Lander Project. The mission was to sample the lunar surface, which would later to be used in support of the Apollo program.

The Centaur was the world's first high energy LOX/LH2 upper stage, utilizing the RL-10 engines and propellant tank insulation panels deployable during ascent to minimize propellant boil off. The structure was patterned after the Atlas thin walled stainless steel pressure propellant tanks. The first launch failed in May 1962, which brought about a congressional investigation. The committee's findings and recommendations resulted in a complete restructuring of the Centaur Project, and ultimately, the transfer of the project from Marshall to the Lewis Research Center in Cleveland, Ohio. The failure did not dampen the project sense of humor, however. Shortly after the launch, there was a cartoon in the hallway leading to the Resident Office, which read, "I launched a Centaur into the air and it fell to earth everywhere." The project transfer to Lewis included reassigning the NASA representatives in the Resident Office at General Dynamics to the newly established Lewis Centaur Project in Cleveland. One positive thing that came out of the transition process was that I had a chance to meet Lee James, Saturn 1/1B program manager, who I met on a number of occasions to familiarize him on details of the Centaur Project. Our paths would cross again.

After eight months with Lewis, I was pleasantly surprised to receive a phone call from Bob Pace at Marshall who offered me the resident manager's position on the Pegasus Meteoroid Detection Satellite Project. The decision came easy, when factoring in the positives of returning to MSFC and gaining management experience vs. climate change and the uncertainty of career opportunities at Lewis.

The meteoroid measurement effort originally resided in the MSFC Research Projects Laboratory under the direction of Dr. William Johnson, and later reassigned to the Saturn 1/1B Program Office, where Dr. Johnson became the Pegasus project office manager.

The Pegasus satellites were initially designated as secondary missions to be flown on the last three Saturn 1 launch vehicles, and as secondary missions, the project organization was staffed with very few personnel as compared to other MSFC projects. Fortunately, even though they were few in number, the competence, experience, and dedication of engineers, both NASA and contractors, assigned to the project were, in my opinion, well above average in all the required technical disciplines.

Originally, the Saturn 1 vehicle development program required 10 flights; however, the first seven flights proved to be successful in satisfying the planned objectives, which raised the question on the need for the last three flights. In one case, a Washington, D.C. newspaper printed an article that stated the last three flight tests "was an exercise in futility". This prompted NASA to change the Pegasus Project to be a primary mission. One of the fall-outs of the decision was a deluge of MSFC laboratory engineers, who had not previously been involved in the Pegasus development, to initiate technical reviews of the satellite design and test program.

By this time, we had made considerable progress and began building flight hardware, and in some cases were well into component qualification. Even though the satellite was being built and qualified to the specifications and requirements imposed on the contractor, the review teams findings led some of the MSFC laboratory directors to propose that the project be moved in house to MSFC and to increase the funding considerably, both of which would have a major impact on the program. It was obvious these recommendations would bring about a confrontation between the project office and the laboratory directors that would require the decision to be made by WvB himself.

A meeting was scheduled with WvB for each side to present their respective positions. Lee James led the project presentation, and even though he was well prepared with facts justifying staying the present course, I personally had some reservations about whether WvB would make a decision contrary to the desires of some of the laboratories. The meeting was held in the 10th floor conference room with the project representatives on one side of the table and the laboratory representatives on the other. The primary issue was whether the spacecraft components had been designed and qualified to MSFC requirements. One example presented was the vibration requirements for the electronic boxes. The project had qualified each box to withstand 10 Gs. The labs believed it should be 25 Gs. After hearing this, WvB, the master at asking the key question at the right time, asked if during the first seven flights of the Saturn 1 were there any vibration measurements made in the payload area. Without hesitation, someone in the back of the room answered "yes" and the maximum level recorded was 3Gs. After the meeting, the decision was made to stay with the project office recommendations, and; ultimately all three Pegasus missions were totally successful and exceeded their design life on orbit. As I look back, the Pegasus project was one of the most rewarding and satisfying programs in my career. I was able to be directly involved in all aspects of the design, development, test and launch operations.

My involvement in the launch operations is particularly noteworthy because it prepared me to become directly involved in the Apollo Saturn Program. During pre-launch and launch operations of the three satellites in 1964 and 1965, the highest priority for the work force at KSC was the preparation for the future manned lunar Apollo program. Therefore, the decision was made to deviate from the way it had been done in the past. The Pegasus NASA project personnel and the Fairchild contractors would deliver the satellites along with the necessary pre launch and launch GSE and conduct all necessary tests including the launch operations with KSC personnel oversight. For each of the three launches, I was designated as the lead for the project and the test conductor, which involved running the last few minutes of the count with "go/no go" authority.

After the completion of the project, Lee James asked me to move to the Cape as the Saturn1/1B resident manager to represent his office in preparation of the Apollo Program. The Saturn 1/1B and Sat V offices were later combined when Lee became Saturn V program manager. My first reaction was not favorable. I felt I had been in residence for the past six years and gained a lot of experience that should qualify me for a position in the Saturn Program Office located in Huntsville. After further consideration, it was easy to convince myself the experience would be beneficial and besides, it fit my objective for progressing to achieve my goal of someday managing a program.

One of my first observations after taking residence at KSC was that Rocco Petrone had been named the director of launch operations, and the discipline he was introducing was different from what I had witnessed during the Pegasus launches. In the past, daily planning meetings were held rather informally and often resulted in delays due to improper or lack of critical inputs. Under Rocco, these meetings became formal, where every technical discipline and organization that had a direct or indirect input necessary to maintain schedule had to be present. Smoking and drinking coffee in the firing room were not permitted. The ban on coffee drinking was the most effective in eliminating potential systems problems during operations. Before this was introduced, I do not remember ever being in the firing room during a test when at least one cup of coffee was not accidentally spilled into a console while in operation.

Even though KSC became a separate NASA field center in 1962, it appeared to me that some of the technical and management personnel were still in a mode of becoming independent of the MSFC control that existed when they were a part of the MSFC organization. Initially, the relationship was somewhat strained, partially because they viewed the Resident Office as an extension of the MSFC control. It took a while working with people I had been associated with previously on the Centaur and Pegasus Projects, to convince everyone my charter did not include directing the operations at KSC. Even so, I had to be careful not to be given credit for any suggestions I might have made that could be interpreted as having a direct effect on improving the launch vehicle launch operations. My first such suggestion came when I recognized the way the Saturn 1 vehicles were being processed after arrival at KSC. Even though they had been inspected and acceptance-tested by NASA and the contractor at Michoud, KSC's procedure included opening electrical boxes and inspecting for workmanship and reworking areas they felt needed repair. To me this was an unacceptable practice that could lead to lost of configuration control and the risk of flight failures. Apparently, this was an interpretation on the part of the KSC technicians of the alleged Debus policy, that when the launch vehicle arrived at KSC it belonged to him. The corrective action I proposed was to have the KSC technicians become a part of the acceptance team at Michoud, where they could provide their inputs and, therefore, eliminate the inspection procedure at KSC. I prepared a memo for Lee James stating the request, which he sent to Rocco Petrone, who complied without question.

During my tenure at KSC, I had the opportunity to be involved in the day-to-day planning meetings representing the program including launch readiness reviews of the first five Saturn 1/B launches and all the Saturn V launches through Apollo 11.

One major objective of the Saturn 1B pre-launch and launch operations was to introduce the concept of controlling the launch by communicating between the pad and the blockhouse, utilizing the RCA 110A computer system. The intent was to perfect the concept, which was required for the Saturn V, due to the long distance between the Saturn V pad and firing room. In essence, it meant the elimination of hardwires between the pad and blockhouse familiar to both the KSC civil servants and their support contractors. Their lack of experience utilizing computers in this manner and continual computer glitches made for some very long hours during prelaunch operations.

One interesting event while the first Saturn 1B was on the pad occured when one of the fuel tanks experienced a negative pressure caused by the inadvertent opening of the fuel pre-valve while the vent valve was closed. The result was that the fuel tank bulkhead tended to reverse. The solution was to replace the failed tank while the vehicle was on the pad, an operation that probably would not be acceptable today. Another interesting event occurred during the first Saturn 1B launch. There was an indication the S-IVB upper stage fuel tank pressure fell below the allowed limits and the computer aborted the launch; however, KSC and MSFC engineers determined the pressure level was acceptable and reinstated the launch. The net effect was that, after it was announced the launch had been scrubbed, the press, which had been stationed on the causeway, packed their cameras and equipment only to see the vehicle liftoff. Both Lee James and Dr. Mueller, the NASA associate administrator for manned space flight, witnessed the launch from the blockhouse and Lee, who felt satisfied he had made a good decision, turned to Dr. Mueller and asked what he thought of the decision. Dr. Mueller realized he would have to face the press, and, without hesitation, told Lee under the circumstances he should have not launched.

During the launch preparations of the first Saturn V, there was a fuel tank incident on the pad similar to the one which occurred on the first Saturn 1B. The fuel tank was loaded with propellant when the pre-valve was accidently opened, which caused a negative pressure on the tank. Even though the bulkhead was not buckled as in the Saturn 1B, it had been stressed. The structural integrity of the tank was verified by first, an internal inspection by lowering an inspector into the tank where no visual damage was evident, then filling the tank with de-ionized water and proof testing. Roy Godfrey, the Saturn V deputy manager at the time, witnessed the operation.

It was during the Saturn V launch processing that it became evident the strained relationship with some of the KSC personnel, which I had experienced earlier, no longer existed. Proof of this came when Dr. Hans Gruene, the KSC Saturn V operations director, asked Lee James if I could be assigned to his office for a period to assist him with the Saturn V launch processing. Lee agreed, and for approximately eight months, I reported to Dr. Gruene.

I found my tenure at KSC to be extremely interesting and beneficial with the chance to be a part of the processing and launches of all the Saturn1B and Saturn Vs launches leading up to the lunar landing of Apollo 11. Witnessing the AS 204 fire from the Complex 34 blockhouse, which took the lives of Grissom, White and Chaffee, was the only bad experience and very sad time for me during my four years at KSC. However I was impressed with the way the agency recovered and maintained the lunar landing schedule, unlike the recovery from the Challenger flight failure.

While at KSC, I established a close working relationship with Rocco Petrone, which later worked to my advantage when he became the MSFC center director.

After the successful Apollo 11 lunar landing, Lee James decided it was time for me to return to Huntsville. The question was in what position? I had put a lot of thought into this, and through some research I found that in the past Dr. Eberhard Rees, deputy MSFC center director, had employed a technical assistant. At the time, the position was vacant. The attraction was the fact that it appeared that after serving as Dr. Rees' assistant for a short period, one had the opportunity to advance to a more responsible position. As an example, two individuals that had held the job in the past were Lee James and Jack Balch. James became Saturn 1/1B program manager and Balch the director of the Mississippi Test Facility.

This appeared to me to be very attractive, and I ask James if he would recommend me to Dr. Rees. I think James had something else in mind that he did not reveal, but agreed to support me.

As the assistant to the deputy director, first with Rees and later with Dr. Lucas when Lucas became deputy director and Rees became center director, my knowledge of how the center was run from the ninth floor was greatly expanded. The exposure to WvB and his senior staff during staff luncheons and program/project reviews was invaluable. The staff luncheon gave WvB the time to informally expose the senior staff to new ideas and innovations to get their reactions, and to discuss any new program issues that may have surfaced. When WvB was relocated to NASA Headquarters, to me the luncheon became less beneficial. Contrary to the facts some employees believed that we were served special gourmet food, in reality it was the same food from the cafeteria. I would later decide it could be a morale problem and I terminated the staff luncheons. Unlike Rees, Lucas gave me more specific tasks to perform that exposed me more to the science community and other development projects, which I had not previously been exposed to.

My first window of opportunity to become a program manager came with the advent of Dr. Rees' retiring and Rocco Petrone becoming the MSFC center director. In a very candid discussion, Rocco informed me that I had been an assistant long enough, and if I wanted to progress in NASA I had to be responsible for something. Just the words I wanted to hear. Fortunately, while working with Petrone at KSC, I was well aware of his management style. In my opinion, he was one of the most competent managers in NASA. He was extremely sharp technically, a rigid task master and demanded efficiency from the entire organization.

At the time there were two potential future programs to be assigned to MSFC. An upper stage designated TUG and Sortie Can, later renamed Spacelab, a manned laboratory both to be flown in the Shuttle bay. Both programs were later tied to the study performed by the Space Task Group headed by Vice President Spiro, which recommended that NASA "internationalize space". Rocco gave me the chance to pick one of the two and I chose the Spacelab.

The Program Development organization at MSFC had the responsibility to perform the concept and definition phases of all new potential programs assigned to MSFC; therefore, a Sortie Can task team had already been established to perform the concept phase for a manned laboratory as a payload for the Shuttle Program. My first duty was to manage the definition phase. Significant progress had been made when the decision was made that the European Space Agency (ESA) would assume the responsibility for the design, development and production of the Spacelab with NASA oversight.

The Europeans had no past experience in developing human rated space systems and NASA was not permitted to impose our specifications. The major challenges for NASA were, first, to apply sufficient NASA technical resources to review the ESA-generated specifications and detail design to insure the NASA requirements were met and, second, to understand the concept where 10 different countries provided a portion of the required funding and the distribution of hardware development assignments between the different countries was equal to each country's contribution to the program. Initial concerns relative to the different cultures, language barriers, and whether some of the contractors selected by ESA had the capability to develop and build safe human-rated flight hardware were soon mitigated.

I served as Spacelab Level II program manager for seven years. At that point the design had been finalized, an engineering model had been built, and all critical hardware had been tested or was in the final test phase. I credit the Spacelab success to the high quality of the NASA and ESA technical and management individuals involved in the program, specifically JSC, KSC and MSFC and the outstanding working and personal relationships we had with our European counterparts.

When Rocco was relocated to NASA HQ, the logical replacement as MSFC center director was William R. Lucas, and in late 1980 I was selected as the deputy center director of MSFC, the third person to officially hold that position since it was created. Lucas and I worked well together in part due to my earlier time as his technical assistant, where I became familiar with his management style and philosophy and what was expected of me and the center. He considered me a confidant, and one of the things he stressed early on was that, when he asked for my opinion, he wanted to know what I thought, not what I thought he wanted to hear. I found him to be sincere in that statement and from that day on, I never hesitated to respond accordingly.

At the time I became deputy, the Shuttle program was only a few months away from the first flight and during the development, my focus had been on Spacelab and my involvement during the design and development phases was limited to insuring the Shuttle provided the required services in the payload bay to the Spacelab and the interfaces were compatible. I did participate in all redesign and the majority of the operations reviews, and was selected to chair the STS-8 Failure Review Board established to determine the cause of the recovery parachutes on both solid rocket booster to fail. We were able to isolate the root cause even though all the data necessary to make the determination was on recorders 3,000 feet below the surface in the Atlantic Ocean.

Initially, in the position as deputy, I concentrated more on overseeing the Spacelab progress thru the first flight and other programs in development at MSFC at the time, specifically the Hubble telescope and early concept and definition of the Space Station.

Dr. Lucas recognized early in the definition phase that the center roles and missions for space station needed to be defined, and I think the JSC management also felt the same. Lucas and I met with Chris Kraft while he was JSC Center Director, and later Gerry Griffin came to a basic agreement that the primary center roles and missions should be between JSC, KSC and MSFC. The concept was patterned after the way the Skylab program had been managed. However, this was contrary to the way the NASA Administrator, Jim Beggs, planned the center roles and missions. His plan was for every NASA center to have specific responsibilities in the development and/or operations. This led to establishment of the control of the program to be located at Reston, VA reporting to NASA Headquarter. It was clear to many at the offset, the concept would be inefficient and much more difficult to manage than the Lead Center approach which, had been successful on past multi-center programs. The case was made, when ultimately the space station organization evolved to one similar to that identified in the early meetings between MSFC and JSC.

What had been an interesting, and I believe a productive, six years as deputy, came to very unpleasant climax in January 1986 with the failure of the Shuttle Challenger. The four months prior to the launch, I was at Harvard participating in the Advanced Management Program. William Graham was appointed as the NASA deputy administrator to Jim Beggs. Shortly after Graham's appointment, Beggs was indicted for fraud, which occurred during his previous employment in industry, and resulted in his taking a leave of absence to prepare his defense. Graham then became the interim Administrator, with little NASA experience in dealing with the field centers or the congressional space committees. Unlike the Apollo fire, where James Webb, the NASA Administrator at the time, took direct control and got approval to let NASA solve the problem, with Challenger, the Rogers Commission was formed by President Reagan; NASA in reality was not in control of the situation.

Since I had spent the winter quarter of 1985 participating in the Advanced Management Program, and I did not participate in any of the Challenger launch readiness reviews, it was decided that I observe the launch in the HOSC in Huntsville. This was beneficial, since all the MSFC Shuttle Contingency Plan Committee chairmen were also located in the HOSC during the launch. Within minutes they were able to convene their respective team members and to begin collecting data and start investigating the cause of the failure.

In all previous anomaly/failure investigations conducted on flight hardware for which MSFC had responsibility, none included loss of life during flight. The NASA and contractor teams joined forces to determine the cause and recommend fixes, and if human error was determined to be directly or indirectly related to the failure, then this along with recommended corrective actions would be cited. The Challenger investigation process seemed to take on a different approach, one of "we have a failure and we will determine who is responsible". This approach caused me concern that, in future anomaly/failure investigations on NASA programs would result in too much focus being placed on the liability aspect of the investigation rather than properly addressing the problem, finding the cause, and recommending solutions in a timely manner. Fortunately, I never found any evidence to substantiate my concern.

In addition, NASA and contractor review teams had always adhered to the policy that the individual team members would refrain from releasing opinions on conclusions to the press prematurely. Not so with some of the Rogers Commission members. This gave the media a field day; every person involved in the Shuttle Program was hounded day and night by the media until the Rogers Commission Report was published. This had a very adverse effect on the morale of all the Marshall employees and their families. The Challenger failure dominated the news and the only news the Marshall employees and their families were exposed to blamed Marshall as the cause of the accident. We attempted to mitigate this with periodic updates where the facts were presented to the first, second and third line supervision, with instructions they should keep their employees informed.

I was not satisfied with the way the investigation was conducted, and I was offended by the criticism of the MSFC management and culture. However, in hindsight, given the time and effort the commission put into the investigation and reviewing the details and facts they had to deal with, I believe they produced a good report with correct findings and conclusions along with proper recommendations.

When Dr. Lucas retired in the July 1986, I assumed the duties as interim MSFC director until September of that year when J.R. Thompson was named as Lucas' replacement. It goes without saying that I was personally disappointed that I was not selected. I felt I deserved it. I had been heavily involved in supporting the Rogers Commission in providing all data they required for their investigation including special studies in support of their findings. I testified before the commission, and managed the center for three months after Lucas retirement all without any NASA administration criticism, as far as I knew.

After considering all the facts, I concluded bringing J.R back to NASA from Princeton where he had served as technical deputy director of operations at Princeton University's Plasma Physics Laboratory, was probably one of the best decisions the agency could make in the interest of the future of MSFC. He had been well respected in NASA, had been away from NASA for three years, had no involvement with the shuttle processing that carried the Challenger, and in the past while at NASA, was not associated with the design, testing or operations of the solid rocket motors, which caused the flight failure of Challenger.

For the next three years until he was selected as deputy NASA administrator, I concentrated first on supporting J.R. in the center's primary objective of returning the Shuttle to flight and center institutional matters. The major return to flight effort included initiating the start of the Advanced Solid Rocket Motor development program intended for future shuttle flights. My relationship with J. R. could not have been better while he was director.

J. R. left the center in good shape when I became director. The Shuttle Program was on track and the other major programs in development at the center

Hubble and Chandra, were progressing well while encountering the typical cost and schedule concerns of most complex development programs. The controversy over locating the Spacelab Missions Control Center at MSFC was still being contested by our sister center. Some of the management at JSC interpreted it to be their responsibility. MSFC on the other hand had, since the beginning of the Spacelab program worked closely with the science community in integrating and testing their experiments in the Spacelab and to expand that into science mission support was both practical and efficient. One point of interest that we did not reveal openly at the time; strategically we at MSFC recognized the potential of evolving the control center to support Space Station.

Under J.R.'s leadership, the culture at Marshall was beginning to change for the better. I attempted to continue this trend and found little resistance from center management or the employees. My contribution was to establish good working relations with the other centers and NASA Headquarters and develop a Strategic Plan that each employee could feel a part of, and introduce a mentality of continual improvement for all processes and procedures to make Marshall more productive.

I am satisfied with the progress at the center during my five years as director. Successful accomplishments include: Shuttle flights, first United States Microgravity Spacelab mission, Hubble Telescope, Chandra X-Ray telescope, the Spacelab Mission Control facility, Astro-1, Burst and Transient Source Experiment (BATSE), and X-Ray Calibration Facility.

The Intellectual Footprint

By Mark McCarter

There's one thing you need to understand about WvB. He didn't just gaze at the Moon. He didn't expend all that infinite imagination of his staring into the heavens. WvB saw promise and opportunities everywhere. He especially saw it in the vast cotton fields that stretched west toward the horizon from what was then the modest city limits of Huntsville. He saw the need for an excellent research-centric university that would support the mission of the United States' space program. He saw the need to erase the boundaries of segregation that sullied the state's image in the 1960s.

In all that, WvB saw what he referred to as the potential for "an intellectual footprint" to be established in north Alabama, a foundation that would resonate long after Neil Armstrong put man's literal footprint on the Moon with no small amount of assistance from Huntsville. WvB's leadership, charisma and intelligence inspired hundreds of team members, from the German scientists who accompanied him to Alabama to the Americans who eagerly joined hands. Together, they designed, built and tested the rockets that would take man to the Moon. Indeed, Huntsville will inexorably be tied to NASA and the space program. The seismic changes in the 1950s and 1960s transformed what was once a small town known as the Watercress Capital of the World, with its economy built on the cotton industry and agriculture, into a player on the international stage, a role the area maintains.

This was decades before the word "synergy" became a corporate buzzword, but it was a philosophy that WvB put to work. He devoutly believed in the need to couple NASA and the U.S. government resources with a larger campus of supporting businesses and feed it with graduates and researchers from a robust local university.

Thus WvB's role as a pioneer in the development of the University of Alabama in Huntsville and the ever-expanding Cummings Research Park in that "intellectual footprint" that includes WvB's home base of Redstone Arsenal and Marshall Space Flight Center. The former would develop the brainpower. The latter two would put it to work.

UAH received its charter as extension center of the University of Alabama in 1949. It began in a building that would later become a middle school and eventually house a brewery, perhaps a bit of symmetry considering the city's ambitious launching of craft beers (many of which took on space-related titles, such as Monkeynaut).

As the school rapidly grew, a larger building was needed, but by the time it was completed in 1961, it was insufficient for the demand. WvB traveled to Montgomery to plead for a larger campus and expanded role. A bond issue was passed that funded more buildings. In 1969, the same year the Huntsville-created Saturn V rocket sent man to the Moon, UAH earned its status as an autonomous university.

An octogenarian retired rocket scientist who continues as historian, archivist and instructor in Von Braun Research Hall is Dr. Charles Lundquist, who worked alongside the building's namesake. "Quite simply," Lundquist says, "the university would not exist without him." "He learned to love (Huntsville), and he helped it grow," WvB's former aide Frank Williams once said. "He had his own city that surrounded him and loved him. Anything he wanted to do, he saw it done."

Now, let's be clear. There was no small amount of pragmatism in WvB's plans, especially when it came to Cummings Research Park. Many of the corporations supporting the work at Redstone were headquartered in California. It would make perfect fiscal sense to have the contractors locate operations in Huntsville, saving the government money in access costs and, frankly, because manpower here would be cheaper with a more palatable cost of living.

Huntsville businessman Milton Cummings had been offered the opportunity to invest in the late 1950s in a struggling company called Brown Engineering. He was friends with WvB and recognized an opportunity for Brown Engineering to become involved with the space program. Cummings' company began to expand, and WvB raised the notion of a research park that would be convenient to Redstone. Cummings and Joseph Moquin acquired the 3,000 acres not far from the UAH campus and joined with WvB in convincing the city to zone it as a research center. Brown Engineering (now Teledyne Brown) became the first tenant, quickly followed by a number of prestigious names in the technical and aerospace field, such as Lockheed, Northrop and Boeing. Appropriately, the area would be named Cummings Research Park in 1973.

It wasn't an era without its challenges for WvB and the business leaders. It wasn't so much a financial challenge, as the spigot from Washington flowed freely, but it was a social challenge. It is almost impossible to have any narrative regarding Alabama history in the 1960s and ignore the topic of racial relations, and indeed the embryonic research park and the work at Redstone itself were not immune.

WvB was supportive of civil rights, but had to maintain a delicate balancing act because of the unwelcoming racial policies of then-Alabama Gov. George Wallace that dominated the state. "It's OK to disagree," WvB once said, "but I wonder if we have to be so disagreeable." After WvB's friend and great ally, President John F. Kennedy, was assassinated on Nov. 22, 1963, President Lyndon Johnson assumed the White House. Though a Southerner himself, Johnson's quest for civil rights became a major priority. WvB supported that effort in Huntsville, again alongside Cummings and, again, it was pragmatic.

Cummings was given a message from Alabama's U.S. Sen. John Sparkman, who said "Johnson is sincere enough about this thing, (that) unless we can find some way to bring some black people into the community to work at (Marshall), it is going to cease to be," according to journalist and author Bob Ward.

Civic leaders such as Cummings and Woody Anderson led the way for race relations in Huntsville, including organizing a strong but subtle integration event at a previously all-white restaurant. Meanwhile, federal officials made clear to Wallace the importance of equal opportunity in the city and its impact on NASA's presence here. It wasn't seamless, but Huntsville as a whole embraced integration much more easily than most Southern states. Perhaps it was because, only a decade earlier, there had been the cultural integration of WvB and his team of more than 100 German scientists into the community. Even WvB might not have seen the "Operation Paperclip" move as a something of a prelude to racial harmony.

As for that "intellectual footprint" and where it stands now? Cummings Research Park is the second-largest of its kind in the United States. More than 300 companies have a presence there. It has expanded its horizons, going from outer space to inner space. It is an incubator of sorts for the growing field of biotechnology and genomic research, for which HudsonAlpha Institute of Biotechnology is an international leader. And just as a hub of space-race business lured its supporting contractors in the synergy of the 1960s, HudsonAlpha has done the same, with myriad companies related to biotechnology establishing a presence within Cummings Research Park.

On the occasion of Cummings Research Park's 50th anniversary in 2012, the park's then-director John Southerland was left to muse what WvB might perceive of what emerged. "I can't tell you how many times in various groups, with park companies and leaders, and they've talked about 'Wouldn't he be proud of what this evolved into?'" Southerland said. "He had very high standards. But he'd be pretty impressed."

WvB was quoted in *Time* magazine in 1958 saying, "Don't tell me man doesn't belong out there. Man belongs wherever he wants to go—and he'll do plenty well when he gets there." He was talking, obviously, about outer space. He could have been talking about the flat landscape of old cotton fields that stands now as another of his grand Huntsville legacies.

Intellectual Critical Mass

1950 arrival of engineers, scientists is genesis leading to one of the most educated workforces on the planet

By Robert A. Altenkirch, President &
Ray Garner, Chief of Staff
The University of Alabama in Huntsville

It started with the U.S. Army's transfer of scientists and engineers to Huntsville, Alabama, from Ft. Bliss, Texas, in 1950. The group arrived in Huntsville to advance the nation's ballistic missile program, an effort deemed crucial for America's defense during the emerging Atomic Age.

Those 119 engineers and scientists were a team assembled by rocket pioneer WvB, and their presence planted an important seed in this small Southern town.

Although this program would continue to grow at a moderate pace during the 1950s, as did the number of scientists and engineers moving to Huntsville, it would be two events that put America's efforts into overdrive to establish superiority in space: the Soviet Union's launch of the Sputnik satellite, and cosmonaut Yuri Gagarin's venture into space. It would not be long before the Redstone team would launch Explorer 1, and Alan Shepard would fly into space aboard a Huntsville-developed rocket in 1961.

Soon, a young and vibrant president would announce an ambitious goal for the nation—transport astronauts to the Moon and return them safely to the Earth before the end of that decade.

That commitment would bring an exponential growth of scientists and engineers to North Alabama. The best minds from every corner of Alabama and from across the United States descended on Huntsville.

The city's workforce would never be the same again.

That massive influx of intellectual talent over the years has created a critical mass of technological capabilities that exist in few places on planet Earth. The greater Huntsville area workforce boasts one of the smartest workforces in America today.

The offices occupied by most of this critical mass of brainpower are located on just 14 square miles of Alabama soil designated with names such as Redstone Arsenal, Cummings Research Park, and The University of Alabama in Huntsville.

This concentration of knowledge and intelligence is why *Forbes* magazine called Huntsville one of the top 10 smartest cities in the world.

Numerous other sources have recognized the city's intellectual prowess. Huntsville was named one of the nation's most educated cities by WalletHub, a personal finance information services company. The accolades don't stop there. NerdWallet proclaimed Huntsville to be America's top engineering city. That last point is confirmed with U.S. Census data. Huntsville is ranked first in the nation on a per capita basis for having the most engineers, computer scientists and mathematicians, according to 2010 data.

Information provided by the U.S. Army at Redstone Arsenal reveals that 69 percent of their 39,000 employees have at least a bachelor's degree, and many of those have earned advanced degrees.

Although NASA's Marshall Space Flight Center workforce isn't as large, figures from the agency reveal the educational levels of Marshall's more than 2,000 civil service employees is among the highest in the region with more than 90 percent possessing at least a bachelor's degree. Of that number, 27.4 percent have a master's degree and 7.8 percent earned a Ph.D.

The corporations in Cummings Research Park also host very high concentrations of intellectual talent. A non-scientific survey among companies in the park reveal that many firms report higher than 80 percent of their employees have a bachelor's or advanced degree. For some of them the number exceeds 90 percent.

Meanwhile, at UAH, the campus has more than 1,300 full-time, permanent employees. As expected with an institution of higher learning there is a higher concentration of employees with a Ph.D.—more than 21 percent. Another 16.7 percent of those employees holds a master's degree, and add to those numbers 34.7 percent of employees with a bachelor's degree.

This strong foundation of highly educated individuals has a marked impact on the community's overall education levels. For example, the number of people in Huntsville age 25 or older who hold a bachelor's or advanced degree is 39 percent. This compares with the Alabama average of 22.6 and the United States' average of 28.8 percent.[1]

This accumulation of brainpower is an outcome of the vision of WvB. He promoted the importance of education if this community was to play a role in America's space program's successes. In the early 1960s, WvB addressed the Huntsville community to kick off a fund-raising drive for an institution of higher learning that eventually became The University of Alabama in Huntsville. He told the crowd: "I am persuaded that we who make our homes in this community believe that this area is destined to continue to grow and become a great and permanent scientific, educational and industrial center."[2]

WvB carried that message to the state capital as well. He spoke to a joint session of the Alabama Legislature in 1961 to encourage the state to invest in a research institute on what is now the UAH campus. During his talk, WvB told the assembled legislators: "Opportunity goes where the best people go, and the best people go where good education goes. To make Huntsville more attractive to technical and scientific people across the country—and to further develop the people we have now—the academic and research environment of Huntsville and Alabama must be improved."[3]

He specifically cited the need for a research university, saying: "It's the university climate that brings the business. Let's be honest with ourselves. It's not water, or real estate, or labor or cheap taxes that brings industry to a state or city. It's brainpower."[4]

It's a testament to WvB's vision that Huntsville's intellectual strength has become a critical national asset since he made those remarks more than a half century ago.

Today, world-class educated workers have made Redstone Arsenal a strategically important installation for the U.S. Army, and crucial to the nation's defense and security. Marshall Space Flight Center is one of NASA's largest field centers[5] and supports America's space science and exploration missions. Cummings Research Park is America's second largest university related research park, and The University of Alabama in Huntsville has become a highly accomplished institution of higher learning and ranked nationally in vital academic and research disciplines that support Redstone Arsenal, Cummings Research Park and Huntsville's economy, including fourth in the nation in aeronautical and astronautical engineering in 2013, and then fifth in 2014. Half of UAH's graduates earn a degree in science and engineering.

This aggregation of human capital is the key factor in Huntsville being able to navigate a smooth and successful transition from an industrial economy into today's knowledge and innovation economy, the latest broad-reaching development in a global economic restructuring.

Because Huntsville's workforce has developed and focused on skills in STEM careers, such as computer scientists, engineers, chemists, and mathematicians, the knowledge of those workers can be easily adapted to other emerging economic and technological trends.

The corporate base within Cummings Research Park has become more diverse, moving beyond aerospace and defense to include software design, engineering services, computers and electronics, advanced manufacturing, cybersecurity, and biotechnology. One key development in that evolution is the constant renewal of the academic programs and research capabilities at UAH.

The knowledge economy is critical in a society that is rapidly changing with more focus on innovation developed through the interaction of research and laboratory environments at government agencies, corporations, and UAH.

Well-developed industrial clusters, particularly those based on knowledge, are important to the continued advancement of economies. They are an important factor in the greater Huntsville area's success in developing its transcending economy.

Research universities, corporate research parks, and federal laboratories are crucial not only for creating and maintaining employment in today's high technology sectors, but also for serving as a location where discoveries take place and where the economy of the future will be born.

Research and education are the first links in the value chain of prosperity, and knowledge developed from those factors will serve as the catalyst for Huntsville's economy long into the future – just as predicted by Dr. von Braun a half century ago.

THE HUNTSVILLE RENAISSANCE
A STUDY OF A COMMUNITY AND ITS LEADERSHIP, 1950-1970

By Zack Wilson
Auburn University History Student, 2013

In the second half of the 20th century, the city of Huntsville, Alabama, underwent a period of growth unlike anything it had previously witnessed. During the 1950s and 1960s the town experienced tremendous growth, transforming nearly every facet of local society. In the short term, the "Huntsville Renaissance" that rapidly blossomed resulted from federal funding and directives propelled by the Cold War. In the long run, however, its growth can be traced to the efforts of the Huntsville community and its leadership on a number of fronts during these years. Community leaders played a crucial role in facilitating desegregation during the height of the Civil Rights Movement, for example. They also had the foresight and necessary planning to bring economic diversification to the area, which liberated Huntsville's economy from utter dependence on federal programs and spending.

Historically, Huntsville's economy was centered on the cotton and textile industry, but when World War II broke out the military built chemical and munitions plants, which brought a plethora of jobs to the area.[1] Unfortunately, the end of the war also meant the end of the need for chemical weapons, and, consequently, the plants shut down. Economic salvation, however, arrived in Huntsville in 1950 in the form of the U.S. Army's rudimentary rocket program. The old "Redstone Arsenal," home to chemical and munitions plants just a few years earlier, became the cradle of the American rocket program, and by the end of the decade the cradle of the American space program as well. Popular opinion regarding the new developments in Huntsville was initially skeptical,[2] and to direct the project the U.S. Army utilized the German rocket engineers and scientists who surrendered following the end of World War II. In 1950, 120 German rocket experts and their families were transferred to this small, still largely rural cotton town.

Among the Germans was a young man named Dr. Wernher von Braun, who had played a chief role in developing and managing the revolutionary V-2 for the Nazis along with his colleagues. The arrival of the rocket program and the WvB team in 1950 ushered in a period of tremendous growth and prosperity for the city, and this alleviated the initial skepticism of many locals. WvB and his fellow Germans would ultimately adopt Huntsville as their home, and they took a keen interest in the development of the community as a whole. The Huntsville residents in turn embraced the Germans as their own, and the echoes of their actions reverberate throughout much of Huntsville today.

Historiography

The story of WvB and his fellow German engineers has been a popular topic for historians interested in the development of rocketry, space flight, and the systems management of complex and large-scale projects.[3] These sources, however, focus less on who WvB was as a person and more on the aforementioned themes. They were written during WvB's lifetime and are often hagiographic in their approach. Former colleagues of WvB have also written on the topic, but they expand the story to capture the experiences and impact of the team of engineers and other personnel who worked under him, with an emphasis on the Saturn project.[4] In, *Von Braun: Dreamer of Space, Engineer of War,* for example, Michael J. Neufeld analyses WvB's history and legacy from the ethical standpoint of rocketry. He covers in detail WvB's life in Germany, including his time working for the Nazis. He compares him to Goethe's Faust, claiming: "In von Braun's case, he accepted the ample resources offered by the Third Reich to build rockets, believing that it would lead to a glorious new future for himself, and humankind, in space."[5] Additionally, an important theme of the work is the moral complexities inherent in rocket development and space travel in the twentieth century.

Other scholars and colleagues of WvB have written more recent narratives about him and have approached him from a humanistic perspective.[6] They shed more insight into the involvement of the Germans in community affairs and note their important contributions outside rockets and space. In particular, Bob Ward argues that the social, cultural, and economic changes that Huntsville experienced are directly related to the arrival of the Germans and their continued influence in the area.[7] This research expands on this point by portraying how WvB was essential to promoting social change and economic diversification during the "Huntsville Renaissance."

Erik Bergaust's book, *Rocket City USA*, published in 1963, is a narrative history of Huntsville that focuses on the town's transformation from cotton to rockets. It gives insight into the cultural changes that Huntsville experienced during the 1950s and into the early '60s, but it is incomplete and concludes with speculations about the future of the "Rocket City." Coinciding with the history of Huntsville is that of the Marshall Space Flight Center (MFSC). It was established in 1960 and became the heart of the U.S. space program as well as an integral and lasting part of the community. Much of the history surrounding the MSFC is dedicated to its space projects and evolution. *Power to Explore*, however, addresses the impact the Civil Rights Movement had on the MSFC and inversely the role of the MSFC in dealing with segregation. The authors argue that NASA helped reconstruct the region's economic, demographic, social, and educational landscape.[8] *Historic Huntsville: A City of New Beginnings*[9] is a narrative account of the city's history into the 1980s. It discusses aspects of the community and its leadership during the 1950s and 1960s. This research expands on the importance of the community, to demonstrate that they were essential in laying a foundation for growth to which Huntsville benefits today.

The goal of this research is to broaden our understanding of the factors that contributed to the "Huntsville Renaissance" evident in all facets of Huntsville society from 1950 to 1970. The ostensible reason for this growth can be traced to the arrival of the Army's rocket program in 1950 and the subsequent creation of NASA in 1958. While these two programs were seen at the time as the driving factor in the "Huntsville Renaissance," historical hindsight illustrates that these programs began to decline by 1970. However, at that moment in time the duration of the Army's presence was unknown and Huntsville was once again dependent on one source of power to fuel its economy. City leaders had the foresight to see the importance of a diversified economy as well as promoting social and cultural changes throughout the community. The number of programs and institutions established during the aforementioned period are too numerous to mention in this paper, however, there are some worth noting. These include the creation of the UAH Foundation (1965) and the establishment of UAH as an autonomous campus (1969); and a vast expansion of the city's cultural life such as the Huntsville Civic Symphony (1954), The Huntsville Youth Orchestra (1962), The Huntsville Civic Ballet (1963), and the Fantasy Playhouse Theater (1961). This research explores the following examples of the city's economic, social, and cultural growth to illustrate how important strong leadership and community cooperation laid the foundation for future generations to continue to expand upon: the administration of Mayor Robert Searcy and the remarketing of Huntsville's image; the Huntsville Industrial Expansion Committee and the Huntsville Madison County Airport Authority and their success in diversifying the local economy; the economic and social impact of local businessman and philanthropist, Milton Cummings; the importance of the community in facilitating desegregation; and the establishment of the Von Braun Astronomical Society. It should be noted that this research does not separate the German community from the rest of the Huntsville community. Many of the Germans arrived in Huntsville in 1950 when the city's population was slightly over 16,000. By 1960 it had increased to over 70,000 due to the influx of skilled labor that was brought to Huntsville. In fact, the Germans had been settled in Huntsville for eight years when NASA was created, and were made citizens of Huntsville in a 1955 ceremony at Huntsville High School. The point being, it was no single group of people — the German community, the old Huntsville community, or the those who arrived post 1950 — that were responsible for the Renaissance, but rather all are cumulatively responsible and came together to form a new and distinct community within Huntsville that stood out from the radically conservative state of the 1960s in which it resided.

The bulk of this paper will draw from a variety of primary documents, such as local, regional, and national newspapers, magazine articles, scholarly journals, local Chamber of Commerce material, and interviews. The goal of this research is not simply to document Huntsville's history, but to illustrate that strong community involvement and leadership were as vital to the city's long term growth as the federal programs were. In sum, the 'Huntsville Renaissance' was set in motion by the arrival of the Army's rocket program, but it was not the reason for the town's long term growth. Rather, the social, economic, and cultural changes experienced during the "Renaissance" were promoted and advanced by the cumulative effort of city leaders, such as WvB, Milton Cummings, the administration of Mayor Robert Searcy, as well as from Huntsville's diverse community. Massive federal spending on rocket programs and NASA were temporary enterprises that enabled the town to catapult into the mainstream society of the 1950s and 1960s, but did not provide the longevity necessary to claim that Huntsville's continued success today can be attributed to their presence. Indeed, these programs altered Huntsville demographics and society to the point that it could have potentially been devastating had the municipality and business leaders not been prepared to address them properly. Not only did they face the dilemma of an economy dependent on federal spending, but the municipality faced the problem of how to rapidly expand Huntsville in order to successfully cater to its population boom. Furthermore, the city leaders and community cooperated together to promote desegregation during the height of the Civil Rights struggle. They welcomed the task and succeeded in laying a foundation for long-term growth that has enabled Huntsville to continue to prosper and expand today.

From a Cotton Town to Rocket City USA

In 1957, Mayor Robert Searcy and the City Council released to the citizens of Huntsville its first annual "report to its citizens." The cover depicted a rocket soaring over an aerial view of the city, and the opening statement read: "This is to be the first of a series of annual reports from your city officials to you, the citizens of Huntsville. But this is more than a mere report. It was laid out to be a colorful, attractive glimpse of your city—America's Rocket City."[10] The community and its leadership capitalized on its new image. Their objective was to promote Huntsville as a modern technological mecca, and to replace the city's association with cotton and textiles to that of space exploration and rockets. Local newspapers also contributed to the effort. An article in the *Huntsville News* on June 24, 1969 ran the title: "From a Textile Mill Town to 'Rocket City' USA." Additionally, Huntsville's new image was promoted on a national level. Publications, such as Erik Bergaust's 1963 book on Huntsville, titled *Rocket City USA*, helped spread national attention and awareness for the city's new image. It further reinforced the image of Huntsville as a modern, progressive "Rocket City." The remarketing of the city's image was also an important step in the towns strive towards economic diversity, because it made the town more appealing to prospective employees and their families. Ed Buckbee, first director of the U.S. Space and Rocket Center (USSRC) in Huntsville, said:

> "What attracted engineers from other parts of the country and foreign countries to Huntsville was the exciting and challenging work. The problem was the engineer's wife and family. How would they adjust to the small southern town? Once they arrived and engaged in the community, they realized that the majority of the citizens were from somewhere else and were newcomers as well."

In fact, at the height of the space race in the mid-1960s, rockets and space travel generated tremendous national and international interest, which drew many tourists to Huntsville to see the self-proclaimed space capital of the world. In 1964 the city received an estimated 148,000 visitors from all over the nation and 20 different foreign countries, eager to visit the MSFC and witness history in the making.[11] Huntsville capitalized on the emerging tourism industry that resulted from the presence of rockets and NASA to establish a viable source of long-term revenue. Through the efforts of WvB, the USSRC was established in Huntsville in 1970, and Ed Buckbee served as its first director. It is home a full-scale Saturn V rocket display, as well as Space Camp and Aviation Challenge. The space museum, as it is known, became an integral part of the community, generating revenue through tourism as well as stimulating enthusiasm for space and science in the youth. Since opening in 1970 the USSRC has attracted 15.5 million tourists, and generated $77 million in revenue for the North Alabama economy in 2012.[12] Its inception and continued success today can be traced to a romantic notion of rockets and spaceflight that became associated with Huntsville during the 1950s and 1960s. During the 1950s and 1960s, city leaders revamped the image of Huntsville from textile to rockets, and created for the town a source of long term revenue through the tourism industry.

The reopening of Redstone Arsenal and the subsequent creation of NASA by President Eisenhower in 1958 drastically altered the demographics of the city. In 1950 the population of Huntsville was roughly 16,000, but by 1967 it had soared to over 143,000.[13] A shortage of housing was an initial problem the community had to overcome. A 1950 *Huntsville Times* article encouraged local residents to offer their support, advertising: "Need is Urgent for Temporary Homes in the Area." It was estimated that 362 families were to move to Huntsville to work on the rocket program, and only 20 homes were available for sale.[14] The lifespan of the program was unknown at the time, and consequently many newcomers were more interested in renting than owning homes. In 1965, however, the city constructed 495 new homes.[15] The initial housing shortage required city leaders to galvanize the community to act in order to meet the rapidly growing demand for housing. Additionally, successful construction of permanent housing facilitated Huntsville's economic diversification process as well.

Milton Cummings and the Establishment of Cummings Research Park

Huntsville's socio-economic transition into mainstream society was not without its opponents and hurdles. Resistance came from "Old Huntsville," or more specifically, those who resided in Huntsville prior to its adoption of rockets in 1950. They believed and held onto a mode of conservative thinking prevalent throughout the South in the 1950s and 1960s. One historian described the party as one that "was not a serious, long-term political movement but rather a transitory phenomenon led by irrational, paranoid people who were angry at the changes taking place in America."[16] The Civil Rights movement and desegregation were among these changes they resisted. In his 1963 book *Rocket City U.S.A.*, Erik Bergaust alludes to tension between the old Huntsville order, seeking to hold onto its traditional ways, and the new Huntsville order, seeking to establish Huntsville as a prominent community that had embraced the technological advancements of the post-World War II era.[17] Not all of "Old Huntsville," however, proved to be resistant to the tremendous and rapid socio-economic expansion of a once small cotton town. Notable among them was

Milton K. Cummings, local business tycoon and philanthropist whose legacy in Huntsville is a testament to the positive impact business leaders can have on a community. He was born in 1911 in Gadsden, Ala., and as a child he lost a leg to osteomyelitis. He credited his disability with instilling in him a sense of what it means to be disadvantaged. In a 1969 interview he reluctantly responded to why his vast wealth had not altered his humble perspective on life, stating:

> "I guess it's because I've been a cripple all my life. I guess it's because I know what it is to be handicapped. Losing a leg is nothing. But think what it means to lose your sight. Think what it means to have no hearing. 'Handicapped' can mean a lot of things. It can mean too little food, too little opportunity…" he said, and then his voice trailed off. He sat quietly alternately smoking non-filtered Camels and menthol Salems.[18]

Cummings initially made a successful living as a cotton merchant, but in 1953 he left this business to venture into the world of the stock market. In 1958, he was asked by board members of a local company, Brown Engineering, to temporarily assume presidency of the company until a more suitable candidate with engineering skills could be found. Cummings, however, quickly turned the dying company around, investing $1 million of his own money and applying a business philosophy that became a cornerstone of Huntsville business in the 1960s. It is best summed up in his own words: "There never has been a time that I put public interest above my own that I did not gain…Dollars should be the by-products of good service."[19] Not surprisingly, he was asked to serve as full time president of the company. Under the leadership of Cummings, Brown Engineering became one of NASA's biggest and most important contractors in the region, employing 3,300 people and generating $33 million in sales in 1962.[20] Other contractors that catered to the aerospace demands of the Marshall Space Flight Center included Chrysler, IBM, Boeing, Lockheed, and Northrop. These companies generated tremendous research and development activity in the region, but prior to 1962 they had headquarters in other areas throughout the country. This made communication between Marshall and its contractors inefficient. In order to remedy this, an industrial research park was proposed by Cummings, and other city leaders such as WvB promoted the endeavor. In 1962 the Huntsville Research Park was established as an industrial R&D center to facilitate interaction between NASA and its contractors, and the aforementioned companies opened facilities in the park. It was later renamed Cummings Research Park in 1973 in honor of Milton Cummings, and continues to be an important pillar of Huntsville economy today. In 1997, the research park received the highest national recognition from the Association of University Related Research Parks (AURRP) for its technological and educational contributions.[21] Today it has expanded to include other companies whose expertise is outside of aerospace, such as biotechnology and telecommunications, and is projected for continued growth and expansion this year.[22]

City Leadership and the Community during the Civil Rights Movement:

During the 1960s Alabama became the heart of the struggle for Civil Rights, and consequently the impact was felt in Huntsville. The city attracted a great deal of outside labor, given its drastic population rise from the 1950s to the 1960s. The result of such a large influx of people from various parts of the country and world created a unique, diverse and culturally expanding town in the heart of Dixie. Additionally, the federal government was heavily invested in Huntsville through NASA and its Army rocket program. These two factors made public opinion in Huntsville more liberal in regards to race when compared to other cities throughout Alabama during the Civil Rights era. Consequently, the city clashed with zealously conservative Gov. George Wallace, who claimed he "would call out the National Guard if the Federal Government forced integration of schools at Redstone Arsenal in Huntsville."[23] In 1964 the federal government passed the Civil Rights Act, which deemed segregation unconstitutional. However, Wallace refused to allow Alabama to desegregate without a struggle. Pressure from the federal level and efforts by the residents of Huntsville helped facilitate the city's transition towards integration more smoothly than other cities throughout the state. The community and its leadership had already taken steps toward desegregation prior to the passage of the Civil Rights Act in 1964. In 1963 local business leaders established the Association of Huntsville Area Contractors (AHAC), with Cummings at the helm.[24] Their objective was to strive for progress in racial relations and to remove the barrier of segregation. It had become apparent to local business leaders, such as Cummings, that their ties to the federal government and continued economic success were dependent upon equal employment opportunities. The AHAC was successful, along with the Huntsville area school system, in acquiring a $2.7 million grant to finance schooling and home improvement for black neighborhoods. They have been credited with facilitating the integration of places such as motels, restaurants, and even promoted desegregated housing, and by 1967 approximately 35 black families were living in white neighborhoods.[25]

Although desegregation went more smoothly in Huntsville than in other parts of Alabama, it was not without its opponents and hurdles. Early in 1962 a lunch counter sit-in had led to the arrest of several locals. In response, Huntsville Mayor R.B. Searcy, the Huntsville Chamber of Commerce, and the Huntsville Industrial Expansion Committee appointed a committee to collaborate with local black leaders to foster integration.[26] The objective of these meetings was to promote nonviolent desegregation, and on July 10, 1962 blacks were served at six downtown restaurants without any issue. Local store managers and owners also contributed to the planning and implementation of the sit-ins, and its success led to further non-violent sit-ins. The success of the endeavor is a testament to the cooperation and progress made in racial relations throughout the community at a time when the struggle for civil rights had taken on a more violent form in other parts of the state. Furthermore, on May 16, 1963 the city hired its first black police officer, Robert Carl Bailey, through a unanimous vote of the City Council and the support of Mayor Searcy.[27] He was hired one year prior to the passing of the Civil Rights Act, demonstrating a proactive approach to desegregation on the part of Huntsville residents and its leaders. Indeed, the prospect of hiring a black police officer had been discussed in City Council meetings as early as 1960.[28]

The community rallied together to promote desegregation on moral grounds. In August 1962, famed evangelist preacher Billy Graham arrived in Huntsville to endorse hits topics of the day, including "integration by love instead of law, space exploration and birth control."[29] He held a rally at Redstone Arsenal that an estimated 100,000 people attended. His message of combating racial inequality with love and religion resonated strongly with a community located in a region commonly referred to as the "Bible Belt." In fact, Huntsville boasted a vast number of churches of various denominations. In 1966, there were more than 100 churches in the city, and since 1957 more than 50 new churches spanning 35 denominations were constructed in the city.[30]

Huntsville citizens were also active in promoting civil rights and helping to break the barrier of segregation. Myrna Copeland serves as an example of a Huntsville resident who embraced the notion of desegregation on moral grounds. She moved from New York to Huntsville in 1957 and brought with her a passion for minorities' rights and social justice.[31] Prior to her arrival she had been involved with the NAACP and American Civil Liberties Union. She was agitated and galvanized by the separate facilities for blacks and whites that she witnessed in Huntsville. Throughout her tenure in the city she participated in a wide array of socially uplifting programs that facilitated social transformation throughout the community. These included involvement in the Civil Rights Movement, anti-poverty programs, voter registration drives, women's rights groups, as well as advocating gay rights.[32] Copeland represents an important example of how social change and prosperity were implemented in Huntsville, not as a result of the missile or space programs, but as a result of the efforts of individuals throughout the community.

Economic Diversification:

Huntsville's initial economic boom can be attributed to the arrival of the Army's rocket program in 1950 and the subsequent opening of the Marshall Space Flight Center in 1960. By 1961 the MSFC employed 5,688 permanent employees, and indirectly employed 43,100 residents of Madison County.[33] The number peaked in 1965 at 7,327 employees, but in subsequent years it saw a decline, and by 1993 it housed 3,626 employees.[34] Huntsville experienced economic stagnation, but not turmoil in the late 1960s due to diminished government funding, particularly in the NASA department. The country had achieved its objective of beating the Soviets to the Moon and popular interest in the "Space Race" began to decline. The city experienced the ramifications of having an economy dependent on one commodity in the past. Prior to World War II it was cotton and textiles, during the war it was the munitions and chemicals plants at the Arsenal, and after the war it was rockets. Huntsville leaders recognized the potential hazards inherent in such a situation and in 1944 they established the Industrial Expansion Committee (IEC) to promote economic diversification. In 1961 the IEC evolved into the Huntsville–Madison County Industrial Expansion Committee, and Guy Nerren became its first permanent, paid director.[35] The process of economically diversifying Huntsville and alleviating its reliance on government contracts began long before the reality of space cuts became an issue, as its roots in 1944 demonstrate. The committee, comprised of city leaders and members of the community, was instrumental in bringing new industry to the area. It was not funded by the municipality, but rather it was funded by the businessmen themselves. They set out on a campaign to market Huntsville as a viable community to bring businesses. In August 1967 the committee made its first big breakthrough in diversification when Automatic Electric decided to locate in Huntsville, providing 2,500 non–government jobs.[36]

In order to attract diversified industry to Huntsville, city leaders took note of the fact that there were a number of factors that compelled a business to locate in an area. One of these factors that this research addresses is the inception of the Port of Huntsville in 1967. In the early 1960s it became apparent to city leaders that Huntsville's airport was inadequately suited to cater to the city's booming population and commercial activity. In 1954 the airport had over 19,000 passengers, but by 1965 the number drastically increased to over 336,000.[37] The Huntsville Madison County Airport Authority was conceived and tasked with the creation of a new jetport, as it was called. Under the leadership of Chairman J.E. Mitchell Jr., the authority achieved its goal and on October 29, 1967 the Port of Huntsville was opened. It served as Huntsville's inland port and provided the growing city with a viable means of transporting its commercial activity. It also facilitated transportation between NASA and its west coast contractors, but more importantly for long term growth, it made Huntsville more appealing to prospective industry. Additionally, city leaders took the opportunity to further promote Huntsville's image a rocket city. In a "Special Report to the City of Huntsville," the Airport Authority stated one of its criteria for the new airport was to "Establish a theme of space and/or openness."[38] The airport served as an important point of entry for many travelers and was often their first impression of the city. By displaying Huntsville's rocket city theme throughout the airport, city leadership contributed to the establishment of Huntsville's reputation as a space and rocket city. Coincidentally, the same year the port was opened, the HIEC succeeded in paving the way for economic diversification when it convinced Automatic Electric to locate in the city. The port continues to serve as an important source of economic activity for the city today. Its importance and continued relevance today is a further testament to a legacy of expansion that was established by city leadership during the city's renaissance.

Huntsville's success at avoiding an economic crisis following the decline of federal funding led other towns and municipalities to emulate its approach to diversification. The city of Tuscaloosa sent its leaders to learn from Huntsville's experience. The *Tuscaloosa News* attributed the success of the IEC to the following factors: collaboration between city officials; collaboration between city officials and citizens; the efforts of a business community not solely motivated profit, but for the cultivation of the community; collaboration between civilian and military officials; and separation of the IEC from the Chamber of Commerce.[39] Huntsville's economic success was further facilitated by inexpensive electrical power derived from the Tennessee River.[40]

While other municipalities across the country were feeling the economic ramifications of diminished government funding, Huntsville's leaders managed to avoid such a crisis. Brevard County, Fla. is another example of an area whose economic livelihood and prosperity was dependent on aerospace and government funding. The county, home to the Cape Canaveral launch pad, experienced economic woes in the late 1960s and 1970s as a result of the federal budget cuts, and accordingly, to a lack of cooperation between municipalities and local organizations.[41] It went from the nation's fastest growing county to one with an unemployment rate of 6.2% by 1970.[42] The economic plight of Brevard County in the 1970s is a similar tale to Huntsville's experience in the mid-1940s when the military closed the munitions and chemical plants.

The Establishment of the Astronomical Society as an Example of Cultural Expansion

Huntsville experienced tremendous cultural expansion during the 1950s and 1960s due, once again, to the efforts of the community. Among these cultural developments was the establishment of the Astronomical Society. The "Rocket City Astronomical Association" was founded in 1955 and engineers and scientists from Redstone Arsenal took a keen interest in the endeavor. The idea for the observatory was proposed to scientists at Redstone by 16-year-old Huntsville High student, Sammy Pruitt, and some of his classmates in 1954.[43] The observatory was built on Monte Santo Mountain by the scientist, engineers, and other members of the community, and among them was WvB. Additionally, they raised funding for a telescope, and when they initially encountered pecuniary troubles they received funding from business firms and individuals that allowed them to construct the observatory.[44] The opening of the Huntsville's first astronomical society is a testament to its evolution from a cotton town to a mainstream, culturally expanding city, united by a vision of the cosmos. The observatory served a scientific purpose, but it also served as a source of scientific inspiration for the youth of the community. The observatory is still exists today, but it was subsequently renamed the Von Braun Astronomical Society, after the man who served as its first president.

Conclusion

In sum, the start of the "Huntsville Renaissance" can be traced to the arrival of the Army's rocket program and German engineers and scientists, such as WvB. Initially these programs bonded Huntsville's economic livelihood solely to the federal government. But they also served as a catalyst for change that was felt in all facets of Huntsville society. The town's story in the 1950s and 1960s demonstrates that the plight of a community encouraging mainstream growth is its constant need for adaptation. In each instance of growth, whether it was culturally, socially, or economically, the common factor can be traced to the efforts of the community and its leaders. City leaders, such as Milton Cummings and WvB, were prudent enough to foresee the importance of a diversified economy, and their success continues to be built upon today. They were also wise enough to see that Huntsville's long term success and continual growth were dependent upon racial integration. They took a proactive approach to ending segregation, and along with support from a geographically diverse community, they succeeded in quietly and non-violently bringing segregation to an end at a time when it had taken on a more radical and violent nature in other parts of the state. The municipality and newspapers marketed Huntsville as a modern, progressive city and helped bring international attention to "America's Rocket City." Cooperation and communication between the municipality, city leaders, and the citizens were imperative to their success. Non-profit organization, such as the Huntsville Industrial Expansion Committee and the Airport Authority were separate entities from the local government, and generated their own funding. However, they shared a common goal: to devote the time, energy, and resources necessary to ensure Huntsville could continue mainstream growth into the future. In order to do this the city also had to develop itself culturally in order to attract prospective industry and residents. The decline of federal spending and budget cuts at the end of the 1960s did not affect Huntsville to the same extent as communities, as the example of Brevard County illustrated, and this can be directly attributed to the efforts of the community itself. Huntsville's continued growth and prosperity today is a byproduct of successful practices, both economically and socioculturally, that were implemented during the 1950s and 1960s. In 2008 Forbes Magazine ranked Huntsville as one of the top five cities in the US to ride out the recession.[45] Its success is a testament to the prosperity generated through community cooperation, and its legacy continues to be relevant today. One of the major themes of today's world is the notion of "globalization," and Huntsville represented a microcosm of cooperation among a variety of cultures and ideas, not for the enrichment of the individual, but for the enrichment of the community as whole, which could be applied to the modern world in terms of global interaction. New industries are constantly emerging and inevitably other towns and cities will experience a growth scenario similar to Huntsville's. The past is a prudent teacher, and the story of Huntsville during the renaissance holds many important lessons for a community seeking to generate long term growth.

From the First to the Future - Space Camp®

By Holly Ralston

Serendipity. There are few occasions when the word is sincerely and perfectly placed. But every now and then, an idea is given voice and a chorus of, "Why not!" stirs the air in a way that becomes a collective breath. Circumstances and people merge to provide sustenance and an idea takes on a life beyond words. Just so, Space Camp and serendipity are perfectly paired.

Excerpt, 1982 Space Camp trainee logbook:
Was your mission successful? "Yes, everybody lived."

It was 1975. Three years earlier, in 1972, Dr. Wernher von Braun had left his work at NASA headquarters and taken an executive role with Fairchild Industries in Germantown, Maryland. The decision evidenced a parting of ways between von Braun and U.S. leadership regarding forward strategy in manned spaceflight. In January of the same year, then U.S. President Richard Nixon announced development of the shuttle program featuring a reusable craft and launch system that would take more than nine years to reach fruition. Simultaneously, discussions regarding the longevity of the Apollo program had grown increasingly heated. In December, after Apollo 17 launched and returned astronauts safely to Earth, the program experienced what many considered to be a premature demise.

This particular day in 1975 found von Braun visiting Huntsville and the U.S. Space & Rocket Center. According to the center's founding executive director, Edward O. Buckbee, it was a favorite pastime for the rocket scientist. "He loved walking through the facility to see what we had," said Buckbee. It wasn't unusual, he added, to find von Braun opening a cockpit to play pilot.

"Then we walked around the Rocket Park and there was a group of children, what we called a yellow school bus tour." Buckbee explained that the students would be at the center most of the day to tour and then climb back on the bus for the ride home. When asked by von Braun whether he thought the students were gaining appreciable knowledge, Buckbee recalling saying, "I think they are, but I bet they would love to climb into that spacecraft over there, put on space suits or snack on space food."

That's when it happened. It happened in 1975, in Rocket Park amid a yellow school bus tour, in a time following cancelled Moon missions and a yet-to-be-realized shuttle program. It emerged from concern about keeping manned spaceflight alive and moving ahead. According to Buckbee, von Braun next commented, "We have all kinds of camps in this country, football camps, cheerleading camps and so forth, why not a space science camp?"

A flurry of letters commenced. "He liked to say, 'Let's give them a glimpse of their future,'" said Buckbee. They decided Space Camp would focus on technology, an imperative in von Braun's words to "open up their minds and let them think about what they can contribute to the future." Tours of nearby Marshall Space Flight Center, talks by NASA scientists and engineers and eventual participation by aerospace companies would ultimately lead to the creation and launch of a program that counts among its alumni several astronauts, numerous entrepreneurs, children of national and international leaders, celebrities, special needs populations and gifted, as well as average students.

With such a pivotal role in the creation of Space Camp, it seems a travesty that von Braun never saw its introduction. In 1977, Wernher von Braun died of cancer. His work lives on, however, through the hundreds of thousands of students who have experienced Space Camp and its sister programs Space Academy, Aviation Challenge and Robotics Camp.

Following the 1981 launch of Space Shuttle Columbia and the first mission for NASA's space transportation system, Buckbee was close to having Space Camp (for many years U.S. Space Camp but later changed to Space Camp) ready as a multi-day, immersive program. This was a time of resurgence in manned spaceflight. The community and nation were once again excited about space and NASA Marshall Space Flight Center was buzzing with activity. An almost daylong tour at Marshall could include a plethora of happenings: astronaut training at the neutral buoyancy tank, shuttle engine test firings, engineering projects and a variety of other projects.

Space Camp had a cadence of days one through five, with a shuttle mission as the culminating event. The youth program was vetted with academic personnel at the University of Alabama in Huntsville, as well as Auburn University.

"One of the things we found out immediately was these kids just finished school tests and the last thing they want to do is walk in here and be tested again. We threw out the pretest and posttest before we ever opened Space Camp," said Buckbee. This was counter to advice from higher education but the decision stood.

The absence of testing in no way reflected a lessening of standards or content. Remnants of von Braun's rocket team joined early, providing hands-on expertise and passion that would have found purchase in university halls. Scientists including Ernst Stuhlinger, Konrad Dannenberg, Willie Prasthofer and Georg von Tiesenhausen, representing original and later members of von Braun's rocket team, carried the fire. Oscar Holderer, another member of the rocket team, designed immersive astronaut-training simulators to complement Space Camp's signature character. Engineers and scientists from industry and throughout the Marshall ranks would eventually populate presentations and special guest lectures. Their presentations on propulsion, engine design, lunar rover, guidance systems, future space and a wide variety of topics found rapt audiences.

"By the second or third day, we had assigned Space Camp trainees their position, so they were beginning to learn what a mission specialist is supposed to do, what a flight director was supposed to do. By Friday, they were tuned into the mission," said Buckbee. Training instilled the importance of teamwork and leadership, as well as technology know-how, communication and process. "The mission truly was the test. That was the smartest thing we ever did."

Was your mission successful? The question has been a constant at Space Camp since the very beginning. Life is not a given for crewmembers of either an actual manned mission or a Space Camp mission. This lesson was never more deeply felt than during the 1986 Space Camp season. The year began with a sobering event. On January 28, 1986, Space Shuttle Challenger exploded 73 seconds after launch, killing all seven of its crewmembers.

Somewhat counter to the disaster, about five months later, the upbeat movie *SpaceCamp* was released in theaters. With such a mix of media exposure and news events, what would be the impact to the Huntsville-based Space Camp?

Going into the 1987 season, Space Camp experienced a phenomenal increase in attendance and demand. As a brand, Space Camp became firmly established. News coverage and guest appearances on popular television programming mounted. According to Buckbee, exploding interest prompted on-campus growth that established the Space Camp Habitat and Training Center. These capital projects, running into millions of dollars, gave rise to corporate sponsorships ranging from Coca-Cola to the world's largest aerospace and defense companies. Interest in additional domestic and international Space Camp locations soon emerged.

Today, Space Camp is recognized worldwide as a pioneer in workforce development programs emphasizing science, technology, engineering and math. Cultivating awareness and talent for high technology industries is among primary program goals.

Space Camp enhances lives. It's true. The science and romance of manned space flight resonates with young and old, those near and far and among a variety of populations. Space Camp grew to include a wider age range with Space Academy®, plus educator, adult, corporate and family camps. Enhanced options for multiple-year attendance emerged at the *Top Gun*-like Aviation Challenge® and Robotics Camp.

Hands-on, immersive and relevant: It's the way Space Camp transforms lives according to Dr. Deborah Barnhart, current chief executive officer of the U.S. Space & Rocket Center. "Space Camp isn't about an electronic, handheld flat screen. We are putting their entire being into the simulator, into the cockpit, into the 5DF*, into the MAT*, into the water, into the tank, into the centrifuge. It's a physical, macro immersion into technology."

According to Buckbee, Barnhart was largely responsible for putting the sizzle in Space Camp. She served as Space Camp director from 1986 to 1990 and wrote the curriculum for advanced Space Academy, the teachers' program and Aviation Challenge. She exited campus to advance her education and join the ranks of industry executives. In late 2010, she accepted the offer to become chief executive over the Rocket Center and Camp.

"I wanted to devote my life to life-changing work, to a place where you can see the results of your efforts in the elevated aspirations of our graduates, young and older," Barnhart said.

Barnhart believes one of the most impactful elements of Space Camp is what she calls the global clash. "We literally have students from all 50 states and around the world." She notes that a full 20 percent of attendance is international. "So when they come here, they enter into a global exchange, a global clash of ideas and ideologies." She explained attendees begin to sense the world as both a smaller place and as a unique entity providing home to many different kinds of people. "I believe that global interface wakes them up to the idea we must all work together on this planet, both to save the planet and to leave the planet."

"Regardless of the outcome of their Space Camp mission, we believe our programs encourage trainees to live more fully, more purposefully and more knowledgeably," added Barnhart.

**Space Camp simulators used to teach physics and human response associated with manned spaceflight. Originally designed by Oscar Holderer.*

How the Next Generation Can Become the Mars Generation

By Dr. Deborah Barnhart

Space Exploration is in our DNA. As a child of the Wernher von Braun moon landing generation, I want to see humans set foot on Mars.

Most days after school, the windows of our homes would rattle with powerful thunder for 35 miles in every direction from Huntsville. NASA was test-firing the F-1 moon rocket engines in preparation for the lunar landing mission. We put our fingers on the big picture windows to feel the rumble. That thunder made our young chests pound, and it has never stopped. Harnessing 1.5 million pounds of thrust in full fire for almost four minutes just to escape gravity, the F-1 Saturn boosters were so remarkable that never since have we launched a more powerful engine.

Our parents put man on the moon, but how can we ensure that our children reach Mars? How do we take that second wobbly step off the planet? Can the next generation become the greatest space generation? It's now been almost 50 years since America landed on the moon. Even today's retirees were in their teens and younger when their grainy black and white TVs transmitted mankind's giant leap. Where will this generation, weaned on satellite TV, internet and cell phones, get its motivation to explore further?

Each generation bemoans the quality of the next: "How can these Twinkie eating kids raised on credit cards, iPods and Wikipedia possibly get to Mars!?" What generation wouldn't pale compared to the Moon Generation? We, their hippie-Boomer children, slugged through 135 missions of the Space Transportation System, better known as Space Shuttle. We took for granted this magnificent space truck that built the International Space Station (ISS). The achievements of Apollo were our birthright, but our mission, by definition, set our sights lower, literally low Earth orbit. The moon is 235,000 miles away, ISS less than one-tenth of that figure. Getting us to the moon was an unlikely successful enterprise. At more than 36 million miles away, the task of going to Mars, by any measure, is far more daunting.

The Mars Generation

The Mars Generation is living in both the best and worst of times. At base, their mortal survival is in question. Shockingly, this is the first generation in history to have a shorter projected lifespan than their parents. But they also have exponentially increasing access to knowledge and information. They are the most informed, the most tech savvy, and the most aware of others around them. And there are more of them. The Mars Generation population in America (born between 1980 and 2000) boasts 82 million souls versus Gen X (1965-1980) at 60 million. They even exceed the 78 million Boomers (1946-1964).

Whatever the cause, this generation appears to have lower physical vitality and more health issues at an earlier age. They are fatter, sicker and more allergic than their parents. Forty percent of American and European children are overweight, and another 30 percent are clinically obese. They are seen by physicians for asthma, allergy and mental conditions five times more frequently than children of the 1950s. At the same time, they have the opportunity to live much longer than previous generations, given the advances in genomic research, organic prosthetics and nutritional information. Can they thrive on Earth long enough to achieve the journey to Mars?

They are hardwired almost non-stop to devices. While they are electronically mainstreamed, they are more personally insular. Many require basic face-to-face social training just to interview and qualify for jobs. Seldom do they move between social strata. Directed instead to groups, schools and activities only in their socioeconomic coteries, they have little interface or consciousness of other cultures and peoples. the exception is in pop culture ecosystems that surround things likemusic, gaming and myriad fandom cultures.

Fearing modern threats, their helicopter parents are excessively sheltering the younger cohort of this generation as well. With less self-sufficiency at early ages, they will have a hard time looking to the future and seeing themselves without external filters. They believe in the American dream, but are allowed less unchaperoned time and travel and are homeschooled. They don't have chores and they have fewer jobs in their adolescence, and they join the military at less than half the rate of even the all-volunteer military of the 1980s. These characteristics, along with other features of our post-Cold-War culture, lead to isolation from other peoples of the nation and planet.

As for their merits, the potential Mars Generation is fearless and unapologetic. Born with little to fear – no World War II, no real nuclear threat - they are confident of the promise of food and a future. They lay claim to every adult privilege without hesitation, impatient to access the whole of life, rather than taking it sequentially. They dive into technology, media, ideology, and the pursuit of a meaningful life with aplomb and without apology. They do things that would have been unimaginable when Neil Armstrong stepped outside to hop on the moon just 50 years ago. They build robots, program computers to connect the globe, make movies with their phones and play electronic games with people around the planet they've never even met.

And they can do it all at once. Mars Gens growing up in the age of technology can multi task more skillfully than any preceding generation. They don't use technology as a tool requiring the deep concentration of sequential logic; they use technology as an extension of their communications. Their immersion in gaming and social media has created a strengthened natural capability to do two or more tasks at the same time. They can use technology without concentrating on the mechanics of typing, navigating or game console control, including the special moves that take multiple button actions to implement. This skill will serve them well in problem solving for space missions.

While they don't yet travel by spaceship, they are without boundaries. Because of popular culture, ubiquitous travel and the fusion of music, the emancipated Mars Generation is conscious of other nations, languages and ethnicities to an unprecedented level. The internet, YouTube and WeChat bring scenes from every corner of the developed world, unceasingly. Their intellectual itch can be constantly scratched. There is no hurdle to learning anything, anytime, almost anywhere.

Despite this and other litanies of pros and cons, generations are defined by their accomplishments, not their aspirations. There will only be a Mars Generation when we step foot on Mars. Space Campers I know are already thinking about what their first words from Mars will be. Probably in 140 characters or less!

The Recipe

So will this dichotomous generation excel beyond even their own ambitions, achieving new human records or will they teeter toward extinction due to declining physical health or hubris? Are they doomed to the fate of humans in Wall-E, to be large and phlegmatic? If they rally their strengths in the recipe that follows, the fate of Mars, as well as Earth, is theirs to determine. Here's how:

Determine to Go

The first step is to determine to go to Mars. Capability is not enough. Just because you own a car doesn't mean you're going to drive from Alaska to Chile. When it comes to conquering Mars, set the destination and stay with it. There may be building-block destinations along the path – Lagrange points, the moon, Phobos, Deimos – but Mars itself is the big goal. The journey itself will yield untold benefits in its wake, just as the Apollo program did in communications, telemetry, medicine and computing.

JFK's galvanizing speech was a perfect example of setting a sizeable goal with a solid determination. "Putting a man on the moon and returning him safely to Earth, within this decade" became both a mission and a mantra.

The NASA budget Kennedy's Congress had wasn't appropriated to match the task, but the team stayed on mission and rallied taxpayers to the dream. Public and congressional will is critical to success. In one example, a grade-school Space Camp trainee was interviewed by a reporter following his graduation. "Well, young man, if you were head of NASA, what would you do?" Immediately, the beaming graduate confidently stated his missions: "I'd ensure that International Space Station was permanently manned by the partner nations, send men back to the moon to begin regolith operations, and send a manned mission to Mars." The reporter harrumphed, "That's pretty expensive. I doubt Congress would support all that!" Undaunted, the youngster replied, "Your Congress might not, but mine will."

Work like hell and advertise

Dr. von Braun was keenly sensitive to such public, media – and congressional – opinion. He constantly directed naysayers about the moon mission to the source. "Go talk to the man who pays the bills," he suggested, directing critics to the nation's taxpayers. His confidence in their enthusiasm, combined with his work ethic, undoubtedly saved the program.

One of his favorite sayings, and his work mantra, was tried and true: "Late to bed and early to rise, work like hell and advertise." Imagine the courage and boldness of a war refugee, widely mistrusted for his role in World War II (with English as a second language) who was engaging, compelling and confident enough to convince the U.S. Congress, a young JFK and the American people, to go to the moon.

Listen to the old thunder

While touring a NASA Center this past summer, an intern cheerfully toured us around the propulsion laboratory. "Here's a Saturn V F-1 engine! We are taking it apart to see how it worked," she chimed. I immediately had two thoughts: 1. Good that you're able to have hands-on, real hardware, and 2. WvB would cry if he saw that 45 years later, we haven't yet moved exponentially past the mighty F-1.

When you realize that the Apollo moon rocket's launch vehicle digital computer had less RAM than a recordable Hallmark greeting card, you understand there was value in the old ways of accomplishment. Take the lessons of those who have gone before, who did it with only a solid math education and a slide rule. Listen to the lessons of the robust simplicity of the Russian program with its uncomplicated but rugged engineering. Build and test. Build and fire. Blow up or go up. Risk aversion because of safety and cost can be fatal to innovation, but foolhardiness can stop progress in its tracks. The Moon Generation was gifted with courage and prudence in addition to their technical skill. Most importantly, don't sit still and don't lose the legacy you have inherited from Wernher von Braun and his contemporaries, and the progress that has been made during Shuttle and Space Launch System programs.

Get serious about ISS and Commercial Space

It's time to leverage the learning from ISS. Much has been learned from living and working in space for the past 15 years since ISS was fully manned, and we still have 15 more years to learn from it. But its scientific contributions have been unheralded and sub-optimized. Our American arrogance pushed us to begin what would eventually become ISS alone as Space Station Freedom. We soon learned that it takes the planet, not only for the money, but also for the collective human will, that it takes to commit to and maintain such a substantive enterprise.

ISS was positioned to be a groundbreaking oasis for manufacturing, medicine and technology development. Crystal growth and pharmaceutical manufacturing were supposed to be transformed by the microgravity environment of an orbiting space station. While there have been technical setbacks and economies of scale that have hampered the original vision, we must leverage commercial promise into reality on this sophisticated satellite. Mars explorers should get serious about exploiting the unique station environment by proposing and funding X Prizes, broader commercial transportation and utility access, and a hospitality mat to tourism. With the capabilities of Bigelow Aerospace's inflatable modules and new spacecraft by Lockheed, Boeing, Blue Origin, Sierra Nevada Corporation, Virgin Galactic and SpaceX, many more short term visitors and experiments should be accommodated, inspiring and exploiting the practice and promise of low Earth orbit.

Gather the globe

One lesson that ISS teaches us is that multi national beats going it alone when it comes to major tasks in space. The laboratories provided by the Europeans, Russians and the Japanese multiplied science on ISS three-fold. It doesn't just take a village, it takes a planet to go to Mars. The oft-quoted adage holds true to space: "If you want to go fast, go alone, if you want to go far, go together."

A generation that colonizes Mars should get serious today about gender and cultures. Haven't we had enough proof that men and women can accomplish different but equally valuable goals? Isn't our understanding of the peoples of the Earth broad enough in the 21st century to capitalize on ALL cultures and dispense with the idea that achievement is limited? To get to Mars, get serious about women and all cultures, with their diverse capabilities.

Exploit technology to revolutionize education

The sad tech education joke of the past generation was that it took 50 years to get the overhead projector from the bowling alley into the classroom. A Mars generation will need to translate the use of technology for education on the exponential curve facilitated by introduction of the internet and the best devices it can enable.

With laptops replacing textbooks, and new research challenging the traditions of didacticism, stay on the path to accelerated learning and push the envelope. Challenge the system that takes 12 years to imbue what might be absorbed in eight. Now that we are mastering homeschooling, virtual school, online classes and alternatives like Khan Academy, it is time to press further. The future is for doers and makers as well as learners.

Get serious about math education. The Program for International Student Assessment, ranks the US at 26th in math out of the 34 Organization for Economic Cooperation and Development countries. It's not good enough for American achievement, and it's not good enough to lead the world to Mars.

Mars Gens need to leverage their incredible electronic access, and their moxie in using it, to slingshot themselves to the top of the math and science ranks. Do whatever it takes in terms of access, school structures and motivation to change this one sour fact: Americans are behind in math.

Save the planet...and its children

One of the potential reasons for accessing Mars is to create a future home for Earthlings, whether out of necessity or for the expansion of human understanding. The Mars Generation should rage against this possibility of the former scenario and vigorously embrace the latter. They should go all out to preserve the Earth, rather than to despoil and abandon it. Before focusing on terraforming Mars, we need to focus on preservation and sustainability of our planet.

Space travel is fundamentally about environmental control and life support. As we learn what is required to sustain life in space for long and difficult journeys, we should immediately apply that knowledge to improving Earth's environment. Survival on Earth and in space presents similar resource management problems of scarcity, adaptation and conservation. As we increase our skill in space travel, let us first apply that knowledge to preservation of our planet so that leaving it is our choice, and never a necessity.

Regardless of what we accomplish in space, or in which generation, there is no higher duty that taking the best possible care of our children. That doesn't mean over parenting. Overindulgence fosters environmental deterioration, a poor diet, inequitable opportunity and a reduced life expectancy. We must endeavor to compensate and remediate, and instill in the Mars Generation this responsibility. No culture can rise without the strength and willpower of its young.

There's plenty of room for parenting advice but for the Mars Gen the best may be the "roots and wings" legacy. Give them a frontier upbringing; make room for them to benefit from mistakes and false starts. Inspire those who follow you but let them take the helm into their own hands.

Perhaps the Mars Generation would also do well to consider the implementation of compulsory national service for Americans, both male and female. Taking a year or even two at age 18 or 19 would yield maturation and understanding of the world that no other experience could provide. National service, not just military, but also perhaps reading, teaching, firefighting, law enforcement or elder care, would introduce young people to potential careers, provide meaningful, paying jobs early in life, and increase cultural exposure and acceptance. It would broaden their horizons and make the world a smaller place at the same time.

Work at the gritty business of governance

The unifying experience of the European Space Agency will prove to be a major tool in the toolbox of conquering space. The formation of the ESA's roles preceding the birth of the truly international orbiting space station took decades. Their funding processes and decision-making was notoriously deliberate (i.e. slow). These may be precisely the skills we will need for establishing governance in the space diaspora.

Let space be the high ground, not a battlefield. As the high ground, each space faring nation will naturally desire to have a strategic advantage. The dilemma plays out in a chess game of trust: cooperate to get there, but watch your back and keep your edge.

While we won't be settling the moon or Mars in the near future, issues of sovereignty and economic impact may quickly come into play. If China lands on the moon and begins to mine helium-3, is it theirs alone to take? If India establishes an independent moon base habitat, are they like Sooners who have staked a claim? What if a meteor is inbound to planet Earth on a trajectory to hit a helpless developing country, which space-faring nation's responsibility is it to take aggressive action to prevent disaster?

A possible forum for this activity is the establishment of a U.S. or even International Space Academy. The U.N. Committee on the Peaceful Uses of Outer Space comes as close to a space-governing body as any organization, but it has no provisions for operational response. There are erudite policy groups such as Georgetown University, and there is International Space University and its groundbreaking work in space agency policy. But there is no international body that enforces policy on the space frontier. Current space maritime laws are poorly defined and unenforced. Stratospheric air space is ruled by multi national agreement, but beyond that, there is an infinitude of space that has no jurisdiction, or laws. Due to the proliferation of low Earth orbit satellites in our lifetime, space junk has become dangerously voluminous and virtually unregulated.

The Mars Generation will have the task of setting the political and economic conditions for opening the frontiers of space in their lifetime. If they do it right, it will serve as a template for peaceful, pan-galactic jurisdiction that can optimize the utilization and eventual occupation of space for the betterment of all humanity. If not, Star Wars may well become their destiny.

Persist, and don't procrastinate

Wernher von Braun planned for at least 20 Apollo missions. When Congress truncated the program, he just went ahead with America's first space station, the Skylab. Never really approved by NASA headquarters or Congress, the von Braun team simply took the third stage of the Saturn V and built an orbital workshop and living quarters that launched as Skylab I. The launches of Skylab, and all three missions including the longest time on orbit to date at the time, were accomplished in just one year.

Even while Apollo was in full swing, WvB had the team working on a space station. Although the funding and glory days of moon landings were long gone, Space Station Freedom proceeded. Freedom had an original projected completion date of 1986. Politics, funding gaps, new partners and Shuttle losses did not stop the program. Even when it passed by just one vote in Congress in 1992, the people behind the program rallied support of the nation and persisted, finally ending with the complete assembly of ISS in 2007.

Persistence on Constellation or Shuttle programs would have prevented us being solely dependent on the Russians to access ISS since the retirement of the Shuttle in 2011. While Shuttle was not a wholly premature ending (it finished its mission as a truck to build a space station), terminating the program left America without a self-provided ride into space for the first time in the history – never a strong position.

At all costs, Mars Gens must learn the lessons of the unreliability of funding in the Congress. No mission to change mankind can be accomplished in one year of funding. Consider the pointless change of Constellation program to Space Launch System. At the end of the day, the programs have essentially the same goals, with new names after the starts and stops that cost millions and needless delay.

Set a goal and go. Don't concern yourself too much too soon with how you are going to get there – you will invariably find yourself in unfamiliar territory. There will be obstacles that seem insurmountable. A clear and strong vision and goal will allow adaptability and ingenuity to find the best course, while aversion and fear will allow only stagnation and atrophy. Oliver Wendell Holmes said it best, "It matters not the course we sail. We sail at times with the wind, at times against the wind. But sail we must, and never drift, nor anchor."

Carry and Pass the Torch

Whether ours, or the next, or the one after that, is the actual generation to occupy Mars, it is up to each to carry the torch in this direction if we are to continue as a space faring people.

Until we can make the next giant leap, we must continue to make the small steps on our watch: Continue to live and work in near-Earth orbit, or at the Lagrange points, return to the moon, occupy Mars and its environs, establish peaceful oceans in space, capture and control asteroids. Continue to encourage the imaginations of the nascent Mars Generation. Engage them about their futures in their schools, their jobs, and their social lives. The most grievous sin they can commit is not dreaming big enough. In order for the Mars Generation to realize its full potential, we must, as Wernher von Braun encouraged, carry and pass the torch.

"When the craft is ready and the oceans of space are calm… the space-age Columbus and Magellan are presently sitting somewhere today in a public school house preparing for an adventure. Here are the people to who we shall pass the baton. But the first lap of the race is ours. And we shall not falter."

BIBLIOGRAPHY

PART I: BEGINNING

Charles L. Bradshaw. *Rockets, Reactors and Computers Define the 20th Century.* Providence House Publishers, 2007.

The Visionaries—Setting the Stage for the Greatest Space Generation

Tom Brokaw, *The Greatest Generation*, Random House, New York, New York, 1998.

Esther C. Goddard, Editor and G. Edward Pendray, Associate Editor, *The Papers of Robert H. Goddard*, Three Volumes, McGraw-Hill, New York, New York, 1970.

Hermann Oberth, *The Rocket Into Planetary Space*, English Translation by Trevor C. Sorensen, *et al.*, Walter de Gruyter GmbH, Berlin, Germany, 2014.

Konstantin E. Tsiolkovskiy, *Collected Works of K.E. Tsiolkovskiy: Volume II–Reactive Flying Machines*, A. A. Blagonravov, Editor-in-Chief for the Academy of Sciences of the USSR, Moscow, 1954; English translation for the United States National Aeronautics and Space Administration, as NASA TT F-237, Washington, D.C., by Faraday Translations, New York, 1965.

Where It All Started

[1]Bergaust 1960

[2]Dunar-Waring 1999

The First Redstone Ballistic Missle Launch Team

First Redstone Ballistic Missile Launch Team Roundtable Discussion, video recording, Tim Hall (Moderator), U. S. Space and Rocket Center (USSRC), October 2015. This and subsequent quotations are from this source.

Jones, *Raymond B., Citizen Soldier: Carl T. Jones*, 2015.

David Newby–Marshall's First Employee Transfer of the WvB Rocket Team

[1]David H. Newby Oral History, Space History Interviews, University of Alabama Huntsville Archives and Special Collections.

[2]Memorandum from the Administrator, 29 April 1959. ABMA Collection, MSFC Historical Reference Collection.

[3]ABMA-NASA Transfer Chronology. October 30, 1959. ABMA Collection, Marshall Historical Reference Collection.

[4]Memorandum. November 3, 1959. ABMA Collection, Marshall Historical Reference Collection.

[5]Summary and Concepts of Transfer Plan, December 11, 1959. ABMA Collection, Marshall Historical Reference Collection.

[6]Andrew J. Dunar and Stephen P. Waring. *Power to Explore: A History of Marshall Space Flight Center 1960-1990.* Washington DC: NASA, 1999. 30.

[7]Michael Neufeld. *Von Braun: Dreamer of Space, Engineer of War.* New York: Knopf, 2007. 346.

[8]*50 Years of Rockets and Spacecraft in the Rocket City.* Paducah: Turner Publishing Company, 2002. (15).

[9]*Power to Explore*, 60-61.

[10]*Marshall Star,* September 29, 1965. Digital Media. Marshall Historical Reference Collection.

Part II: Organization

Space Sciences

The history of the transition of the von Braun team from Fort Bliss to Huntsville and the events leading up to the launch of Explorer I are well-documented in a number of books. Most of the historical material in this Chapter was taken from the following:

Power to Explore – History of the MSFC, 1969-1990, Chapter 1 Origins of the MSFC Authors A.J. Dunar and S.P. Waring, MSFC History office, US Gov't Printing Office. Jan 1, 1999 ISBN 0160589924

David S. Akens, "Historical Origins of the George C. Marshall Space Flight Center," MSFC Historical Monograph No. 1 (Huntsville: MSFC, 1960).

Von Braun: Dreamer of Space, Engineer of War, Michael J. Neufeld, Vintage, Nov. 2008, ISBN:0307389375

(A good review of the political situation facing the von Braun team leading up to the launching of Explorer I)

Major General John B. Medaris with Arthur Gordon, Countdown for Decision (New York: G. P. Putnam's Sons, 1960).

(Gives the Army's perspective of the politics at the time as well as a firsthand account of the Nickerson affair.)

Dr. Space, The Life of Wernher von Braun, Bob Ward, Naval Institute Press, Annapolis Md. 2005, ISBN 1-59114-926-6

(Shows a more personal side of von Braun and the members of his team.)

Accounts of memories and events that took place in the old Research Projects Office are taken from personal communications with Chuck Lundquist, Billy Jones, Bill Snoddy, Don Cochran, Landa Thornton, Stan Fields, John Bensko and from the Donald Tartar Interviews (University of Alabama Library Archives) with George Bucher, Jim Downey, Jim Kingsbury, Chuck Lundquist, Bob Naumann, Bill Snoddy and Russ Shelton.

Most of the papers pertaining to the work carried out by the personnel in the Research Project Office are available either in the Redstone Scientific Information Center or in the University of Alabama Library Archives.

Part III: Management

Management

Arthur L. Slotkin. *Doing the Impossible: George E. Mueller and the Management of NASA's Human Spaceflight Program.* Springer-Praxis Publishing, 2012.

Unless otherwise noted, all quotes compiled by Ed Buckbee, 2009-10.

Part V: Reflections

Intellectual Critical Mass

[1]http://quickfacts.census.gov/qfd/states/01000.html

[2]The Voice of Dr. Wernher von Braun, page 140.

[3]Address by Dr. Wernher von Bruan to a joint session of the Alabama Legislature, June 20, 1961, 11 a.m. – pages 16-17.

[4]Address by Dr. Wernher von Bruan to a joint session of the Alabama Legislature, June 20, 1961, 11 a.m. – page 18.

[5]http://www.nasa.gov/centers/marshall/pdf/159998main_MSFC_Fact_sheet.pdf

The Huntsville Renaissance

[1]Luther J. Carter, "Huntsville: AL Cotton Town Takes off into the Space Age," *American Association for the Advancement of Science* 155 (March 1967): 1224.

[2]Huntsville had tasted the fruits of economic prosperity brought about by federal government contracts during WWII and it had a keen desire for more. Senator John Sparkman was sent to lobby Washington to bring an air force wind tunnel to the area, but lost it to Tullahoma, Tennessee. He settled for the Army's new rocket program and the 120 German scientist and engineers that were to helm the project. Many in Huntsville initially saw it as a consolation prize. For more detail see: Bob Ward, *Dr. Space: The Life of Wernher von Braun* (Annapolis: Naval Institute, 2005), 75–76.

[3]See, for example, Helen B. Walters, *Von Braun: Rocket Engineer* (New York: Macmillan, 1964); David M. Heather, *Wernher von Braun* (New York: Putnam, 1967); John C. Goodrum, *Wernher von Braun: Space Pioneer* (Huntsville: Strode Publishers, 1969).

[4]See, for example, Frederick I. Ordway III and Mitchell R. Sharpe, *The Rocket Team* (New York: Crowell Publishing 1979).

[5]Michael J. Neufeld, *Von Braun* (New York: Random House Inc., 2007) 5.

[6]See, for example, Ernst Stuhlinger and Frederick I. Ordway III, *Wernher von Braun: Crusader for Space* (Florida: Krieger Publishing Company, 1994); Bob Ward, *Dr. Space: The Life of Wernher von Braun* (Annapolis: Naval Institute Press, 2005).

[7]See, for example, Bob Ward, "Rocket City Legacy," in *Dr. Space*, 168-174.

[8]Andrew J. Dunar and Stephen P. Waring, "Marshall Reconstruction," in *Power to Explore: A History of Marshall Space Flight Center 1960-1990* (Washington DC: NASA History Series, 1999); see also, Ed Buckbee, *50 Years of Rockets and Spacecraft: NASA Marshall Space Flight Center* (Missouri: Acclaim Press, 2009).

[9]See Elise Hopkins Stephens, "We Have Lift Off," in *Historic Huntsville: A City of New Beginnings*, (California: Windsor Publications Inc., 1984), 102-128.

[10]Ed Buckbee, e-mail message to author, April 21, 2013.

[11]"A Showcase for the Space Age in Alabama," *New York Times*, October 24, 1965.

[12]Ed Buckbee, email message to author, April 21, 2013.

[13]Luther J. Carter, "Huntsville: AL Cotton Town Takes off into the Space Age," *American Association for the Advancement of Science* 155 (March 1967): 1224.

[14]Pat McCauley, "Shortage of Rental Housing is Critical for Next 4 Months," *The Huntsville Times*, May 14, 1950

[15]Bob Ward, "Construction Booms Over Record High," *The Huntsville Times,* March 1965.

[16]Matthew Dallek, "The Conservative 1960s," The Atlantic Online, entry posted December 1995, http://www.theatlantic.com/past/docs/issues/95dec/conbook/conbook.htm (accessed April 19, 2013).

[17]Erik Bergaust, *Rocket City U.S.A.*, 205-212.

[18]Hugh Merrill, "Some U.S Industrialist Just Don't Fit the Mold," *The Huntsville Times*, October 5, 1969.

[19]Louise Davis, "New Throne For a Cotton King," *The Nashville Tennessean Magazine,* Jan. 19,1964.

[20]Roger Honkanen, "Huntsville Ex-Clerk is One of NASA's Top Contractors," *The Atlanta Journal and Constitution Magazine*, May 31, 1964.

[21]"Research Park Receives Award," *Times Daily,* July 13, 1997.

[22]Martin Swant, "Cummings Research Park: Still Growing Despite Federal Budget Struggles (Outlook 2013)," *The Times*, March 1, 2013, http://www.al.com/business/index.ssf/2013/03/cummings_research_park_no_grow.html (accessed March 8, 2013).

[23]"Warning in Alabama: Governor Says Integration Will Bring Use of Troops," *The New York Times*, November 14, 1959.

[24]Carter, "Huntsville," 1228.

[25]Ibid., 128.

[26]"Huntsville Integrates Lunch Counters," *The Florence Times*, July 10, 1962.

[27]Don Cox, "City Hires First Negro Policeman," *The Huntsville Times,* May 17, 1963.

[28]Ibid.

[29]Pat Houtz, "Graham Gives Views, Readies for a Rally," *The Huntsville Times*, August 25, 1962.

[30]Huntsville Industrial Expansion Committee, *Information Kit: Facts & Figures on living in Huntsville Madison County Alabama* (Huntsville, 1966).

[31]Melinda Gorham, *The Huntsville Times*, June 6, 1999.

[32]Ibid.

[33]Andrew J. Dunar and Stephen P. Waring, *Power to Explore: A History of Marshall Space Flight Center 1960-1990* (Washington DC: NASA History Series, 1999) 616.

[34]Ibid., 616.

[35]Jack Hartsfield, "Huntsville Beat Aerospace Doldrums, Now Boasts a Diversified Employment," *The Huntsville Times*, March 26, 1978.

[36]Ibid.

[37]Huntsville Industrial Expansion Committee, *Information Kit: Facts & Figures on living in Huntsville Madison County Alabama* (Huntsville, 1966).

[38]Huntsville Madison County Airport Authority, *Special Report to City of Huntsville & County of Madison Terminal Complex Plans* (Huntsville, 1966), 2.

[39]"When Citizens are Active," *The Tuscaloosa News,* April 30, 1964.

[40]Edward C. Burks, "Industry Mushrooms in North Alabama," *The New York Times*, May 30, 1965.

[41]Hartsfield, "Huntsville Beat Aerospace Doldrums."

[42]Elizabeth Whitney, "No Golden Eggs in Brevard, But Silver Lining?" *The St. Petersburg Times*, September 15, 1970.

[43]Monte Sano Observatory Built by Employees," *The Marshall Star*, January 18, 1961.

[44]Ibid.

[45]Helen Coster, "Affordable Places to Weather the Downturn," *Forbes*, November 11, 2008, http://www.forbes.com/2008/11/12/cheap-cities-affordable-forbeslife-cx_hc_1112realestate.html (accessed March 14, 2013).

Secondary Sources

Buckbee, Ed. *50 Years of Rockets and Spacecraft: NASA Marshall Space Flight Center*. Missouri: Acclaim Press, 2009.

Coster, Helen. "Affordable Places to Weather the Downturn." *Forbes*. November 11, 2008. http://www.forbes.com/2008/11/12/cheap-cities-affordable-forbeslife-cx_hc_1112realestate.html (accessed March 14, 2013).

Dallek, Matthew. "The Conservative 1960s." The Atlantic Online, entry posted December 1995. http://www.theatlantic.com/past/docs/issues/95dec/conbook/conbook.htm (accessed April 19, 2013).

Dunar, Andrew, and Stephen P. Waring. *Power to Explore: A History of Marshall Space Flight Center 1960-1990*. Washington DC: NASA History Series, 1999.

Goodrum, John C. *Wernher von Braun: Space Pioneer.* Huntsville: Strode Publishers, 1969.

Heather, David M. *Wernher von Braun*. New York: Putnam, 1967.

Neufeld, Michael J. *Von Braun*. New York: Random House Inc., 2007.

Ordway III, Frederick, and Mitchell R. Sharpe. *The Rocket Team*. New York: Crowell Publishing 1979.

Stephens, Elise Hopkins. *Historic Huntsville: A City of New Beginnings*. California: Windsor Publications Inc., 1984.

Stuhlinger, Ernst, and Frederick I. Ordway III. *Wernher von Braun: Crusader for Space*. Florida: Krieger Publishing Company, 1994.

Swant, Martin. "Cummings Research Park: Still Growing Despite Federal Budget Struggles (Outlook 2013)." *The Times*. March 1, 2013. http://www.al.com/business/index.ssf/2013/03/cummings_research_park_no_grow.html (accessed March 8, 2013).

"Research Park Receives Award." *Times Daily*. July 13, 1997.

Ward, Bob. *Dr. Space: The Life of Wernher von Braun*. Annapolis: Naval Institute, 2005.

Walters, Helen B. *Von Braun: Rocket Engineer.* New York: Macmillan, 1964.

Primary Sources

"A Showcase for the Space Age in Alabama." *New York Times*. October 24, 1965.

Carter, Luther J. "Huntsville: AL Cotton Town Takes off into the Space Age." *American Association for the Advancement of Science* 155 (March 1967): 1224.

Cox, Don. "City Hires First Negro Policeman." *The Huntsville Times*. May 17, 1963.

Davis, Louise. "New Throne For a Cotton King." *The Nashville Tennessean Magazine.* Jan. 19, 1964.

Edward C. Burks. "Industry Mushrooms in North Alabama." *The New York Times.* May 30, 1965.

Elizabeth Whitney. "No Golden Eggs in Brevard, But Silver Lining?" *The St. Petersburg Times.* September 15, 1970.

Gorham, Melinda. *The Huntsville Times.* June 6, 1999.

Hartsfield, Jack. "Huntsville Beat Aerospace Doldrums, Now Boasts a Diversified Employment." *The Huntsville Times.* March 26, 1978.

Honkanen, Roger. "Huntsville Ex-Clerk is One of NASA's Top Contractors." *The Atlanta Journal and Constitution Magazine.* May 31, 1964.

Houtz, Pat. "Graham Gives Views, Readies for a Rally." *The Huntsville Times.* August 25, 1962.

Huntsville Industrial Expansion Committee. *Information Kit: Facts & Figures on living in Huntsville Madison County Alabama* (Huntsville, 1966).

"Huntsville Integrates Lunch Counters." *The Florence Times.* July 10, 1962.

Huntsville Madison County Airport Authority. *Special Report to City of Huntsville & County of Madison Terminal Complex Plans* (Huntsville, 1966).

McCauley, Pat. "Shortage of Rental Housing is Critical for Next 4 Months." *The Huntsville Times.* May 14, 1950.

Merrill, Hugh. "Some U.S Industrialist Just Don't Fit the Mold." *The Huntsville Times.* October 5, 1969.

"Monte Sano Observatory Built by Employees." *The Marshall Star.* January 18, 1961.

Ward, Bob. "Construction Booms Over Record High." *The Huntsville Times.* March 1965.

"When Citizens are Active." *The Tuscaloosa News.* April 30, 1964.

Index

A

Abernathy, Ralph – 257
Able, *monkey* – 58, 164
ABMA Guidance and Control Laboratory – 59, 61
Adams, Bill – 44, 45
Aden, Bob – 25, 70, 77, 80
Advanced Research Projects Agency (ARPA) – 45, 55, 63
Advanced X-ray Astrophysics Facility (AXAF) – 128
Aeroballistics Laboratory – 65
Air Research and Development Command – 42
Alabama Space & Rocket Center – 166, 175, 188, 204, 240
Albert, Carl – 198, 242
Albright, Eddie – 78
Alcott, Russ – 70
Aldrin, Edwin "Buzz" – 83, 96, 114, 181, 192, 257
Alford, Lionel – 92
Allen, Harry Julian – 41
Allen, Jim – 204
Allen, William – 81
Alsup, Bryce – 81
Altenkirch, Robert – 5, 279
Amalavage, Al – 70
Ames Research Center (ARC) – 58
Anderson, Dorrance – 70, 81
Anderson, Woody – 278
Angele, Wilhelm – 67, 71
Apollo – 6, 7, 41, 211, 256
Apollo 1 – 133, 134
Apollo 4 – 83
Apollo 6 – 83
Apollo 8 – 6, 179
Apollo 11 – 13, 21, 83, 96, 181, 182, 188, 192, 228, 240, 249, 272
Apollo 12 – 83, 181
Apollo 13 – 82
Apollo 14 – 181
Apollo 15 – 181
Apollo 16 – 181
Apollo 17 – 6, 8, 171, 181, 193, 293
Apollo Program – 7, 14, 61, 63, 94, 118, 122, 128, 235, 269
Apollo Saturn 501 – 178
Apollo Saturn Program – 165, 269
Apollo Telescope Mount (ATM) – 104, 105, 106, 107, 108, 109, 114, 115, 116, 117, 118, 119, 120, 121
Apollo VIII – 257
Apollo XI – 257
Applegate, Fred – 74
Armed Forces Special Weapons Project (AFSWP) – 55
Armstrong, Neil – 33, 83, 96, 114, 139, 163, 181, 182, 192, 257, 276, 297
Army Ballistic Missile Agency (ABMA) – 28, 31, 34, 43, 44, 45,47, 48, 51, 55, 57, 60, 62, 63, 69, 71, 72, 91, 152, 165, 166, 234, 267
Army Ordnance Missile Command (AOMC) – 28
Arsement, Leo – 69, 80
Artley, Gordon – 64
Ashton, Al – 102
Asquith, Bob – 70
Astrionics Laboratory – 74, 76, 77, 78, 79, 80, 81, 176
Atherton, Jim – 80
Atlas – 41, 42, 43, 267
Atomic Energy Commission – 18
Aviation Challenge – 216, 217, 294

B

Baggs, Ellis – 70, 77, 81
Bailey, G. – 81
Bailey, Robert Carl – 288
Baker, *monkey* – 58, 164
Balch, Jack – 272
Baldwin, R.B. – 61
Barbree, Jay – 238
Barnard, Christiaan – 156
Barnett, Reed – 22, 23, 26
Barnhart, Deborah – 5, 213, 215, 295, 296
Barr, Glen – 81
Barr, Tom – 69, 80
Barsky, *Russian Interpreter* – 114
Bean, Alan – 83, 84, 112, 181
Becher, Rudy – 69
Beckham, Richard – 78
Beggs, Jim – 274
Beichel, Margot – 253
Belew, Leland (Lee) – 136, 193, 266
Bell, Alexander Graham – 18
Bell, Lucian – 69, 91
Benedict, Howard – 238, 247

Benham, Ted – 231
Bennett, Michael – 102
Bensko, John – 44, 45, 62, 63
Bergaust, Erik – 247, 283, 285, 286
Berger, Ed – 177
Bergman, Jules – 238
Bethay, Woody – 30
Beyer, Rudy – 70
Biggs, Bob – 231
Bizarth, Elmer – 101
Blackstone, John – 70
Blanche, Jim – 81
Blanton, Jim – 70, 74, 80
Blount Brothers – 22
Blumrich, Josef – 62
Body, W. – 70
Boehm, Josef – 67, 70, 81
Bolton, Tom – 70
Bonestell, Chesley – 243
Borman, Frank – 163
Borrelli, Mike – 176
Bossart, Karel – 267
Bowden, Ann – 228
Boykin, Claude – 78
Bracker, A. L. – 69
Bradshaw, Charles – 5, 35, 37
Bramlet, Jim – 91, 235
Brandner, Frederick – 69
Bridges, Jack – 70, 81
Bright, Frazier – 71
Bright, Lillian – 228
Brokaw, Tom – 14
Bromberg, Jack – 92
Brooksbank, Bill – 127
Brooks, Melvin – 70, 74, 80
Brothers, Johnny – 78
Broussard, Pete – 81
Brown, Harold – 74
Brucker, William – 43, 249
Bucher, George – 44, 46, 63, 190, 266
Buchhold, Theodore – 67
Buckbee, Edward (Ed) – 1, 6, 8, 15, 31, 85, 96, 132, 188, 204, 206, 213, 218, 231, 238, 266, 285, 286, 293, 294, 295
Burdine, Bill – 70
Burke, Harlan – 69
Burns, Howard – 91
Bush, Eddie – 177
Bush, George H.W. – 148, 213
Buzbee, Claude – 229

C

Cagle, Eugene (Gene) – 70, 115
Camp, Joe – 102
Cape Canaveral – 16, 22, 30, 132, 160, 185, 246, 247, 265, 291
Carlson, Norm – 185
Carpenter, Liz – 160
Carpenter, Scott – 162, 163, 227
Carr, Gerald (Jerry) – 112, 113
Carroll, Stan – 176
Casey, Charles – 70
Cataldo, Gene – 44
Caudle, John – 70, 81
Centaur – 24, 267
Centaur Project – 268
Central Systems Engineering Laboratory – 79, 80
Cernan, Gene – 181
Chaffee, John – 229
Chaffee, Roger – 133, 271
Challenger – 274, 275
Chambers, Charley – 69
Chandley, Curley – 24, 25, 26
Chandra X-ray Observatory – 60, 129
Chandra X-Ray Telescope – 276
Chapman, Harvey – 71
Chase, John – 69
Chassay, Roger – 127
Chatfield, Miles – 40
Christensen, Dave – 266
Christofilos, Nicholas – 55
Christopher, Ruth – 44
Christy, Jim – 78
Chubb, Bill – 70, 74
Churchwell, John – 185
Clarke, Arthur C. – 147, 182, 247
Clark, George – 60
Cline, Fred – 136
Cloud, Bo – 24
Cochran, Don – 46, 52, 58
Cole, Roger – 70
Collier, Heidi Weber – 5, 252
Collins, Michael (Mike) – 96, 114, 192, 246
Computation Laboratory – 65
Conard, Gertrude – 124, 157, 160
Conner, Jack – 5, 21, 26, 27, 85, 177
Conrad, Charles "Pete", Jr. – 83, 84, 107, 110, 111, 163, 174, 181
Cook, Dick – 124
Cook, Don – 74
Cook, Lewis – 70
Cook, Thomas – 29
Cooper, Charlie – 98, 99, 100, 101
Cooper, L. Gordon – 101, 162, 163, 227
Copeland, Myrna – 289
Coppock, Huey – 81
Coppock, Hugh – 70
Cornelius, Charley – 176
Costes, Nick – 63
Counter, Duane – 81
Counts, Parker – 229
Cox, John – 69, 80
Craft, Harry – 125
Crain, Bob – 231
Crisp, Amos – 239
Cristoforetti, Samantha – 220
Cronkite, Walter – 156, 238, 247

Crossfield, Scott – 33
Crumpton, Walt – 65
Cruse, Bill – 102
Cruse, Clarence – 102
Crutcher, Andy – 71
Cummings, Milton – 277, 284, 287, 292
Cummings Research Park – 15, 277, 278, 279, 280, 281, 286, 287

D

Dannenberg, Konrad – 211, 294
Darity, Martin – 204
Darwin, Charles – 124
Davis, Gene – 65
Davis, Wilbur – 18
Debus, Kurt – 7, 18, 22, 30, 62, 89, 134, 143, 168, 185
DeNeen, Carl – 136
Denver, John – 156
Derrington, Jim – 69, 80
DeSanctis, Carmine – 124
Development Operations Division – 29, 153, 154
Devine, Paul – 177
Digesu, Fred – 68, 69, 176
Disney, Walt – 113, 197, 240, 247
Doane, George – 81
Dobbs, Glenn – 102
Dobbs, Steve – 70
Dodds, Ian – 147, 148
Dornberger, Walter – 15
Downey, Jim – 44, 49, 63, 124, 128, 129, 266
Downs, Hugh – 247
Downs, Sanford – 69
Drawe, Gerhard – 67, 70, 74, 262
Duerr, Frederick – 91
Duggan, Jack – 69, 80
Duke, Charlie – 181

E

Earls, Tom – 102
Eberle, Erv – 231
Edison, Thomas – 18
Edwards, Billy – 101
Ehricke, Kraft – 267
Eichelberger, Bob – 69, 80
Eifler, Charles – 204
Eisenhardt, Otto – 64, 105
Eisenhower, Dwight D. – 20, 28, 29, 43, 44, 55, 85, 155, 286
Ek, Matt – 231
Elliott, Neil – 163
Emens, Frank – 69
Emme, Eugene – 203
Engler, Erich – 62
Erwin, Bently – 177
Escue, Tom – 80
Explorer – 24, 41, 48
Explorer I – 14, 20, 44, 50, 51, 69, 162, 164, 227, 248, 254, 267, 279
Explorer II – 51
Explorer III – 51
Explorer IV – 51, 52, 53, 54, 55, 56, 60
Explorer V – 53, 55
Explorer VI – 59
Explorer VII – 59
Explorer VIII – 60
Explorer XI – 60

F

Fabrication and Assembly Engineering Laboratory (Fab Lab) – 63, 64, 65, 66, 67, 98
Faget, Max – 7
Farmer, John – 74
Felch, Jim – 81
Fellers, Charles – 102
Ferguson, Bob – 69
Ferrell, George – 70
Fesler, Vernon – 177
Fichtner, Hans – 70, 77
Field, Elmer – 91
Fields, Stan – 54
Figurola, Tulio – 102
Fikes, Gene – 70
Fisher, Ken – 229
Flack, Ray – 70
Fletcher, James Dr. – 147
Flynn, Alexander – 124
Flynn, Frank – 102
Foster, Betty – 258
Foster, Carol – 5, 252, 255, 258
Foster, J.N. (Jay) – 5, 63, 64, 65, 66, 123, 141, 255, 258
Fox, Tom – 74
Franklin, Jack – 65
Frary, Spencer – 44, 45
Freedman, Jack – 229
Friedrich, Hans – 36, 37
Frost, Walt – 69, 80
Fuller, Paul – 231

G

Gagarin, Yuri – 14, 279
Gage, Claude – 234
Galileo – 18
Gandhi, Mrs. – 148
Gant, Roger – 102
Garner, Ray – 5, 279
Garrett, Harrison – 81
Garriott, Owen – 112
Gassaway, Gilbert – 70, 80
Gates, Dan – 45, 48
Geiger, Hans – 50
Geissler, Ernst – 18, 143, 168
Gemini – 32, 41, 163, 211, 256
George C. Marshall Space Flight Center – 27, 30, 60

George, Jack – 70
Gerow, Herman – 63
Gibson, Ed – 110, 111, 112, 113
Gierow, Herman – 128
Gilbert, G.K. – 61
Giles, Jack – 204
Gilino, Norman – 70, 80
Gillespie, Charlie – 177
Gilruth, Robert (Bob) – 7, 89, 134, 139, 140
Ginn, Nathan – 69
Glass, Bill – 70, 77, 80
Glass, Charles – 70
Glennan, Keith – 28, 138
Glenn, John H. – 162, 163, 227
Goddard, Robert H. – 10, 12, 46, 161
Goddard Space Flight Center – 128
Godfrey, Roy – 91, 126, 271
Goetz, Otto – 231
Goldberg, Leo – 121
Golley, Paul – 74
Goodhue, Walt – 70
Gordon, Dick – 83
Gordo, *squirrel monkey* – 58
Gore, Senator – 233
Gorman, Harry – 18, 124, 138, 144
Gould, John – 70, 81
Grace, Clinton – 92
Graff, Charley – 81
Grafton, Bill – 21, 22, 23, 177
Graham, Billy – 289
Graham, Clyde – 102
Graham, William – 274
Grau, Dieter – 67, 68, 138, 143
Greaver, Bill – 61
Greer, Robert – 92
Griffin, Gerry – 274
Grissom, Virgil. I. "Gus" – 133, 162, 163, 227, 260, 271
Gruber, Pat – 102
Gruene, Hans – 133, 185, 262, 271
Guidance and Control Laboratory – 66, 67, 69, 71
Guided Missile Development Division (GMDD) – 35, 40, 43, 44, 67
Guilian, William – 18
Guillebeau, W. W. – 70
Guire, Nancy – 5, 153, 189

H

Haeussermann, Walter – 18, 67, 69, 73, 80, 84, 168, 176, 253
Haire, Jack – 65
Hale, Jim – 231
Haley, Foster – 18, 239
Hall, Tim – 5, 25, 27
Hamby, Red – 177
Hamill, James P. – 18
Hammer, Don – 102
Harbert Management Corporation – 22
Hardin, Ernie – 102
Hardy, George – 173, 233
Harper, John – 70
Harper, Warren – 69, 80
Harris, Gordon – 18
Hart, Johnny – 156
Hauff, Chris – 78
Hearn, Glenn – 204
Heaussermann, Walter – 136, 137, 143
Heckman, Dick – 103, 173
Heflin, Howell – 233
Heimburg, Karl – 18, 86, 87, 133, 137, 138, 143, 168, 244, 262, 263
Heimburg, Klaus – 263
Heimburg, Ruth – 263
Heimburg, Stephan – 263
Hellebrand, Emil – 150
Heller, Gerhard – 40, 44, 48, 50, 51, 52, 53, 61, 190
Helmut Hoelzer – 37
Hembree, Ray – 47
Hendricks, John – 56
Hendrix, Doug – 74
Hermann, Adolph – 70
Hight, Hermon – 74
Hines, Bill – 238
Hitler, Adolf – 246
Hjornevik, Wesley L. – 29
Hoberg, Otto – 67, 69
Hoelzer, Helmut – 18, 136, 137, 143, 168
Hoffman, Helmut – 70
Holderer, Oscar – 210, 294, 296
Holladay, Alvis – 81
Holland, Bob – 54
Holland, Paul – 176
Holmes, Bonnie – 86, 101, 139, 153, 154, 155, 156, 158, 159, 160, 187, 188, 189, 225, 226, 241, 247, 248
Holmes, Brainerd – 89
Holmes, Emern – 156
Holmes, Kenny – 156
Holmes, Ray – 156
Holweger, Denny – 229
Hoover, George W. – 42
Hopkins, Hop – 177
Hopson, George – 231, 232
Horton, Bill – 80
Horton, William – 115
Hosenthien, Hans – 68, 70, 80, 253
Hovik, Oscar – 70
Hovis, Howard – 70
Howard, Bill – 70, 176
Hubble Space Telescope – 128, 129
Hubble Telescope – 174, 276
Huber, Bill – 125, 138
Hueter, Eike – 261
Hueter, Hans – 18, 23, 168, 261, 267
Hueter, Ruth – 261
Hueter, Uwe – 5, 252, 261

Hueter, Wendula – 261
Huggins, Carl – 69, 80
Hughes, David W. – 5
Humphery, Hubert – 155
Humphrey, Hubert – 166
Humphrey, Jack – 185
Huntsville International Airport – 22
Huntsville Rocket Team – 69, 74, 76, 84
Huntsville's International Airport – 15
Huntsville's Von Braun Center – 27
Huth, Chauncey – 18
Huzel, Dieter – 232

I

Illinois Institute of Technology Research Institute (IITRI) – 48
International Geophysical Year (IGY) – 13, 42, 50, 57
International Space Camp – 212
International Space Station – 48, 210, 211, 214, 220, 222, 296
Irwin, James (Jim) – 33, 181
Ise, Rein – 5, 151, 165, 233

J

Jacks, Bennie – 27
Jackson, Guy – 239
James, Lee – 268, 269, 270, 271, 272
Jastrow, Robert – 62
Javins, Dave – 102
Jean, O.C. – 125, 127
Jefferies, Bill – 78
Jenke, Richard – 78
Jennings, Charles – 71
Jet Propulsion Laboratory (JLP) – 43, 44, 50, 57, 58
Jobe, Sherman – 78, 81
Johnson, Jerry – 58, 231
Johnson, Joe – 71
Johnson, Lady Bird – 160, 197
Johnson, Lyndon B. – 155, 191, 197, 243, 257, 278
Johnson Space Center – 82, 96, 108, 110, 120, 134, 146, 148
Johnson, William – 268
Johnstone, Harry – 193
Jones, Billy – 51, 59
Jones, Bob – 85
Jones, Carl T. – 22
Jones, Charles – 74
Jones, Clyde – 70, 81
Jones, Joe – 239, 248
Juno I – 57
Juno II – 57, 59
Jupiter – 44
Jupiter-A – 43
Jupiter (AM-13 flight) – 58
Jupiter C – 24, 43, 44, 47, 50, 54
Jupiter IRBM AM-18 – 58
Jurgensen, Klause – 81

K

Kalange, Mike – 70, 81
Kampmeier, Heinz – 69, 80
Kapryan, Walt – 7
Kasparek, Walter – 70, 81
Kastanakis, John – 26
Kaufmann, Ralph – 81
Keathley, William – 115
Keithley, William – 129
Kelley, Bill – 71
Kelley, Fred – 70, 81
Kennedy, John F. – 6, 44, 63, 74, 85, 86, 87, 96, 98, 133, 155, 156, 159, 160, 189, 191, 195, 240, 242, 243, 278
Kennedy, Senator – 233
Kennedy Space Center (KSC) – 7, 13, 31, 53, 78, 89, 96, 107, 120, 146, 269, 270, 271
Kennel, Hans – 70, 74
Kent, Marion – 18
Kerry, Senator – 233
Kerwin, Joe – 107, 110, 111, 174
Khrushchev, Nikita – 44
Kidd, Jerry – 228
King, Martin Luther, Jr. – 257
King, Olin – 61
Kingsbury, Jim – 40, 41, 44, 129, 144, 232
Kirby, Cliff – 70, 81
Knott, Don – 70
Koelle, Hermann – 45, 138
Kopera, Frank – 177
Kraft, Chris – 7, 139, 274
Kraushaar, Bill – 60
Kreider, Bill – 70, 81
Kroeger, Hermann – 69
Kroh, Hubert – 67, 69
Krome, Henning – 176
Kruidenier, Bob – 80
Kuebler, Manfred – 70
Kuerschner, Mr. – 67
Kuers, Werner – 135, 136
Kuettner, Joachim (Jack) – 31, 32, 33, 34, 162

L

Lamb, Dale – 70
Lange, Ernst – 67, 70, 81
Lange, Oswald – 66
Langley Research Center – 28
Larson, Ed – 231
Lary, Frank – 231
Launching and Handling Laboratory – 66
Lawrence Berkley National Laboratory – 55
Lawrence Livermore National Laboratory (LLNL) – 45
Lawrence Radiation Laboratory – 57

Lee, Bill – 102
Lee, Charles – 70, 81
Lee, Gene – 70, 81
Lee, Thomas J. "Jack" – 5, 127, 128, 190, 193, 232, 266
Lemay, John – 71
Letterman, David – 218
Lewis, Charles – 103
Lewis, Jim – 70, 77, 81
Lewis Research Center – 49
Lewis, Richard – 238
Ley, Willy – 13
Lincoln, Evelyn – 156, 189
Lindbergh, Charles – 156
Lindner, Anni – 255
Lindner, Kurt – 67, 255
Lindquist, Bob – 177
Lindstrom, Robert (Bob) – 144, 165, 232, 234, 235
Linstead, Bill – 71
Lombardo, Joe – 231
Los Alamos National Laboratory (LANL – 49
Lousma, Jack – 112
Lovelace, Alan – 245
Lovell, Jim – 163
Low, George – 7, 147, 148
Lucas, Jack – 68, 70, 81
Lucas, William R. (Bill) – 5, 121, 122, 136, 137, 144, 157, 193, 229, 232, 272, 273, 274
Ludwig, George – 50
Luehrsen, Hannes – 30
Lunar Module – 171
Lunar Program – 63
Lunar Roving Vehicle (LRV) – 63, 96, 171, 257
Lundquist, Charles (Chuck) – 5, 41, 47, 48, 50, 51, 56, 58, 63, 166, 266, 277

M

Mack, Jerry – 70, 74
Macuch, Marvin – 70
Malone, Lee – 69
Mandel, Carl – 70, 81
Manned Spacecraft Center (MSC) – 7, 61, 85, 118, 139
Manteuffel, Dr. – 67, 68
Manufacturing-Engineering Laboratory (ME Lab) – 98, 99, 105, 108, 110
Mark, Hans – 55
Marshall, Bob – 124, 125
Marshall, George C. – 20, 85, 155
Marshall's Fabrication Laboratory – 170
Marshall Space Flight Center (MSFC) – 7, 14, 17, 44, 49, 60, 61, 63, 66, 69, 72, 74, 76, 78, 89, 90, 92, 94, 101, 108, 111, 115, 116, 118, 120, 121, 122, 123, 124, 126, 127, 128, 129, 141, 143, 144, 146, 147, 148, 154, 157, 167, 211, 213, 235, 252, 260, 277, 280, 281, 283, 287, 289, 294
Martineck, Hans – 71
Martin, Hank – 44, 45
Martin, Jim – 102
Matthews, Frank – 70
Maus, Hans – 18, 66, 139, 143
Maxfield, James R. – 245
McCarter, Mark – 276
McCarty, John – 231
McCool, Alex – 23
McCullogh, James – 91
McDivitt, James – 163
McDonough, Georg – 144
McElroy, Neil – 43
McIllwain, Carl – 50
McLendon, Sam – 101
McMahen, Charles – 70
McMillion, Jim – 144
Medaris, John B. – 16, 18, 19, 29, 43, 44, 47, 48, 248
Medlock, Joe – 78
Mercury – 31, 33, 41, 163, 164, 211, 219, 227
Mercury 7 – 162, 163
Mercury-Redstone – 24, 164, 165, 220
Mercury-Redstone Rocket – 162
Michoud Assembly Facility – 30
Mikulski, Senator – 233
Mikuni, Don – 231
Miles, Gentry – 45, 190
Miller, Ed – 48
Miller, Gene – 177
Milner, Bob – 70, 77
Mink, Harold – 70, 80
Mintz, Ed – 231
Mitchell, Ed – 181
Mitchell, J.E., Jr. – 290
Mitchell, Walt – 231
Mobility Test Article (MTA) – 199
Moody, Jewell – 91
Moore, Brooks – 5, 22, 24, 67, 70, 80, 144, 176, 266
Moore, Claud – 177
Moquin, Joseph – 277
Morea, Saverio "Sonny" – 97, 126
Morgan, Tom – 70, 81
Morris, Bob – 231
Morris, Delmar – 18, 144
Moser, Bob – 185
Moss, Gail – 101
Mrazek – 18
Mrazek, Willi – 136, 143, 168, 235
Mueller, Fritz – 67
Mueller, George – 7, 89, 94, 134, 139, 140, 148, 150, 151, 235, 271
Mueting, Evelyn – 154
Mullane, Robert – 234

Murphy, Clark – 177
Murphy, James – 125

N

NASA (National Aeronautics and Space Administration) – 7, 18, 20, 27, 28, 29, 30, 45, 49, 59, 60, 62, 69, 71, 72, 74, 85, 89, 90, 94, 95, 98, 104, 107, 115, 118, 119, 121, 122, 123, 126, 128, 129, 132, 133, 136, 140, 141, 142, 146, 147, 148, 149, 154, 233, 238, 241, 243, 245, 248, 249, 258, 265, 267, 273, 277, 283, 286, 288, 289
National Academy of Sciences (NAS) – 42
National Advisory Committee for Aeronautics (NACA) – 28
National Advisory Committee on Aeronautics (NACA) – 22
Naumann, Robert "Bob" – 5, 40, 47, 51, 53, 55, 57, 60, 127
Navajo Missile – 42
Naval Research Laboratory (NRL) – 42
Neighbors, Joyce – 81
Nelson, Richard – 92
Nerren, Guy – 290
Neubert – 18
Neubert, Erich – 168, 193, 255
Neufeld, Michael J. – 283
Neutral Buoyancy Simulator (NBS) – 97, 98, 101, 104, 105, 106, 107, 108, 109, 110, 111, 112, 113, 114, 174, 175, 200
Neville, Don – 98, 99, 100, 102
Nevins, Pete – 101
Newby, David (Dave) – 18, 27, 28, 29, 30, 31, 124, 166
Newell, Homer – 126
Nicaise, P.D. – 74
Nickerson, John – 41, 49
Nicks, Ed – 261
Nike X – 17, 20
Nikolayev, Cosmonaut – 114
Nixon, Richard – 147, 293
Noel, John – 70
Novak, Max – 64
Nuclear Engine for Rocket Vehicle Applications (NERVA) – 49, 194
Nurre, Gerald – 74, 129

O

Oberth, Hermann J. – 10, 12, 13, 42, 46, 161, 166
Odell, Bob – 128, 129
Odom, Brian C. – 5, 27
Odom, James – 129
Office of Manned Space Flight (OMSF) – 89, 94, 95
Office of Naval Research (ONR) – 42
Oliver, Jean – 62, 125, 128
Operation Hardtack – 24
Operation Paperclip – 16
Oppenheimer, J. Robert – 50
Orden, Ray Van – 71
Ordnance Missile Laboratories (OML) – 40
Ordway, Fred – 249, 266
Orem, Ernest – 70, 81
Orton, Bert – 71
Ostrander, Donald R. – 72

P

Pace, Bob – 268
Paetz, Robert – 64, 65
Pagenkopf, Benjamin – 229
Paine, Tom – 133, 146, 147, 148
Paludan, Ted – 69, 80
Parker, Bill – 92
Parnell, Tom – 128
Pascal, Etheridge – 70
Patterson, James – 70
Paul, Hans – 232
Paulus, Ron – 211
Payne, Billy – 80
Payne, Molly – 154
Peacock, Max – 185
Pearson, Jim – 23, 24, 177
Peenemunde – 15, 17
Peenemunde East – 16
Peenemunde Historic-Technical Museum – 15
Pegasus Meteoroid Detection Satellite Project – 268
Pegasus Project – 269
Penovitch, Frank – 78
Perry, Guy – 177
Pershing Missile – 20
Pershing Weapons System – 91
Petrone, Rocco – 7, 95, 156, 232, 269, 270, 271, 272, 273
Pfaff, Helmut – 70, 81
Phillips, Samuel – 7, 89
Piccard, Jacques – 113
Pickering, William – 43, 44, 50, 162
Pickett, Andy – 185
Pierce, Charlie – 229
Pioneer III – 57, 58
Pioneer IV – 58
Pogue, Bill – 112, 113
Polaris IRBM – 43
Poppel, Gerda – 254
Potter, Bill – 64, 65
Potter, Dick – 44
Pouliot, Richard – 102
Powell, J.T. – 69, 80
Powell, Luther – 78
Powers, H.C. – 70
Prasthofer, Willie – 294

Pratt, Dick – 239
Preston, G. Matt – 185
Price, Harold – 186
Price, John – 69
Priest, Jack – 74
Project Argus – 55
Project Atlas – 42
Project Horizon – 45
Project Orbiter – 42, 49
Pruitt, Ed – 103
Pruitt, Sammy – 291
Puckett, Alton – 102

Q

Quality Laboratory – 66
Quayle, Dan – 233

R

Rackley, Joe – 81
Ralston, Holly – 5, 292
Randall, Joe – 81, 144
Reagan, President Ronald – 6, 274
Redstone Arsenal – 15, 16, 17, 18, 20, 25, 26, 28, 29, 30, 35, 40, 68, 148, 153, 191, 234, 252, 255, 261, 264, 267, 277, 279, 280, 281, 282, 286, 289, 291
Redstone BMLT – 21, 22, 23, 24, 27
Redstone Missile – 20, 31, 40, 42
Redstone Missile System – 91
Reed, Wanda – 239
Rees, Eberhard – 18, 29, 52, 144, 147, 156, 168, 185, 189, 193, 198, 227, 232, 272
Register, Bob – 78
Reinartz, Stanley R. "Stan" – 5, 148, 165
Reinbolt, Earl – 81
Reinbolt, Jim – 70
Research Projects Laboratory – 66
Research Projects Lunar Working Group – 62
Research Projects Office (RPO) – 44, 45, 46, 49, 51, 53, 57, 58, 59, 60, 61, 63
Richards, Ludie – 70, 76, 77
Richardson, Jerry – 128, 129
Richards, Richard – 70
Riddick, Ed – 239
Rigell, Ike – 78
Riggenbaugh, Dave – 101
Riggs, Ken – 25, 26
Riley, Charles – 80
Ritter, Glen – 74
Robinson, Bill – 62
Robotics Camp – 294
Roe, Fred – 70
Roop, Lee – 5
Rorex, Jim – 24, 25, 69
Rosinski, Werner – 70, 80
Rothe, Henry – 67
Rowan, Arthur – 91
Rudledge, Frank – 177
Rudolph, Arthur – 91, 92, 95, 186, 235
Ryan, Bob – 231
Ryan, Cornelius – 156, 240, 247

S

Saidla, Bob – 87
Salyut I – 14
Sanchini, Dom – 229, 230, 231
Saturn 1B – 271
Saturn Apollo Lunar Program – 66
Saturn Apollo Program – 82, 83, 121, 142, 146
Saturn I – 63, 71, 72, 74, 75, 76, 77, 85, 86, 142, 143, 144, 170, 185, 191, 227, 234
Saturn IB – 76, 78, 79, 86, 134, 143
Saturn Program – 63, 78
Saturn V – 6, 7, 24, 74, 76, 82, 83, 86, 90, 91, 92, 94, 95, 170, 172, 183, 186, 190, 193, 206, 220, 227, 235, 245, 256, 265, 271, 277
Saturn V Flight Program – 82
Saturn V Lunar Launch Vehicle Program – 81
Saturn V Moon Rocket – 31
Saturn V Program – 74, 79, 91, 92, 93, 186
Saunders, Grady – 80
Schick, Bill – 193
Schilling, Martin – 262
Schirra, Walter M. "Wally", Jr. – 34, 162, 163, 227
Schlemmer, Norm – 231
Schlidt, Dorette – 132
Schlidt, Rudy – 44
Schlitt, Dr. – 67
Schmitt, Harrison – 5, 8, 171, 181
Schneider, Bill – 193
Schocken, Klaus – 45
Schorsten, Ed – 122
Schultz, Dave – 74
Schulze, Wilhelm – 23
Schwedetsky, Dr. – 67
Schweickart, Rusty – 110, 111, 114
Schwinghamer, Bob – 64, 65, 109, 231, 232, 233
Scofield, Harold – 80
Scott, Dave – 181
Seamans, Robert – 7
Searcy, Robert – 284, 285, 288
Seifert, Al – 28
Seitz, Bob – 49
Seltzer, Sherman – 176
Sevastyanov, Cosmonaut – 114
Shafer, Irene – 239
Sharp, Terry – 125
Shelton, Edward – 102
Shelton, Harvey – 74

Shelton, Russell (Russ) – 49, 56, 190
Shepard, Alan B., Jr. – 14, 31, 32, 33, 34, 162, 163, 164, 181, 219, 220, 227, 260, 279
Shephard, Jim (J.T.) – 139, 154, 156, 193
Sheppard, Alan – 74
Shields, Bill – 70
Shockley, Wayne – 70
Shuhlinger, Ernst – 264
Shurney, Bob – 103
Siepert, Al – 29
Simac, Clifford – 63
Simpson, Bob – 70
Sims, Clifton – 70
Skaggs, Jim – 147, 148
Skylab – 14, 94, 97, 100, 101, 102, 103, 104, 106, 107, 108, 110, 111, 112, 113, 116, 117, 118, 119, 120, 121, 127, 149, 150, 173, 174, 232, 302
Skylab Program – 101, 102, 113, 121, 150, 228, 274
Skylab Program Office – 103, 104, 105, 116
Slattery, Bart, Jr. – 87, 198, 239, 247, 248
Slayton, Donald K. "Deke" – 162, 163
Slotkin, Arthur L. – 139
Smith, Carlyle – 231
Smith, Dick – 70, 193
Smith, E.C. – 74
Smith, Gerald – 211, 231
Smith, Mary Jo – 53
Smith, Ralph – 78
Smith, "Snuffy" – 52
Smithsonian Astrophysical Observatory (SAO) – 55
Smith, Spencer – 91
Smith, Ted – 92
Smythe, Prof. – 50
Snark – 42
Sneed, Bill – 5, 89, 91, 92, 125, 186
Sneed, Mike – 5, 252, 260
Snellgrove, Jim – 70
Snoddy, Bill – 48, 51, 52, 53, 54, 56, 57, 59, 60, 125, 128
Sorensen, Victor – 18
Southerland, John – 278
Soviet Union – 6, 14, 17
Space Academy – 294
Space Camp – 206, 207, 208, 209, 210, 211, 213, 214, 215, 218, 219, 220, 221, 293, 294, 295, 296
Space Camp Habitat and Training Center – 295
Spacelab – 272, 273, 276
Space Shuttle – 126, 127, 129, 211, 265
Space Shuttle Challenger – 295
Space Shuttle Columbia – 294
Space Technology Laboratory (STL) – 41
Sparkman, John – 85, 249, 278
Speer, Fred – 128, 129
Splawn, Jim – 5, 97, 98, 99, 100, 101, 113, 174, 211, 226, 228
Spraul, James – 102
Spurr, J.E. – 61
Sputnik – 20, 43, 44, 62, 267, 279
Sputnik I – 14, 247
Sputnik II – 14
Stacy, Robert – 81
Stafford, Thomas – 163
Stangeland, Joe – 231
Staples, Evelyn – 225
Starkey, Jim – 185
Starkey, Theo – 71
State University of Iowa (SUI) – 55, 59
Stein, Richard – 44
Stennis Space Center – 30
Steurer, Wolfgang – 45, 46
Stewart, Edward C. II – 5
Stewart, Homer – 42
Stocks, Charlie – 98, 99, 100, 102
Stokes, Jack – 173
Stone, Ed – 266
Strategic Defense Initiative – 6
Striplin, Bill – 70
Stroud, John – 70, 80
Structures and Mechanics Laboratory – 66
Stuerer, Wolfgang – 40, 44
Stuhlinger, Christoph – 5, 252, 264
Stuhlinger, Ernst – 18, 42, 44, 47, 48, 50, 59, 62, 66, 67, 138, 143, 168, 190, 294
Stulting, Jim – 70, 80
Styles, Paul – 18
Suomi, Verner – 59
Sutherland, Brenda – 231
Swaim, Bob – 228
Swanson, Conrad – 63
Swearingen, Charles – 70, 77, 78, 81
Swords, Ben – 70

T

Taylor, Hugh – 70, 77
Taylor, Jim – 70, 81
Teague, Oli E. "Tiger" – 142, 244, 245
Technical Feasibility Study Office (TSFO) – 40, 41, 42, 44, 46
Technical Services Office – 30
Teller, Edward – 57
Tessman, Bernhard – 262
Test Laboratory – 64, 66, 175, 177
Teuber, Dieter – 70
Thermo Physics Branch – 48
Thiel, Adolph (Dolph) – 36, 37, 40, 41
Thomas, Don – 266
Thomas, John – 128, 211
Thomason, Herman – 70
Thompson, Art – 190

Thompson, J.R. – 5, 103, 173, 192, 229, 230, 231, 275
Thompson, Zack – 176, 231
Thomson, Jerry – 231
Thor – 41, 42, 43, 44
Thornton, Landa – 44, 46, 47
Threlkill, Bill – 69
Tingle, Annette – 5, 21, 27, 157, 190
Titan – 41, 42, 43, 239
Tjulander, Ray – 228
Toftoy, Holger – 16, 18
Torstenson, Charlie – 98, 99, 100, 102
Trott, Jack – 65
Troupe, Clifford – 101
Troupe, Jack – 177
Truly, Richard – 233
Tsiolkovskiy, Konstantin E. – 10, 11, 46, 161
Tuggle, Richard – 70, 81
Turner, Gerald – 70, 77, 81
Tussel, Earl – 177
Tutt, Richard – 70
Twigg, John – 185

U

Uptagrafft, Fred – 157
Urey, Harold – 62
Urlaub, Matthew – 91
U.S. Astronaut Hall of Fame – 31
USS *Albermarle* – 56
USS *Norton Sound* – 55, 56
U.S. Space & Rocket Center – 27, 204, 205

V

V-1 Program – 15
V-2 Program – 15
V-2 Rocket – 13, 15, 16
Vallely, Pat – 74
Van Allen, James – 50, 51, 57, 59, 162
Vandersee, Fritz – 87, 104, 175
Vanguard – 43, 50
Vanguard Missile – 20, 42
V-I Buzz Bomb – 16
von Braun, Iris – 254, 258, 259
von Braun, Margrit – 5, 252, 258, 260
von Braun, Maria – 260
von Braun, Pete – 259
von Braun Space Crescent – 169
von Braun Team – 15, 16, 17, 18, 19, 20, 28, 31, 40, 42, 65, 66, 84, 89, 90, 121, 144, 164, 171, 282
von Braun, Wernher – 7, 8, 13, 14, 15, 16, 17, 18, 19, 20, 27, 29, 30, 31, 34, 40, 41, 42, 43, 44, 46, 48, 50, 67, 68, 73, 74, 84, 85, 86, 87, 89, 90, 91, 94, 95, 98, 101, 113, 121, 122, 123, 126, 132, 133, 134, 136, 139, 140, 141, 142, 143, 144, 145, 146, 147, 148, 149, 150, 151, 152, 153, 154, 155, 156, 157, 158, 160, 162, 163, 165, 166, 168, 173, 176, 182, 184, 185, 187, 188, 191, 194, 195, 196, 197, 198, 199, 200, 201, 202, 203, 204, 206, 225, 226, 232, 233, 234, 235, 238, 239, 240, 241, 242, 243, 244, 245, 246, 247, 248, 249, 250, 252, 254, 255, 257, 258, 260, 261, 262, 267, 269, 272, 276, 277, 278, 279, 280, 281, 282, 283, 284, 286, 287, 291, 292, 293, 294, 296, 299, 302, 303
von Neumann – 42
von Pragenau, George – 176
von Saurma, Ruth – 5, 133, 187, 239, 240, 243
von Tiesenhausen, Georg – 62, 294
Vowe, Gisela – 252
Vowe, Ursula – 252
Vruels, Fred – 127

W

Wagner, Bob – 228
Wagner, Herman – 70, 81
Wagner, Ignatius – 71
Waites, Henry – 74
Wallace, George – 278, 288
Walls, Bobby – 81
Ward, Bob – 5, 246, 266, 278, 283
Wear, Larry – 231, 267
Webb, James – 7, 89, 93, 274
Webb, Jim – 150
Weber, Al – 56, 63
Weber, Fritz – 69, 78, 252
Weber, Lissy – 252
Weidner, Herman – 84, 193
Weidner, Hermann – 144, 152, 153
Weil, Jack – 229
Weisskopf, Martin – 128
Weitz, Paul – 107, 110, 111
Westinghouse, George – 40
Whisenant, Roy – 177
White, Bill – 70, 80
White, Ed – 133, 271
White, Paul – 70
White Sands Missile Range – 17
Whiteside, Carl – 78
Whites Sands Missile Range – 16
Wiesenmaier, Bernie – 70, 74
Wilford, John Noble – 250
Wilheim, Willy – 231
Wilkerson, Ken – 102
Williams, F.L. – 147
Williams, Frank – 5, 124, 132, 133, 137, 138, 146, 147, 148, 152, 154, 193, 277
Williams, Frasier – 56, 57
Williams, Herbert – 70
Willis, Al – 70
Wilson, Bill – 65

Wilson, Charles – 41
Wilson, Secretary – 43, 49
Wilson, Zack – 5, 282
Winkler, Carl – 70, 81
Winkler, Forrest – 74
Wojtalik, Fred – 70, 77, 80, 128, 129
Wolfe, Tom – 250
Wolfsberger, John – 78
Woodbridge, Dave – 47
Wood, Byron – 231
Wood, Lelous – 70, 81
Wood, Lewis – 70, 74, 77
Woodruff, Don – 80
Woosley, Al – 70
Wooten, E.C. – 239
Word, James – 71
Worlund, Len – 231
Wyatt, D.D. – 147
Wyman, Chuck – 81

Y

Yarkin, Sam – 91
York, Herbert – 55
Youngblood, Phil – 70, 80
Young, Dick – 58
Young, John – 163, 181
Young, Richard (Dick) – 58
Yount, Pat – 185

Z

Zeiler, Albert – 185
Zeismer, Erick – 70
Zerlaut, Gene – 48
Zucrow, Maurice – 235